INTEGRATED URBAN DRAINAGE

DR OMPAL SINGH

notionpress.com

INDIA • SINGAPORE • MALAYSIA

ISBN

Paperback 979-8-90431-739-3
Hardcase 979-8-90479-864-2

Dedication

To the late **Dr H. S. Chauhan**, my professor, mentor, and steadfast guide. An internationally respected authority in drainage engineering, his wisdom and unwavering support shaped my journey. I remain deeply grateful for his lasting influence.

Contents

Acknowledgements

The author expresses sincere gratitude to **Dr M.K. Chauhan** for the opportunity to undertake and contribute to important drainage projects in different parts of the country. His technical insights into the drainage issues facing Indian cities across varied geographical locations have been instrumental in defining the direction and depth of this book.

The author also wishes to acknowledge **Shri Devendra Singh** for sharing his extensive expertise in Remote Sensing and GIS. His shared expertise in characterising urban drainage systems, particularly within the NCR and Kolkata regions, provided the necessary "food for thought" to give this book its final shape.

Appreciation is extended to **Shri Yagyik Vashisth** for his academic discussions related to the application of different hydrological and hydraulic approaches to develop practical solutions. His inputs were critical in ensuring the technical robustness of the methodologies discussed in the book

The author further extends sincere thanks to the **officers of DLF Gurugram** for providing the opportunity to implement flood control methodologies within their estates. Access to real-world field conditions, paired with their experienced technical feedback, has greatly enriched the practical insights presented in this work.

This publication is a result of the collective knowledge and field exposure provided by these individuals, and their support is deeply valued.

O.P.Singh

Preface

Urban drainage is often treated as a secondary or supporting element of infrastructure, addressed only after roads are built, plots are developed, and services are installed. Yet, as experience across cities and settlements repeatedly shows, drainage is not a peripheral concern—it is a foundational system that determines the functionality, resilience, and livability of the built environment. This book, *Integrated Urban Drainage*, has been written with the central premise that drainage must be understood, planned, and managed as an integral part of urban development rather than as a reactive afterthought.

The motivation behind this work arises from the persistent and widespread challenges observed in Indian cities and towns—ranging from localised waterlogging and recurring flooding to systemic failures linked with rapid urbanisation, loss of natural drainage, infrastructure fragmentation, and inadequate maintenance. These issues are not isolated technical deficiencies; they are the cumulative outcome of decisions taken across sectors such as land development, road construction, sanitation, and utility management. The need, therefore, is not merely for better drain design, but for a more holistic and integrated understanding of how water behaves within evolving urban landscapes.

This book attempts to bridge the gap between theory and practice by presenting urban drainage as an interdisciplinary field that connects hydrology, hydraulics, infrastructure planning, environmental management, and governance. It explains how drainage problems develop progressively—from early disturbances in natural terrain to complex system failures in dense urban environments. Through a structured approach, the text highlights the importance of continuity, outfall adequacy, maintenance, and institutional coordination in achieving effective drainage performance.

A distinguishing feature of this work is its contextual grounding in Indian conditions. Real-world examples from metropolitan cities such as Delhi, Mumbai, Chennai, Bengaluru, Gurugram, and Kolkata, as well as from smaller and diverse urban settings, are used to illustrate how similar processes manifest differently depending on geography, development patterns, and governance structures.

The book is intended for students, practising engineers, planners, administrators, and all those involved in urban infrastructure development. It seeks to provide both foundational knowledge and applied understanding—ranging from basic concepts of runoff generation and flow behaviour to diagnostic frameworks for identifying drainage failures and strategies for developing resilient systems. Emphasis is placed not only on engineering design but also on sustainability, climate responsiveness, and the preservation of natural drainage assets.

It is hoped that this book will contribute to that transition by fostering a deeper understanding of integrated urban drainage and by encouraging more coordinated, informed, and sustainable approaches to managing water in our cities and settlements.

O.P.Singh

Chapter 1

Introduction to Urban Drainage

Learning Objectives

By the end of this chapter, the learner should be able to:

1. Explain the concept and scope of urban drainage within the broader framework of settlement infrastructure.
2. Describe how drainage problems evolve progressively with infrastructure development.
3. Discuss the role of roads, land development, imperviousness, sanitation interaction, utility congestion, and natural drainage loss in urban flooding.
4. Interpret urban drainage challenges using Indian examples from Delhi, Mumbai, Chennai, Bengaluru, and Kolkata.
5. Recognise the need for integrated, sustainable, and climate-resilient drainage planning.

1.1 Introduction

Urban drainage is one of the most fundamental components of urban infrastructure. Every city modifies the natural movement of water through land development, road construction, building activity, utility installation, and surface sealing. As natural land is replaced with roofs, pavements, embankments, and compacted surfaces, rainfall that was once absorbed, delayed, or conveyed by the landscape is converted into concentrated runoff that must be managed through an artificial drainage system.

For this reason, urban drainage is not simply a matter of constructing drains. It is a broader process of understanding how water behaves within a changing urban environment. It concerns the collection, conveyance, detention, storage, disposal, and, in many cases, treatment of stormwater and excess surface water. It is closely related to roads, sanitation, land use planning, public health, environmental protection, and urban resilience.

A major problem in many cities is that infrastructure development proceeds faster than drainage planning. Roads are improved before outfalls are strengthened, plots are raised without coordinated grading, natural channels are narrowed or encroached upon, and drainage systems are overloaded by wastewater and solid waste. As a result, drainage problems emerge in sequence. They begin as localized ponding and gradually develop into repeated waterlogging, flooding, road damage, and environmental degradation.

In urban hydrology, the most significant change is the shape of the hydrograph (the graph of water flow over time). In natural systems, the curve is “broad,” meaning water reaches the stream slowly. In dense urban areas, the curve becomes “sharp” and “tall,” a phenomenon known as flashiness, where a small amount of rain triggers a massive, immediate spike in water levels.

1.2 Urban Drainage as A Foundational Infrastructure System

1.2.1 Meaning of Urban Drainage

Urban drainage may be defined as the planned management of excess water generated in built-up areas through rainfall, runoff, seepage, and related sources. It includes the collection of water from roads, roofs, plots, and public spaces; its movement through drains, channels, culverts, inlets, and pipes; and its disposal to suitable receiving systems such as natural drains, lakes, wetlands, rivers, detention structures, or engineered outfalls.

1.2.2 Drainage as a Basic Public Service

Drainage is a basic public service because every settlement produces runoff during rainfall. If this runoff is not intercepted and conveyed efficiently, it accumulates in low-lying areas, depressions, roads, intersections, and built-up zones. This creates waterlogging, traffic disruption, unsafe pedestrian conditions, deterioration of pavements, property damage, and reduced urban livability.

Unlike some infrastructure systems whose failure may remain localized, drainage failure rapidly affects multiple sectors. It weakens road performance, complicates sanitation, increases maintenance costs, contributes to environmental pollution, and may expose communities to stagnant or contaminated water. The quality of drainage therefore influences both the physical performance and the social quality of the urban environment.

1.2.3 Urban Drainage as an Interdisciplinary Field

Urban drainage does not function in isolation. It interacts with roads, sewers, water supply corridors, utility ducts, buildings, open spaces, natural drainage paths, and environmental systems. Because of these interactions, drainage planning must be interdisciplinary. A hydraulically adequate drain can still fail if a road blocks its outfall, if a utility trench disturbs its grade, or if solid waste management is poor. For this reason, drainage engineering must be understood not only as hydraulic design but also as coordinated urban infrastructure planning.

1.3 Natural Drainage Before Infrastructure Expansion

1.3.1 Hydrologic Function of Natural Terrain

Before urbanization, the landscape performs many drainage functions naturally. Soil permits infiltration, vegetation slows runoff, shallow depressions provide temporary storage, and natural drainage lines convey excess water toward rivers, streams, ponds, canals, wetlands, and low-lying outlets. These natural processes reduce runoff peaks and delay the concentration of stormwater.

Natural terrain therefore acts as a distributed drainage system. It stores, filters, infiltrates, slows, and conveys water through multiple pathways. In many cases, the natural catchment can manage ordinary rainfall without the need for large engineered structures. This hydrologic balance is gradually disturbed when land is urbanized.

1.3.2 Traditional Settlement Adaptation

In early settlements, drainage often develops informally in response to topography. Houses are placed on relatively safe ground, paths follow ridges or stable alignments, and runoff escapes through open shoulders, shallow drains, or natural depressions. Such arrangements are simple but often effective where built density is low and open land is still available.

However, these informal systems depend on continuity of natural flow paths. Once land is subdivided, filled, paved, and enclosed, their function declines. The settlement then becomes dependent on artificial drainage even if no full system has yet been planned.

1.3.3 Loss of Natural Drainage Memory

An important problem in urban development is the gradual disappearance of awareness about how water originally moved across the landscape. Seasonal channels, old ponds, wetlands, tanks, swales, and shallow depressions may be treated as vacant or developable land. Their drainage function becomes evident only after repeated flooding reveals that they once acted as storage or conveyance features.

This loss of natural drainage memory is a major reason why cities repeatedly build over vulnerable land. In urban drainage education, it is therefore essential to emphasize that understanding inherited topography and natural water pathways is a prerequisite for effective planning.

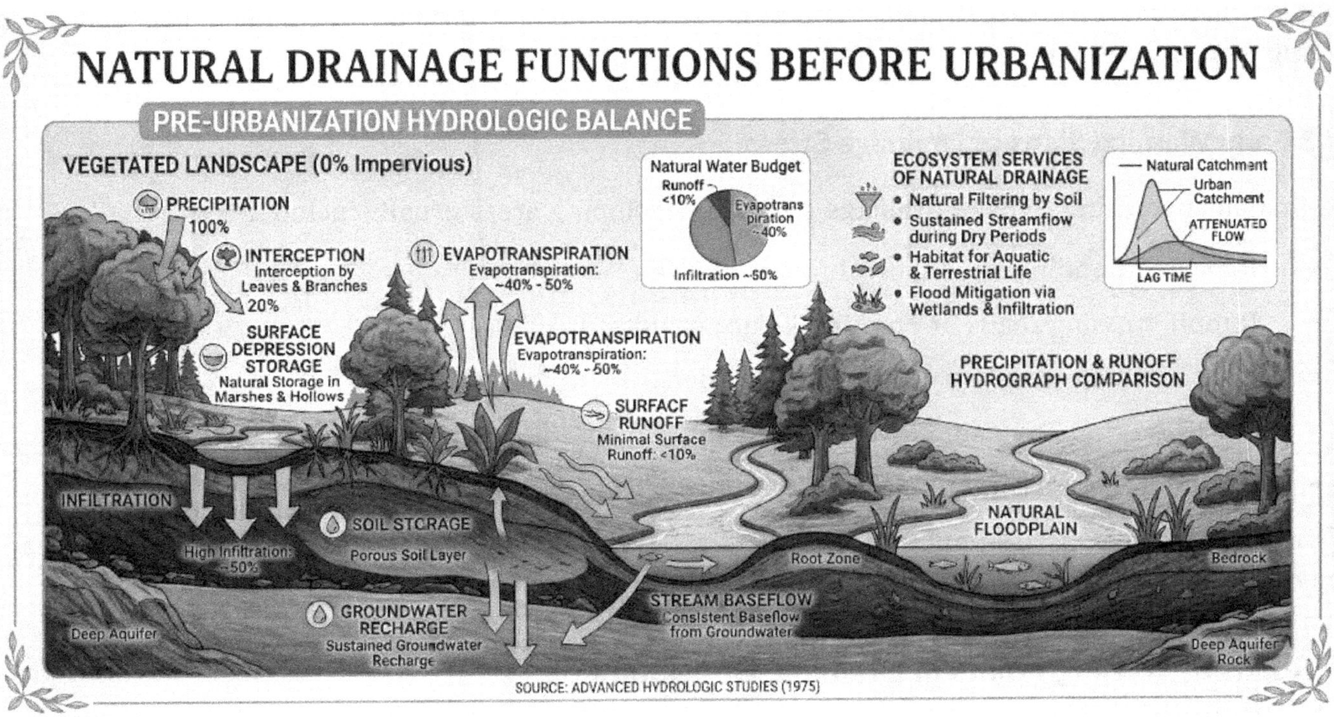

Figure 1.1 Natural drainage functions before urbanization

1.4 Early Land Development and the First Drainage Disturbances

1.4.1 Site Preparation and Altered Runoff

The first stage of urban growth usually includes site clearing, levelling, earth filling, plot demarcation, temporary roads, and early building construction. These activities alter micro-topography, compact the soil, remove vegetation, and disrupt natural overland flow routes. Even before a formal drainage network is established, these changes begin to affect runoff behavior.

Water that once spread gradually across open land begins to pond behind embankments, flow along disturbed grades, erode temporary slopes, and collect at newly created low points. Such local disturbances are often underestimated, but they represent the earliest phase of drainage problem development.

1.4.2 Plot Filling and Independent Grading

In many development areas, individual plots are raised independently to create level construction platforms or to protect against waterlogging. Although this may help a particular site, it often displaces runoff onto neighbouring land or public roads. When many plots are filled independently, the original drainage pattern is fragmented and replaced by a patchwork of barriers and depressions.

This is one of the first major hydrologic consequences of uncoordinated infrastructure development. Instead of runoff following natural slopes, water movement becomes controlled by random local interventions. The resulting drainage condition is unstable and difficult to predict unless area-wide grading control is applied.

1.4.3 Early Warning Signs of Drainage Stress

The first clear warning signs of drainage stress in developing areas usually include:

- Repeated ponding after short-duration rainfall.
- Runoff crossing roads or entering adjacent plots.
- Erosion along temporary slopes or shoulders.
- Localized flooding near culvert approaches or construction zones.
- Sediment deposition in unfinished drains or roadside channels.

These warning signs are important because they appear before full urbanization. If addressed early, they can guide better drainage planning before hydraulic problems become entrenched.

1.5 Roads as Turning Points in Drainage Development

1.5.1 Roads as Hydrologic Modifiers

Roads are among the most powerful modifiers of urban drainage. They increase impervious surface area, alter natural slopes, intercept runoff paths, and concentrate flow along defined alignments. A road therefore does not merely occupy land; it changes the way rainfall is converted into runoff and transmitted through the urban catchment.

Every road requires a drainage logic. This includes pavement drainage, shoulder drainage, side drains, curb and gutter systems where applicable, cross-drainage structures, and adequate outfalls. If these are absent or insufficient, the road becomes both a source of runoff and a barrier to runoff movement.

1.5.2 Road Level Raising and Side Effects

Urban roads are often resurfaced or raised over time. While this may improve ride quality, it may also create drainage problems if adjoining plots and drain levels are not adjusted accordingly. A raised road can become a hydraulic ridge, trapping water on one side or forcing runoff into adjacent low-lying properties.

1.5.3 Drainage-sensitive Road Planning

A road should never be designed independently of drainage. Its longitudinal slope, cross fall, median arrangement, shoulders, inlets, culvert spacing, and outlet relationships must all be considered. In high-density cities, underpasses, flyovers, and depressed corridors are particularly sensitive and require specialized drainage arrangements.

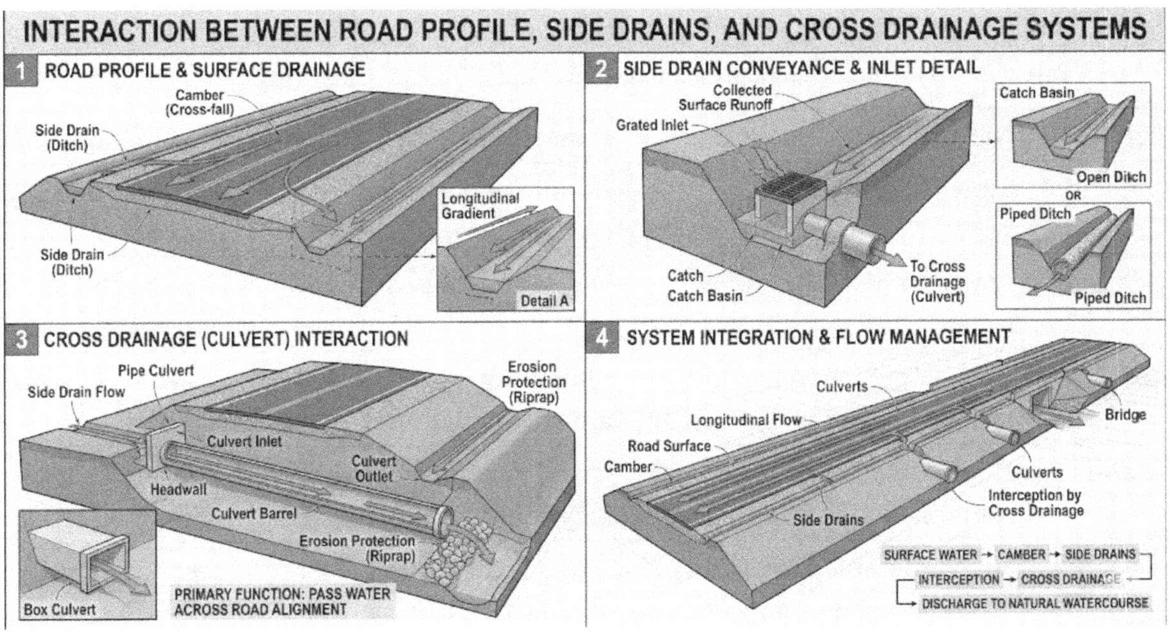

Figure 1.2 Interaction between road profile, side drains, and cross drainage

1.6 Impervious Growth and Hydrologic Transformation

1.6.1 Expansion of Impervious Surfaces

As urban development proceeds, natural ground is increasingly replaced by roofs, roads, courtyards, parking areas, platforms, and paved walkways. These surfaces reduce infiltration and increase direct runoff. Rainfall that was previously absorbed or delayed is converted rapidly into surface flow.

This is one of the defining hydrologic shifts in urbanization. The amount of runoff increases, and the time taken for runoff to reach the drainage system decreases. Together, these changes increase peak flow and put greater stress on drainage infrastructure.

1.6.2 From Local Nuisance to Network Overload

At lower levels of development, increased runoff may produce only local nuisance flooding. However, as imperviousness accumulates across a larger area, the drainage problem becomes systemic. Street inlets surcharge, roadside drains overflow, branch drains lose capacity, and trunk systems become bottlenecked.

This marks a major transition in drainage problem development. The issue is no longer only where water forms on the surface, but whether the entire drainage network can collect, convey, and discharge runoff under peak conditions.

1.6.3 Loss of Informal Storage Spaces

Urban growth also removes informal storage spaces such as vacant plots, roadside depressions, open yards, and low-lying land pockets. These areas often performed valuable stormwater buffering functions even though they were not recognized as part of the drainage system. Once paved or built over, the city loses part of its capacity to delay and spread runoff.

Table 1.1 Hydrologic changes from natural land to dense urban surface conditions

Parameter	Natural Land Conditions	Dense Urban Conditions	Impact of Urbanisation
Runoff Velocity	Slow; hindered by vegetation and soil friction.	Rapid, smooth surfaces and piped drainage.	Increased flood risk and erosion.
Peak Discharge	Low and distributed over time.	High and concentrated.	"Flashier" streams; higher flood peaks.
Lag Time	Long; time between rain and peak flow is extended.	Short; peak flow occurs almost immediately.	Less time for emergency warnings.
Groundwater Recharge	High; consistent replenishment of aquifers.	Very Low; surfaces seal off the water table.	Lowering of the regional water table.
Stream Baseflow	High; sustained by steady groundwater seepage.	Low; streams may dry up during non-rain periods.	Loss of perennial stream flow.
Water Quality	High; natural filtration through soil and roots.	Low; carries metals, oils, and thermal pollution.	"Urban Stream Syndrome" (degraded habitat).
Storage Capacity	High; stored in depression, soil, and vegetation.	Minimal; limited to artificial basins or sewers.	Loss of natural "shock absorbers" for storms.

1.7 Utility Congestion and Underground Competition

1.7.1 Competition within the Urban Corridor

In built-up cities, drainage must coexist with water supply pipes, sewers, electric ducts, telecommunication lines, gas systems, transport structures, and building services. This competition for limited space is one of the defining challenges of urban drainage design and rehabilitation.

1.7.2 Hydraulic Consequences of Utility Conflict

When drains are diverted, narrowed, flattened, or interrupted to accommodate other utilities, their hydraulic performance declines. Even small changes in section or slope may cause sediment deposition, obstruction, and reduced conveyance capacity. In covered systems, such defects are difficult to detect until a major storm exposes them.

1.7.3 Construction Damage and Maintenance Difficulty

Repeated excavation for utility repairs often damages drains or makes maintenance access more difficult. In dense urban corridors, drainage rehabilitation is therefore not only an engineering problem but also a coordination problem involving multiple agencies and service providers.

1.8 Drainage, sanitation, and public health interaction

1.8.1 Misuse of Storm Drains

Where sewerage systems are incomplete or poorly enforced, storm drains are often used to carry greywater or sewage. This introduces solids, grease, organic matter, and sediment into systems meant primarily for intermittent storm runoff.

1.8.2 Effects on Hydraulics and Hygiene

The entry of wastewater reduces carrying efficiency, increases blockage, generates odour, and creates stagnant, polluted conditions. During storms, overflow from such drains spreads contamination into streets and inhabited areas, converting a drainage problem into a public health problem.

1.8.3 Need for Integrated Planning

This interaction shows why stormwater and sanitation systems must be planned together. Effective urban drainage depends not only on physical channel capacity but also on control of wastewater entry, waste disposal practices, and coordinated urban service management.

1.9 Encroachment on Natural Drainage Assets

1.9.1 Natural Drains as Functional Infrastructure

Streams, nallahs, ponds, lakes, wetlands, flood-compatible corridors, and low-lying storage areas are part of the urban drainage system even if they are not engineered structures. They store, slow, and convey stormwater and often provide emergency flood relief during heavy rainfall.

1.9.2 Effects of Encroachment

When natural drains are narrowed, filled, built over, or inadequately culverted, the city loses conveyance and storage simultaneously. Stormwater that was once distributed over a broader natural system is forced into a smaller and more rigid artificial pathway, increasing flood risk.

1.9.3 Loss of Resilience

Natural drainage assets provide flexibility that constructed systems often lack. They can accommodate overflow, variable storm intensity, and temporary storage. Once these features are lost, cities become more dependent on precise functioning of engineered drains and therefore more vulnerable to failure.

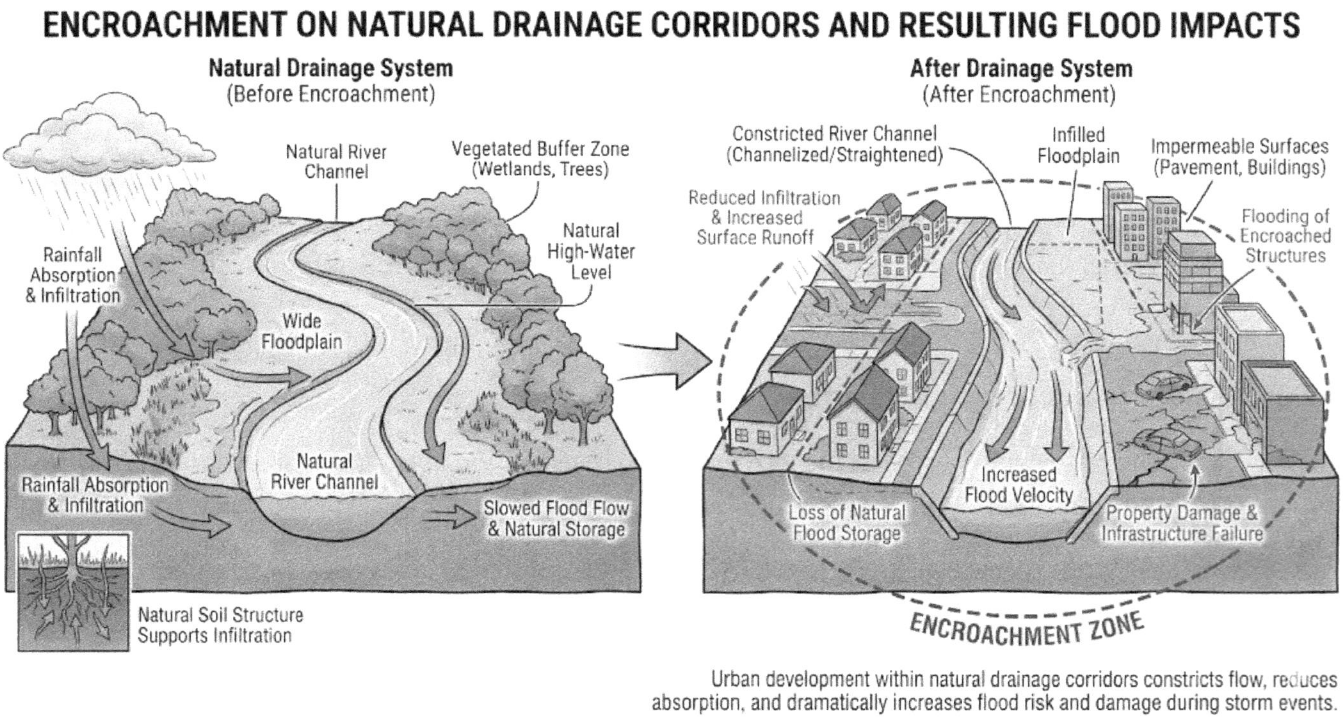

Figure 1.3 Encroachment on the natural drainage corridor and resulting flood impacts

1.10 Incomplete Networks and Outfall Deficiencies

1.10.1 Fragmented Drainage Development

Urban drainage systems often grow in fragments. A road may receive side drains without a downstream connection, a neighbourhood may build local drains without trunk improvement, and branch lines may discharge into constrained outfalls. This creates a local collection without full system performance.

1.10.2 Importance of Continuity

A drainage system functions effectively only when runoff is collected, conveyed, transferred, and finally disposed of through an uninterrupted pathway. If any segment is missing or undersized, water accumulates behind the weakest element.

1.10.3 Outfalls as Controlling Elements

Outfalls are critical because they determine whether the system can actually release water during storms. High tailwater conditions, silted channels, tidal effects, blocked receiving waters, or overloaded trunk drains can all prevent proper discharge even when upstream drains are well designed.

Table 1.2 Hierarchy of urban drainage systems from micro-drainage to final outfall

Component Level	Drainage Category	Description & Scale	Typical Infrastructure	Relative Flow Capacity	Example Function
1	Micro-Drainage (Source Control)	The initial collection of runoff on a single property or lot.	Downspouts, gutters, driveway kerbs, swales, small detention areas (bioretention), rainwater tanks.	Very Low	Directs rainwater from roofs and pavement away from buildings and into the primary system; often incorporates infiltration.
2	Primary/ Lateral Drainage (Collector)	Collects runoff from multiple lots or specific street blocks and conveys it to larger sewers.	Kerbs and gutters, catch basins (gullies), local lateral sewer pipes (e.g., 200mm - 450mm diameter), grass swales along roads.	Low to Medium	Gathers surface flow from streets and lots, channeling it underground.
3	Secondary Drainage (Trunk Collector)	Integrates flow from primary systems across multiple blocks or small sub-catchments.	Main collector sewers, trunk sewers (e.g., >450mm diameter), large culverts, regional swales.	Medium to High	Concentrates the flow from various primary drains and routes it through a neighborhood.

(Contd.)

Component Level	Drainage Category	Description & Scale	Typical Infrastructure	Relative Flow Capacity	Example Function
4	Main Conveyance / Channel (Regional)	Handles the accumulated flow from large urban areas, leading to the ultimate disposal point.	Large concrete channels, massive regional sewer pipes, multi-barrel box culverts, engineered open channels.	Very High	Provides the critical, high-capacity link to move huge volumes of stormwater out of the developed area.
5	Final Outfall	The single point of discharge for the entire system into receiving water or treatment.	Final Outfall Structure, headwalls, energy dissipators, trash racks, potentially a wastewater treatment plant.	Maximum	Releases the collected and conveyed stormwater safely back into natural streams, rivers, lakes, or oceans, or sends it for treatment.

1.11 Maintenance Decline and Operational Failure

1.11.1 Effective Versus Nominal Capacity

Drainage performance in practice depends on effective capacity rather than nominal capacity. A drain partially blocked by sediment, waste, vegetation, or structural damage may operate far below its intended performance during a storm.

1.11.2 Maintenance Becomes Difficult Over Time

As cities grow denser, maintenance becomes harder. Access is restricted, traffic is heavy, drains are covered, encroachments increase, and institutional responsibilities are divided. At the same time, the consequences of maintenance failure become more severe because urban land use is less tolerant of flooding.

1.11.3 Maintenance as Part of Design

Maintainability must therefore be considered during design. Inspection points, access for desilting, realistic section choices, and operational practicality are essential features of a successful drainage system.

1.12 Institutional Fragmentation and Governance Constraints

1.12.1 Divided Responsibilities

Drainage performance depends on coordination among multiple agencies responsible for roads, sewerage, urban development, water supply, natural channels, solid waste, and environmental management. In many cities, no single institution manages the full runoff pathway.

1.12.2 Fragmentation Worsens Flooding

A road project may block drainage, a land project may remove storage, and a utility project may disturb a drain, even when each project appears individually acceptable. Flooding then emerges as a systems failure caused by fragmented governance rather than by one isolated defect.

1.12.3 Need for Integrated Management

Urban drainage governance should therefore be integrated across land development, utility management, road planning, and environmental protection. This is especially important in large cities where infrastructure systems are highly interdependent.

1.13 Climate Variability and the Future of Urban Drainage

1.13.1 Combined Pressure of Urbanisation and climate

Urban drainage systems are increasingly stressed by the joint effects of rapid urbanization and changing rainfall patterns. Systems designed under historical assumptions may become inadequate when runoff generation rises and storm intensity becomes more uncertain.

1.13.2 Amplification of Existing Weaknesses

Climate variability does not create all drainage problems, but it amplifies existing weaknesses such as inadequate outfalls, lost wetlands, overloaded drains, and poor maintenance. Cities with embedded drainage deficiencies are therefore especially vulnerable under more intense rainfall events.

1.13.3 Toward Resilient Systems

Future urban drainage planning must combine conventional drainage works with detention, infiltration, source control, blue-green systems, and preservation of natural drainage assets. Resilience requires both structural capacity and landscape-based adaptation.

1.14 Sequential Model of Drainage Problem Development

The development of drainage problems with infrastructure growth may be summarised as a stage-wise process:

1. Natural terrain provides infiltration, storage, and open flow paths.
2. Early land development disturbs local slopes and creates localized ponding.
3. Road construction modifies runoff pathways and produces barrier effects.
4. Imperviousness increases runoff volume and peak flow.
5. Local drains are introduced, but downstream continuity remains weak.

6. Utility congestion, wastewater entry, and waste deposition reduce efficiency.
7. Natural drains and storage areas are encroached upon or lost.
8. Maintenance failures reduce effective capacity.
9. Climate stress exposes the full inadequacy of the urban drainage system.

This sequence is important because it shows that urban flooding is typically cumulative and path-dependent. The most effective interventions are often those made early, before the city embeds drainage errors into its physical form.

1.15 Indian Metropolitan Case Boxes

Case Example 1.1 Delhi: Metropolitan Growth and Localised Waterlogging

Delhi demonstrates how large road networks, dense development, low-lying pockets, and a complex drain hierarchy can produce repeated waterlogging even in a highly urbanized capital city. Water accumulation at intersections, depressed corridors, and underpasses highlights the importance of local drainage connectivity, inlet efficiency, and coordinated maintenance.

Figure 1.4 Surface flooding (AI-generated to explain the situation)

Key Lessons from Delhi

- Road improvement must be matched by local drainage adjustment.
- Drainage bottlenecks in key urban corridors can disrupt the wider city.
- Agency coordination is essential in flood-prone metropolitan areas.

Case Example 1.2 Mumbai: High-intensity Rainfall in a Dense Coastal City

Mumbai illustrates the combined effect of intense rainfall, dense urban fabric, constrained outfalls, and limited open space. In such conditions, drainage systems must manage large runoff volumes within a short time, while also dealing with receiving-water limitations and extreme urban density.

Figure 1.5 Flooding due to high-intensity rainfall in a dense coastal city

Key lessons from Mumbai

- Dense coastal cities require strong outfall and trunk-drain planning.
- Loss of natural storage and surface permeability increases vulnerability.
- Urban drainage must account for extreme event conditions, not only routine storms.

Case Example 1.3 Chennai: Low-lying terrain and Drainage corridor Pressure

Chennai shows how urban expansion over low-lying areas and drainage-sensitive terrain increases dependence on stormwater connectivity and natural water pathways. The city highlights the importance of preserving drainage corridors, storage areas, and integrated outfall systems under rapid development.

Figure 1.6 Low-lying terrain and drainage corridor pressure

Key lessons from Chennai

- Low-lying urban terrain requires careful preservation of flow routes.
- Development without drainage-sensitive planning increases flood exposure.
- Formal stormwater systems must work with, not against, natural hydrology.

Case Example 1.4 Bengaluru: Urbanisation within a Lake-and-valley System

Bengaluru provides an important example of how drainage problems emerge when urban growth alters hydrologic connectivity among lakes, valleys, and built-up areas. The city's experience shows that interconnected natural systems remain relevant even in highly modern urban settings.

Figure 17 Flooding due to Urbanisation within a lake-and-valley system

Key lessons from Bengaluru

- Hydrologic connectivity is as important as local drain capacity.
- Lake-linked and valley-linked systems should be treated as drainage assets.
- Rapid land conversion increases runoff faster than drainage systems adapt.

Case Example 1.5 Kolkata: Drainage challenges in a flat metropolitan landscape

Kolkata illustrates the importance of canals, lowlands, wetlands, and larger receiving systems in a relatively flat urban environment. In such settings, outfall performance and channel continuity are especially important because small obstructions can significantly increase waterlogging duration.

Figure 1.8 Flooding caused by impeded drainage in a flat metropolitan landscape

Key lessons from Kolkata

- Flat cities depend strongly on maintained outfalls and connected channels.
- Even modest blockages may have large flood consequences.
- Natural and semi-natural receiving systems are critical to urban drainage reliability.

Case Example 1.6 Gurugram: The Concrete Challenge

Gurugram, often dubbed a "Millennium City," exemplifies the urban drainage paradox. Its rapid transformation from a village to a global corporate hub has been largely unplanned and heavily concretized, dramatically reducing the land's natural permeability. This "concrete jungle" effect, combined with inadequate stormwater infrastructure and choked natural water bodies, leads to recurring waterlogging on major arterials like Golf Course Road and rapid flooding in low-lying residential sectors. This image captures how modern infrastructure, like corporate high-rises and the elevated metro, is paralyzed by even a moderate downpour, leaving traffic to navigate deep water and residents stranded.

Figure 1.9 Gurugram flooding due to rapid urbanisation, concrete overload, and paralysed modern roads.

Key lessons from Gurugram

- **Respect Natural Topography:** Urban planning must align with the natural Aravalli slopes and original drainage paths rather than obstructing them with elevated roads and haphazard construction.
- **Adopt "Sponge City" Principles:** Restore lost wetlands and increase permeable surfaces to naturally absorb runoff, reducing the city's total reliance on concrete drainage systems.
- **Sync Density with Infrastructure:** High-density housing approvals (like the "Stilt + 4" policy) must be strictly capped by the actual capacity of the local drainage and sewage networks.
- **Unified Governance and Maintenance:** Consolidate drainage management under a single accountable authority to ensure year-round, tech-monitored desilting and a cohesive master plan

Case Example 1.7. Jodhpur, Rajasthan: Disrupted Traditions and Flash Floods

Jodhpur, the "Blue City," faces drainage problems of a different nature. Located in an arid region with unique rocky hill topography, the city is historically dependent on traditional water harvesting systems. Encroachment upon natural drainage channels and neglect of these ancient stepwells (*baolis*) have made the city highly vulnerable to sudden, intense rainfall events, or "flash floods." This image depicts a narrow, blue-painted alleyway in the walled city during a flash flood. Rushing muddy water, carrying debris, navigates the blue-painted walls of houses. The sun breaks through dramatically after the short, intense burst, illuminating the raw force of the water and the disruption to daily life, as people attempt to navigate the flood and stepwells, now overflowed, create secondary hazards.

Figure 1.10 Jodhpur flooding with arid geography, rocky hills, and flash floods in blue alleys.

Key lessons from Jodhpur

1. The steep, rocky topography of the surrounding hills creates high-velocity runoff that requires dedicated bypass channels to protect the narrow, low-lying alleys of the old city.
2. Restoring and desilting the traditional network of ancient stepwells is essential to provide natural "surge tanks" that can absorb sudden volumes of water during flash floods.
3. Modern drainage infrastructure in the "Blue City" must be specifically scaled and designed to handle intense rain bursts that quickly turn heritage streets into dangerous water funnels.

Case Example 1.8 Vadodara, Overflow and the Urban Squeeze

Vadodara's drainage problems are inherently linked to the Vishwamitri River, which flows through the city. Urbanization has not only encroached upon the river's banks but has also led to the filling of traditional *talavs* (ponds), which once acted as natural reservoirs for excess rainwater. The primary issue is seasonal flooding when the Vishwamitri overflows its banks. A photorealistic image of Vadodara experiencing this riverine flooding is needed. Key landmarks, featuring Indo-Saracenic architectural elements, are partially submerged. Streets are flooded, and people use boats for navigation. The river itself is swollen and brown, carrying sediment. Drainage outlets, visibly submerged, are dysfunctional, with floodwater backing up into residential areas.

Figure 1.11 Vadodara faces problems of Riverine flooding caused by swollen river and submerged dysfunctional drains

Key lessons from Vadodara

1. Encroachment on the Vishwamitri River's floodplains and natural wetlands must be strictly prohibited to restore the river's natural carrying capacity during heavy rains.
2. The city's drainage network requires automated gates and high-capacity pumps to prevent river water from backflowing into residential areas when the Vishwamitri swells.
3. Long-term flood mitigation depends on adopting integrated watershed management that prioritizes the river basin's natural hydrology over piecemeal infrastructure fixes.

Case Example 1.9 Mandi, Himachal Pradesh: Hills, Nullahs, and the Monsoon Threat

Mandi, situated in steep mountainous terrain, is highly susceptible to monsoonal landslides and flash floods, where the primary driver of drainage failure is terrain and rapid runoff. Key drainage problems stem from unplanned construction on unstable slopes and the encroachment of buildings on natural hill-nullahs (streams). This blocks natural drainage paths and funnels intense torrents of water and debris during monsoons. This photorealistic image of Mandi captures a severe weather event, where intense rain triggers a flash flood. A narrow, rocky nullah cuts through the town, surging with a violent mix of water and debris—trees, rocks, and potentially parts of structures—churning in the violent torrent. Buildings on the slopes look threatened, and a partially washed-out road or bridge is visible, emphasizing the raw force of water and the risk of mountain life.

Figure 1.12 The flash flood caused by heavy rain threatened slope structures, and a washed-out road in mountain town Mandi having steep terrain, and construction on slopes.

Key lessons from Mandi

1. The most critical lesson is that natural mountain streams (*nullahs*) must remain sacrosanct. In Mandi, encroaching upon or narrowing these channels for construction has turned manageable runoff into high-velocity, destructive flash flood
2. In hilly terrains like Mandi, poor drainage is the primary cause of slope instability. When water is not efficiently channeled away, it seeps into the soil, increasing pore-water pressure and leading to catastrophic landslides
3. Mountain drainage must account for boulders, uprooted trees, and silt, not just liquid water. During recent floods, many of Mandi's culverts and drains failed because they were physically blocked by debris, causing water to overtop and wash out roads.

Case Example 1.10 Kanpur, Industrial Effluents and complex drainage crisis

Kanpur, with its extensive industrial and combined sewage history, faces a complex drainage crisis marked by pollution and system overload. Decades of unmanaged industrial and domestic waste have clogged the city's extensive sewage and drainage networks. The city's key problem is the "drainage-sewage mix." The combined systems, often in poor condition, overflow during rain, sending mixed sewage and industrial effluents onto the streets. This photorealistic image of an older industrial part of Kanpur shows streets deep in black, putrid water, where garbage and industrial waste float. Drains are visibly overflowing, with dark, foul water bubbling up from manholes, creating a stark visual of severe public health risks, polluted air, and grim environment of a marginalized industrial town.

Figure 1.13 The drainage problem of an industrial town having combined sewers, overflowing sewage, and public health risk.

Key lessons from Kanpur

1. Urban drainage solutions must focus on end-to-end capacity (collection, transport, and specific chemical treatment) rather than just moving the "exit point" of the waste.
2. A segregated drainage network—where industrial waste is pre-treated and handled entirely separately from municipal waste—is not an option, but a necessity for public health.
3. A mart" drainage system is impossible without an equally robust solid waste management system. Without controlling surface litter and ensuring regular desilting, even the most advanced underground sewer lines will eventually succumb to clogging and the accumulation of hazardous gases like methane.

Case Example 1.11 Patna's inadequate, overburdened, and antiquated drainage system

Waterlogging has become a chronic urban crisis in Patna, particularly during the monsoon season. This is driven by an inadequate, overburdened, and antiquated drainage system (*nullahs*) that often dates back to the British era. The primary culprit is severe clogging; drains are frequently blocked by accumulated garbage, plastic waste, and silt due to poor maintenance and disposal practices, preventing rainwater from flowing efficiently to the Ganga River. During intense downpours, this leads to widespread flooding, paralysing transportation, threatening public health, and causing immense suffering for residents.

Figure 1.14 Widespread flooding, paralysing transportation, and causing immense suffering for residents

Key lessons from Patna

1. The city learned that a single point of failure in the main pumping stations (like those in Kankarbagh or Rajendra Nagar) can turn neighbourhoods into permanent lakes within hours.
2. The habit of using *nullahs* as garbage dumps creates "choke points" that lead to backflow.
3. Year-round maintenance and strict enforcement against plastic dumping are far more effective than emergency "pre-monsoon" cleaning, which is often too little, too late.

1.16 Planning implications for resilient urban drainage

The sequence described in this chapter leads to several planning implications.

1. Drainage must be integrated into land development from the beginning. Grading, roads, plot layout, utility placement, and open-space reservation should all reflect runoff pathways and outfall requirements.
2. Natural drainage systems should be preserved as functional infrastructure. Wetlands, ponds, nallahs, tanks, valley lines, and low-lying storage areas should not be treated as expendable land reserves.
3. Drainage should be planned at the catchment scale. Local works are useful only when they fit into a larger hierarchy of micro-drainage, minor drainage, trunk systems, and receiving waters.
4. Maintenance and governance must be built into the system. A technically sound drainage network can still fail if it cannot be inspected, cleaned, repaired, and institutionally coordinated.

5. Future drainage systems should be resilient and adaptive. Conventional gray infrastructure must increasingly be complemented by source control, detention, infiltration, blue-green corridors, and sustainable drainage measures.

1.17 Summary Points

1. Urban drainage is a foundational urban service, not a secondary utility.
2. Drainage problems develop progressively with infrastructure growth.
3. Roads, plot filling, imperviousness, and utility conflict are major drivers of drainage stress.
4. Natural drainage assets provide essential storage and conveyance functions.
5. Outfall adequacy and downstream continuity are critical to system performance.
6. Maintenance and governance are as important as physical design.
7. Indian metropolitan examples show the diversity of drainage contexts and the similarity of developmental causes.
8. Resilient urban drainage requires integrated, catchment-based, and sustainability-oriented planning.

1.18 References

1. Ahiablame, L. M., Engel, B. A., & Chaubey, I. (2012). Effectiveness of low impact development practices: Literature review and suggestions for future research. Water, Air, & Soil Pollution, 223(7), 4253–4273.
2. Butler, D., & Davies, J. W. (2011). Urban drainage (3rd ed.). Spon Press.
3. Fletcher, T. D., Andrieu, H., & Hamel, P. (2013). Understanding, management and modelling of urban hydrology and its consequences for receiving waters: A state of the art. Advances in Water Resources, 51, 261–279.
4. Marsalek, J., Jiménez-Cisneros, B. E., Karamouz, M., Malmquist, P.-A., Goldenfum, J. A., & Chocat, B. (2008). Urban water cycle processes and interactions. CRC Press/UNESCO-IHP.
5. National Disaster Management Authority. (2010). National guidelines: Management of urban flooding. Government of India.
6. Narasimhan, B. (n.d.). Sustainable urban drainage systems. In Urban stormwater management and related practices in India.
7. Rossman, L. A. (2015). Stormwater Management Model User's Manual, Version 5.1. U.S. Environmental Protection Agency.
8. Shuster, W. D., Bonta, J., Thurston, H., Warnemuende, E., & Smith, D. R. (2005). Impacts of impervious surface on watershed hydrology: A review. Urban Water Journal, 2(4), 263–275.
9. Woods Ballard, B., Wilson, S., Udale-Clarke, H., Illman, S., Scott, T., Ashley, R., & Kellagher, R. (2015). The SuDS manual (C753). CIRIA.

10. Yazdanfar, Z., & Sharma, A. (2015). Urban drainage system planning and design: Challenges with climate change and urbanization: A review. Water Science and Technology, 72(2), 165–179.
11. Zhou, Q. (2014). A review of sustainable urban drainage systems considering the climate change and urbanization impacts. Water, 6(4), 976–992.

Chapter 2

Drainage Problems, Failures, and Challenges

Learning Objectives

By the end of this chapter, the learner should be able to:

1. Identify common drainage problems such as flooding, waterlogging, ponding, siltation, and blockage.
2. Distinguish among hydraulic, structural, operational, environmental, and institutional causes of drainage failure.
3. Assess the consequences of inadequate drainage on infrastructure, mobility, health, and environmental quality.
4. Compare drainage challenges in urban, peri-urban, and village settings.
5. Apply a basic framework for diagnosing drainage-related problems in the field.

2.1 Introduction

Drainage systems are intended to remove excess water safely and efficiently, yet in practice many systems perform below expectation because of poor planning, inadequate design, weak construction, insufficient maintenance, encroachment, and changing land use conditions. The consequences of such failure extend beyond hydraulic inconvenience and often include road damage, stagnant water, mobility disruption, environmental degradation, public health risks, and economic loss.

Drainage problems rarely appear as isolated defects. They usually emerge through a chain of linked failures involving runoff generation, obstruction of flow, inadequate conveyance, damaged structures, poor maintenance, and weak institutional response. For example, runoff generated by a paved road may enter a drain that is undersized, partially blocked by waste, and restricted at the outlet, producing overtopping and flooding that appear local but are actually systemic in origin.

This chapter examines the major problems, failures, and challenges that affect drainage systems in urban, peri-urban, and village settings. It explains how such problems appear in the field, what causes them, how they differ by settlement context, and why a systematic diagnostic approach is necessary before design, rehabilitation, or management measures are proposed.

2.2 Nature of Drainage-Related Problems

2.2.1 Meaning of a Drainage Problem

A drainage-related problem arises whenever excess water is not collected, conveyed, stored, or discharged in a controlled manner. In some cases the problem is obvious, such as road flooding, drain overflow, or standing water in built-up areas. In other cases the symptoms are indirect and gradual, including pavement deterioration, damp foundations, foul-smelling stagnant channels, bank erosion, or repeated maintenance difficulty.

A drainage problem should therefore be understood not merely as visible water accumulation, but as any condition in which the intended function of the drainage system is compromised. This broader view is important because many serious failures are first noticed not during storms, but through deterioration in roads, recurring siltation, structural defects, or environmental decline.

2.2.2 Categories of Drainage Failure

The source chapter classifies drainage problems into hydraulic, structural, operational, environmental, and institutional categories. Hydraulic problems arise when the system cannot carry the required discharge. Structural problems involve cracking, collapse, settlement, corrosion, or deformation of drainage components. Operational problems are associated with poor cleaning, blockage, lack of inspection, or malfunctioning control structures. Environmental problems involve sedimentation, contamination, and degradation of receiving waters, while institutional problems arise from unclear responsibilities, weak funding, or fragmented governance.

This classification is useful because it prevents drainage diagnosis from becoming too narrow. A flooded street may not be caused only by insufficient drain size; it may result from poor solid waste management, structural collapse, blocked outfalls, lack of maintenance access, or failure of inter-agency coordination.

2.2.3 System-level Nature of Drainage Problems

Drainage issues often develop as interconnected failures rather than isolated defects. A local depression may pond because road levels are incorrect, because the nearby inlet is blocked, because downstream flow is restricted, or because the outfall is submerged. In many cases, several of these conditions occur together.

For this reason, drainage problems should be interpreted at system level. The visible symptom and the controlling cause are often located in different places, and effective intervention depends on tracing the hydraulic and infrastructural chain from runoff source to final disposal.

2.3 Flooding, Waterlogging, and Ponding

2.3.1 Flooding

Flooding is one of the most visible drainage failures in settlements. It occurs when runoff exceeds the carrying capacity of the drainage network, when flow paths are obstructed, or when receiving waters prevent effective discharge. Flooding may be localised, such as at a road sag, culvert entrance, or market intersection, or more extensive, affecting neighbourhoods, transport corridors, and low-lying urban sectors.

Urban flooding has become more severe in many places because urbanization increases impervious cover, shortens concentration time, and reduces the buffering role of natural drainage and storage spaces. As a result, storms that once produced manageable runoff may now trigger rapid and damaging inundation in built-up areas.

2.3.2 Waterlogging

Waterlogging refers to the prolonged presence of water on or near the ground surface because drainage is slow, the outfall is ineffective, slopes are poor, or discharge continuity is missing. In built-up environments, waterlogging may persist after rainfall has ended, especially where drains are flat, disconnected, or partially blocked. In peri-urban and rural areas, it may also result from road embankments, filled plots, blocked field drains, and poorly managed local outfalls.

Waterlogging is particularly damaging because it combines duration with inconvenience. Even shallow water, when persistent, weakens surfaces, affects access, promotes vector breeding, and creates repeated nuisance in residential and public spaces.

2.3.3 Ponding

Ponding is the temporary accumulation of shallow water in local depressions such as courtyards, roadsides, parking areas, sidewalks, and unpaved lanes. Although ponding is often considered a minor issue compared with flooding, it is an important sign of drainage inefficiency and may represent the early stage of a larger system problem.

Repeated ponding indicates that local runoff is not being intercepted or conveyed properly. Over time, such conditions can damage surfaces, reduce pedestrian accessibility, create unsanitary surroundings, and reveal underlying problems in grading, inlet layout, or local connectivity.

2.3.4 Distinguishing the Three Conditions

Flooding, waterlogging, and ponding differ mainly in scale, duration, and intensity, but all indicate failure of stormwater handling. Flooding is generally larger in spatial extent or depth; waterlogging emphasises prolonged stagnation; and ponding refers to more localised and often shallower water accumulation. From an engineering perspective, all three must be treated seriously because each reflects inadequate drainage performance.

Table 2.1 Distinction between flooding, waterlogging, and ponding by scale, duration, causes, and impacts

DISTINCTION BETWEEN FLOODING, WATERLOGGING, AND PONDING

	FLOODING	WATERLOGGING	PONDING
SCALE Small to Large SCALE	• Regional, large areas, river basins, entire cities. • Examples: Extensive inundation, major displacements.	• Medium to large landscapes, agricultural fields, entire regions with high water tables. • Chronic saturations.	• Localized, small to medium areas, depressions, specific streets, fields. • Examples: Small surface pools.
DURATION TIMELION	• Variable. Can be hours (flash floods) to days, weeks, or even months. • Significant duration for widespread issues.	• Long-term, chronic, semi-permanent to seasonal. Sustained soil saturation over extensive periods.	• Short-term, transient. Temporary surface accumulation, lasting hours to a few days after rainfall. Dissipates.
CAUSES CAUSES	• Heavy and prolonged rainfall, river overflow (banks overflowing), coastal storm surges, tsunamis, rapid snowmelt, dam failure.	• High water table, poor soil drainage (natural or artificial), heavy irrigation, seepage from canals, low-permeability soils.	• Short, intense bursts of rainfall on already wet or impermeable surfaces, flat topography, localized drainage blockages, low infiltration rate.
IMPACTS	• Major infrastructure damage (buildings, roads, bridges), loss of life, displacement of people, massive economic loss, disease outbreaks, crop failure (widespread).	• Reduced crop yields due to root hypoxia (lack of oxygen), soil degradation, salinization, altered plant communities, decline in soil health.	• Temporary transport disruption (puddles on roads), localized crop stress, mosquito breeding, minor property damage (seepage into basements), inconvenience.

2.4 Sedimentation and Blockage in Drains

2.4.1 Sedimentation

Sedimentation is the deposition of soil, silt, sand, and other particulates within drains, culverts, channels, sumps, and chambers. It reduces hydraulic capacity by decreasing effective flow area and increasing roughness. Sediment typically enters drainage systems from exposed soil, unpaved shoulders, construction sites, eroding side slopes, poorly protected embankments, and disturbed catchment surfaces.

In urban and peri-urban areas, sedimentation is especially severe during active construction or where earth surfaces remain unprotected. Construction-related erosion can sharply increase sediment delivery to drains, and the accumulated material may continue to affect drainage performance long after the initial disturbance.

2.4.2 Blockage

Blockage occurs when the flow path is obstructed by solid waste, vegetation, debris, collapsed structural fragments, or accumulated sediment. In urban areas, common blockage materials include plastics, packaging, market waste, road sweepings, and construction debris. In village areas, vegetation growth, loose soil, household refuse, and makeshift crossings often reduce continuity of flow.

Blockage is especially damaging at hydraulic control points such as inlets, bends, culvert entrances, narrow covered sections, and shallow outlets. Even a relatively small obstruction at such locations can sharply reduce effective discharge and produce upstream flooding.

2.4.3 Interaction Between Sedimentation And Blockage

Sedimentation and blockage often reinforce each other. A partially blocked drain slows flow, which encourages further deposition of silt and debris; this in turn reduces velocity and capacity further, causing progressive deterioration. In many practical cases, what begins as minor debris accumulation evolves into chronic stagnation, overtopping, or backwater conditions.

This interaction is important because it shows that sedimentation and blockage are not merely maintenance nuisances. They are also indicators of broader design, planning, and land management problems.

2.4.4 Design and Planning Dimensions

Sedimentation and blockage should not be seen only as maintenance issues. Poor inlet design, erodible drain surfaces, inadequate self-cleansing velocity, missing silt traps, and the absence of debris interception all contribute to recurring problems. Effective drainage design must therefore account for realistic field conditions, not only ideal hydraulic behaviour.

2.5 Hydraulic Inadequacy and Flow Restriction

2.5.1 Meaning of Hydraulic Inadequacy

Hydraulic inadequacy occurs when a drain, culvert, pipe, or channel cannot convey the required flow without overtopping, surcharge, or excessive ponding. This may result from undersized sections, flat slopes, insufficient inlet spacing, poor junction geometry, submerged outlets, or growth in catchment runoff that was never reflected in system upgrading.

In many urban systems, hydraulic inadequacy does not arise because the original design was entirely wrong for its time. Rather, it develops because land use changes, paving, roof expansion, road widening, and settlement growth alter catchment behaviour beyond the capacity of the old system.

2.5.2 Flow Restriction and Bottlenecks

Flow restriction may result from abrupt bends, poor junctions, shallow outlets, reverse slopes, backwater effects, inadequate freeboard, or a localized undersized reach. In many drainage systems, a short restricted segment controls the performance of a much larger upstream network. A narrow outlet, submerged culvert, or poor junction near the downstream end may make an otherwise adequate upstream system ineffective.

This bottleneck effect is one of the most important ideas in drainage diagnosis. It explains why visible flooding may occur far upstream of the actual controlling deficiency.

2.5.3 Diagnosing Hydraulic Problems

Hydraulic inadequacy cannot be diagnosed simply by observing surface water. Proper diagnosis requires examination of drain dimensions, slopes, gradients, continuity, inflow conditions, junction performance, and downstream controls. In practice, recurring failures often reflect hydraulic limitations that have never been systematically assessed after settlement expansion.

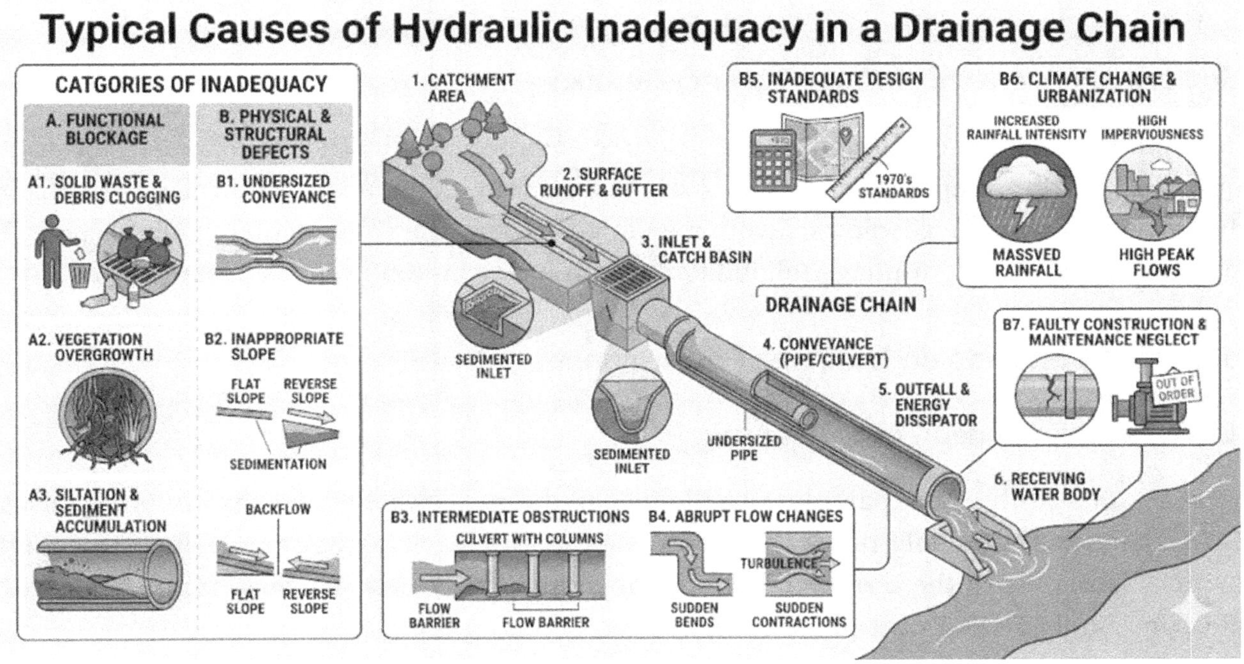

Figure 2.1 Typical causes of hydraulic inadequacy in a drainage chain

2.6 Structural Failure of Drains and Appurtenances

2.6.1 Types of Structural Failure

Structural failure refers to physical deterioration, deformation, settlement, or collapse of drainage elements such as drain walls, channel linings, slabs, culverts, pipes, manholes, chambers, headwalls, and covers. Common defects include cracking, joint separation, wall collapse, cover sinking, pipe misalignment, pipe deformation, scour at foundations, and undermining of aprons or side supports.

Structural distress may develop due to poor design, inadequate wall thickness, weak materials, poor workmanship, lack of curing, poor compaction, overloading, erosion, settlement, corrosion, and age-related degradation. In heavily trafficked urban areas, repeated utility works and vehicle loads may further reduce the life of drainage assets.

2.6.2 Hydraulic Consequences of Structural Defects

Structural and hydraulic failures are closely linked. A cracked drain wall may admit soil and roots, a displaced joint may trap sediment, a collapsed slab may block flow, and a deformed pipe may reduce section area. Thus, a drainage structure may appear physically damaged while simultaneously functioning as a hydraulic restriction.

For this reason, structural inspection must be part of drainage performance assessment. Repair decisions should not focus only on visible damage but also on the effect of that damage on flow continuity and system safety.

2.6.3 Structural Vulnerability in Low-Resource Settings

Structural failure is often accelerated in low-resource environments by thin walls, poor mortar quality, inadequate foundation preparation, lack of curing, and weak construction control. In such contexts, even drains with acceptable geometric layout may fail early because construction quality is poor. This is an important reminder that drainage reliability depends on execution as much as on design.

2.7 Solid Waste Impacts on Drainage Performance

2.7.1 Drains as Unintended Waste Pathways

Solid waste is one of the most persistent challenges in drainage systems. Drains are often misused as disposal channels for household refuse, market waste, plastics, packaging, leaves, silted sweepings, and construction debris. Once these materials enter the system, they obstruct flow, trap sediment, create stagnant pools, and increase overflow frequency.

Research on urban flooding has repeatedly identified waste-related clogging as a major contributor to drainage failure. This is especially critical where drains are open, maintenance access is limited, or waste collection services are irregular.

2.7.2 Hydraulic and Operational Impacts

The effect of solid waste is most severe at control points such as inlets, bends, culvert entrances, screens, and narrow covered sections. A relatively small amount of lodged debris can significantly reduce discharge and trigger upstream flooding. In open drains, waste accumulation may divert water out of the channel and across adjacent land or roads.

Waste also interacts with sedimentation. Debris traps silt, reduces flow velocity, and creates a self-reinforcing cycle of clogging and stagnation. Over time, the effective carrying capacity of the system declines sharply.

2.7.3 Engineering Significance

Solid waste is not purely a behavioral issue; it is also a drainage engineering issue. The problem is linked with lack of regular waste collection, absence of screens or traps, poor awareness, weak coordination between drainage and solid waste departments, and inadequate access for cleaning. As a result, drain design should include realistic provision for trash handling, cleaning access, and inspection.

2.8 Public Health and Environmental Consequences

2.8.1 Public Health Effects

Poor drainage has both direct and indirect public health impacts. Stagnant water in blocked drains, local depressions, and poorly drained plots creates conditions favorable for mosquito breeding and other vectors. Mixed stormwater and wastewater flows may emit foul odors, expose people to contaminated water, and increase pathogen contact through overflow, splash, and stagnant pools.

These impacts are especially severe in dense settlements where drains, roads, markets, and dwellings are close together. Children, elderly persons, and people who must walk through flooded areas face increased health and safety risks.

2.8.2 Environmental Impacts

The environmental effects of poor drainage are also serious. Sediment-laden runoff, polluted drain flows, oil, grease, nutrients, organic matter, and litter may degrade ponds, lakes, streams, wetlands, and receiving channels. Uncontrolled outfalls may also cause erosion and channel instability, increasing sediment loads and damaging downstream aquatic environments.

In this sense, drainage systems are not only flood conveyance systems; they are pathways through which urban land-use impacts are transmitted to the wider environment.

2.8.3 Social and Livelihood Consequences

Inadequate drainage affects access, mobility, safety, and livelihoods. Flooded markets, unusable roads, damaged building edges, and prolonged waterlogging can interrupt work, reduce trade, affect schooling, and increase household inconvenience. Thus, the social burden of poor drainage extends well beyond visible floodwater.

2.9 Settlement-specific Drainage Challenges

2.9.1 Urban Areas

In dense urban areas, drainage challenges often include high runoff concentration, inadequate capacity, encroached drains, utility conflicts, rapid surcharge, and limited space for expansion. Because impervious cover is high and spare hydraulic capacity is low, even minor obstruction can quickly produce major flooding.

2.9.2 Peri-urban and Semi-Urban Areas

In semi-urban and peri-urban areas, drainage problems often arise from the mismatch between old rural drainage patterns and new urban development. Roads are widened, plots are raised, buildings spread into former open land, and natural channels are narrowed or converted, but drainage networks remain fragmented and incomplete. These areas therefore experience transitional drainage stress that is often more complex than either purely urban or purely rural conditions.

2.9.3 Village Areas

In village settings, drains are often discontinuous, unlined, poorly graded, or discharged into unsuitable local depressions. Household access crossings, road embankments, loose soil, and limited maintenance may obstruct flow. Since maintenance often depends heavily on local participation rather than formal institutional systems, performance varies substantially from place to place.

2.9.4 Importance of Context

The same symptom may arise from different causes in different settlement settings. Ponding near a road may reflect inlet failure in a city, incomplete roadside drain continuity in a peri-urban layout, or poor local grading in a village street. Therefore, diagnosis must always consider settlement context.

Table 2.2 Comparison of drainage problems in urban, peri-urban, and village environments

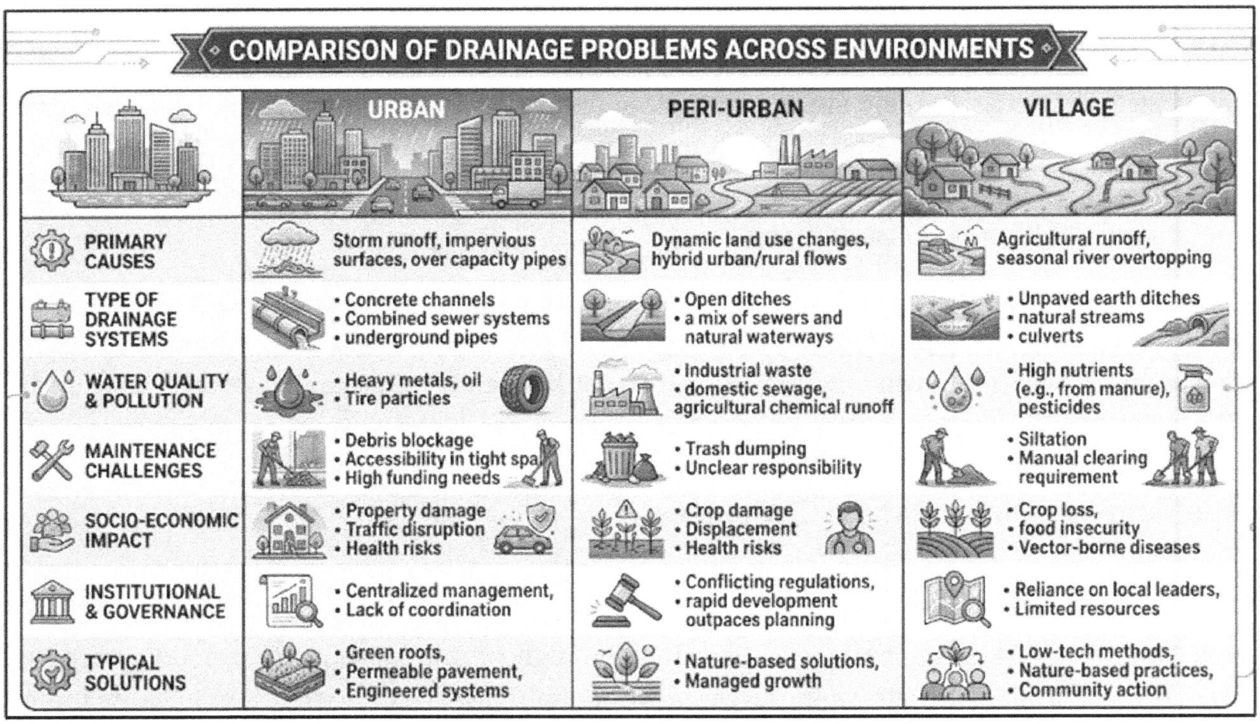

COMPARISON OF DRAINAGE PROBLEMS ACROSS ENVIRONMENTS

	URBAN	PERI-URBAN	VILLAGE
PRIMARY CAUSES	Storm runoff, impervious surfaces, over capacity pipes	Dynamic land use changes, hybrid urban/rural flows	Agricultural runoff, seasonal river overtopping
TYPE OF DRAINAGE SYSTEMS	• Concrete channels • Combined sewer systems • underground pipes	• Open ditches • a mix of sewers and natural waterways	• Unpaved earth ditches • natural streams • culverts
WATER QUALITY & POLLUTION	• Heavy metals, oil • Tire particles	• Industrial waste • domestic sewage, agricultural chemical runoff	• High nutrients (e.g., from manure), pesticides
MAINTENANCE CHALLENGES	• Debris blockage • Accessibility in tight spa • High funding needs	• Trash dumping • Unclear responsibility	• Siltation • Manual clearing requirement
SOCIO-ECONOMIC IMPACT	• Property damage • Traffic disruption • Health risks	• Crop damage • Displacement • Health risks	• Crop loss, • food insecurity • Vector-borne diseases
INSTITUTIONAL & GOVERNANCE	• Centralized management, • Lack of coordination	• Conflicting regulations, • rapid development outpaces planning	• Reliance on local leaders, • Limited resources
TYPICAL SOLUTIONS	• Green roofs, • Permeable pavement, • Engineered systems	• Nature-based solutions, • Managed growth	• Low-tech methods, • Nature-based practices, • Community action

2.10 Diagnostic Framework for Drainage Problems

2.10.1 Need for Systematic Diagnosis

A systematic diagnostic framework is necessary to move from visible symptom to engineering response. Without a structured diagnosis, interventions may treat only the symptom while leaving the controlling cause unresolved. For example, desilting a flooded drain may provide short-term relief if the actual controlling problem is an undersized culvert or submerged outfall.

2.10.2 Basic Diagnostic Sequence

The practical diagnosis sequence involves:

1. Identify the observed problem clearly (e.g., road flooding, persistent waterlogging, foul stagnant flow, culvert overtopping, recurrent blockage).
2. Document its location, duration, rainfall conditions, and spatial extent.
3. Group possible causes into categories: hydraulic, structural, operational, environmental, and institutional.
4. Trace the entire drainage path from inflow area to outfall, examining upstream catchment conditions, intermediate junctions, road crossings, and final outlet performance.
5. Prioritise the causes that control overall system performance rather than those most visible.

2.10.3 Five Guiding Questions

A good diagnosis should answer five key questions:

- What is the exact nature of the problem?
- Where does it occur and under what conditions?
- What factors contribute to it?
- Which cause is primary and which are secondary?
- What intervention is most likely to provide durable improvement?

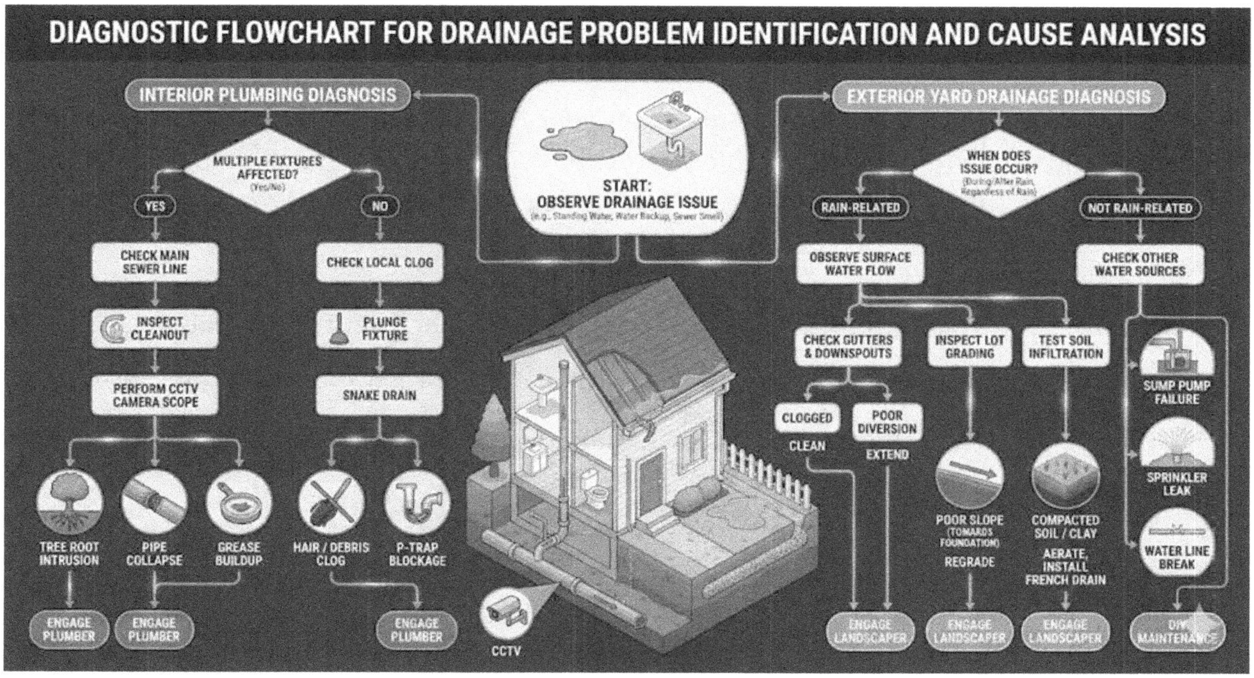

Figure 2.2 Diagnostic flowchart for drainage problem identification and cause analysis

2.11 Indian Case Boxes

Case Box 2.1 Urban inlet blockage and street flooding

In many Indian cities, street flooding develops not only because drains are too small but because inlets and covered sections are blocked by plastic waste, market refuse, road sweepings, and silt. This demonstrates how localized blockage at hydraulic entry points can produce area-wide mobility disruption.

Key lessons:

- Inlet performance is critical to local drainage reliability.
- Solid waste management and drainage maintenance are interdependent.
- Small control-point failures can generate disproportionately large flood effects.

Case Box 2.2 Peri-urban culvert bottleneck

Peri-urban growth often widens roads and raises land levels without proportionate culvert or outfall upgrading. In such areas, former rural flow paths become concentrated at undersized crossings, leading to prolonged upstream waterlogging and repeated overtopping during rainfall events.

Key lessons:

- Transitional areas require drainage reassessment during urban expansion.
- A single undersized crossing may control the performance of a wider drainage network.
- Road development should always be coordinated with catchment drainage continuity.

Case Box 2.3 Village roadside waterlogging

In village settings, roadside waterlogging results from poor local slope, discontinuous drains, household access crossings, and discharge into unsuitable depressions. Though the physical scale may be small, the effect on accessibility, hygiene, and local quality of life may be significant.

Key lessons:

- Simple continuity and grading improvements may solve persistent local problems.
- Community maintenance plays an important role where formal systems are weak.
- Small-scale drainage failures can still have major social effects.

2.12 Planning Implications

Practical implications include:

1. Drainage problems must be treated as multi-causal and system-based rather than isolated local defects.

2. Design, construction quality, maintenance, and solid waste management must all be integrated into drainage planning.
3. Diagnosis should always extend from the inflow area to the outfall.
4. Urban, peri-urban, and village settings require different interpretive frameworks even when visible symptoms appear similar.
5. Drainage engineering should incorporate public health and environmental thinking rather than focusing only on flood conveyance.

2.13 Summary Points

1. Drainage failures may be hydraulic, structural, operational, environmental, or institutional.
2. Flooding, waterlogging, and ponding differ in scale and duration but all indicate inadequate stormwater handling.
3. Sedimentation and blockage reduce effective hydraulic capacity and often reinforce each other.
4. Hydraulic inadequacy may result from undersizing, poor slope, bottlenecks, or changed catchment runoff.
5. Structural defects impair both durability and hydraulic performance.
6. Solid waste is a major drainage engineering issue, not only a behavioral problem.
7. Poor drainage affects public health, mobility, environmental quality, and livelihoods.
8. Diagnosis must be systematic and should trace the entire drainage path.

2.14 References

1. Butler, D., & Davies, J. W. (2017). Urban drainage (4th ed.). CRC Press.
2. Central Public Health and Environmental Engineering Organisation. (2019). Manual on sewerage and sewage treatment systems. Ministry of Housing and Urban Affairs, Government of India.
3. Ministry of Housing and Urban Affairs. (2017). Standard operating procedure (SOP) for urban flooding. Government of India.
4. National Disaster Management Authority. (2010). National guidelines: Management of urban flooding. Government of India.
5. United Nations Environment Programme. (2015). Global waste management outlook.
6. U.S. Environmental Protection Agency. (1999). Preliminary data summary of urban storm water best management practices.
7. World Meteorological Organization, & Global Water Partnership. (2017). Integrated flood management tools series: Management of sediment in floods.
8. Yazdanfar, Z., & Sharma, A. (2015). Urban drainage system planning and design—Challenges with climate change and urbanization: A review. Water Science and Technology, 72(2), 165–179.

Chapter 3

Hydrologic and Hydraulic Fundamentals

Learning Objectives

By the end of this chapter, the learner should be able to:

1. Describe rainfall-runoff processes relevant to drainage engineering.
2. Explain the influence of catchment characteristics on runoff generation.
3. Estimate peak discharge using appropriate basic methods.
4. Apply principles of open channel and closed conduit flow in drainage systems.
5. Interpret the importance of slope, velocity, resistance, and sediment transport in design.

3.1 Introduction

Hydrology and hydraulics form the scientific basis of drainage engineering. Hydrology explains how rainfall is transformed into runoff over a catchment, while hydraulics explains how that runoff moves through drains, channels, pipes, culverts, and storage systems. Together, these two fields allow the engineer to estimate design flows, size drainage components, and evaluate system performance under varying site and storm conditions.

In settlement drainage, hydrologic and hydraulic analysis must always be linked. A runoff estimate is meaningful only if the conveyance system can safely carry it, and a well-sized drain is useful only when the design flow has been estimated realistically. This chapter introduces the principal hydrologic and hydraulic concepts needed for drainage planning and design in urban, semi-urban, and village environments.

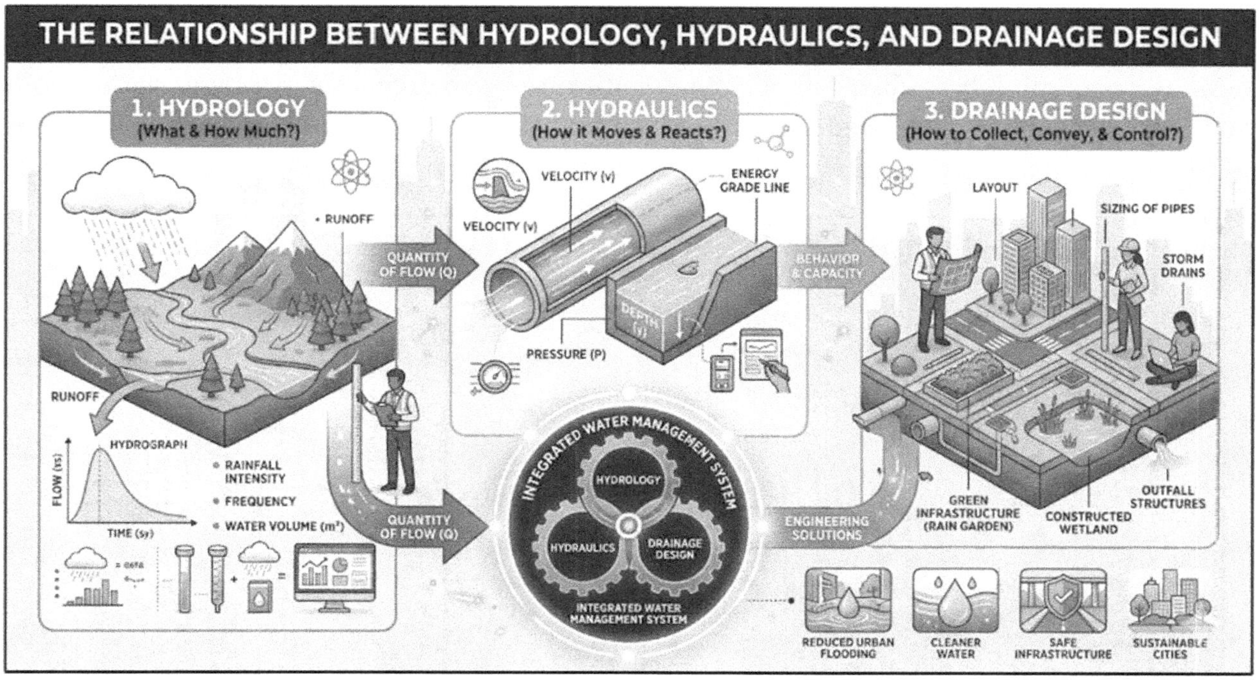

Figure 3.1 Relationship between hydrology, hydraulics, and drainage design

3.2 Hydrologic Cycle and Drainage Relevance

3.2.1 Hydrologic Cycle

The hydrologic cycle is the continuous movement of water through evaporation, condensation, precipitation, infiltration, runoff, storage, and groundwater flow. In drainage engineering, the main concern is the portion of precipitation that appears as excess water on or near the land surface and must be managed to prevent flooding, waterlogging, erosion, and structural damage.

When rainfall reaches the ground, part of it is intercepted by vegetation and evaporates, part infiltrates into the soil, and part is stored temporarily in surface depressions. The remainder flows downslope as overland runoff, enters drains or natural channels, and eventually reaches an outfall or receiving water body.

3.2.2 Relevance to Drainage Engineering

The hydrologic cycle is directly relevant to drainage because it controls the timing, volume, and rate of runoff generation. Any drainage design must account for how much rainfall becomes runoff, how quickly it reaches the drainage system, and how long it remains in transit within the catchment.

Settlement development alters this cycle significantly. Roofs, roads, paved courtyards, and compacted surfaces reduce infiltration and accelerate runoff concentration, while land filling and development may modify natural storage and flow paths. Therefore, drainage engineering requires a practical understanding of how human activities alter the local hydrologic response.

3.2.3 Human Modification of Hydrologic Response

In natural terrain, rainfall is distributed among infiltration, interception, storage, and runoff in a relatively balanced way. In developed areas, that balance shifts toward rapid surface runoff. This is why urban drainage systems require more careful hydrologic estimation than undeveloped catchments.

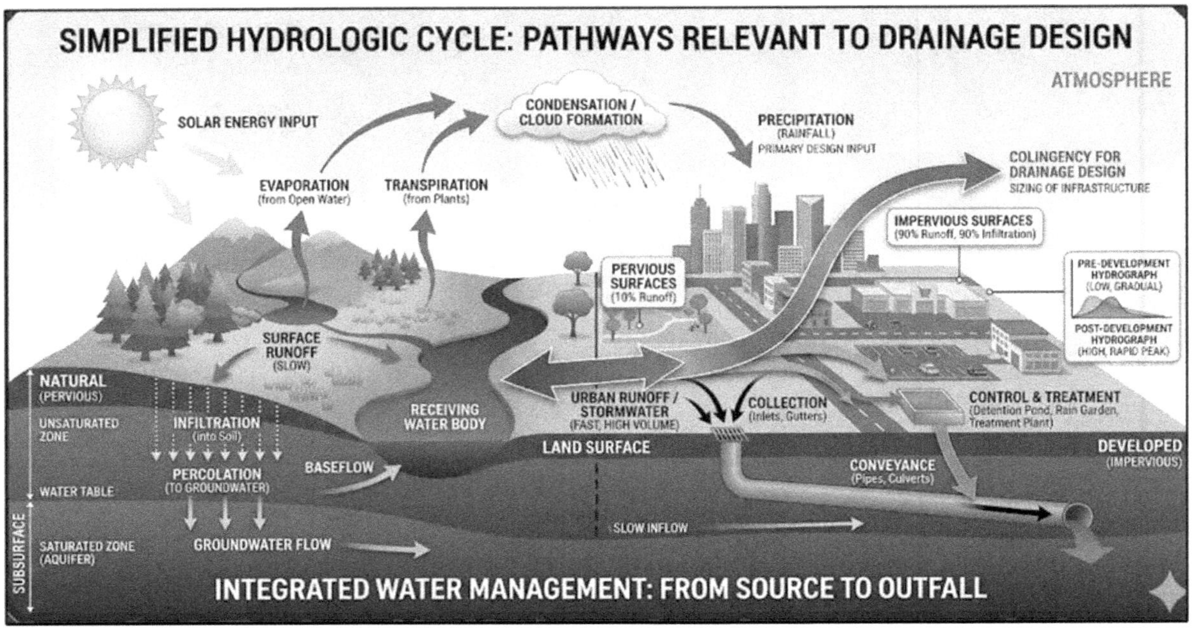

Figure 3.2 Simplified hydrologic cycle showing pathways relevant to drainage design

3.3 Rainfall Characteristics and Storm Events

3.3.1 Rainfall as Design Input

Rainfall is the principal input to most drainage systems. However, not all rainfall events produce the same drainage response. The hydraulic load on a drainage system depends not only on total rainfall depth but also on intensity, duration, temporal distribution, spatial variation, and antecedent moisture condition.

3.3.2 Rainfall Intensity And Duration

Rainfall intensity is especially important in drainage design because short-duration high-intensity storms can generate rapid runoff and high peak flows. A brief, intense storm may create greater peak discharge than a longer but less intense event. Rainfall duration affects both total runoff volume and the extent to which different parts of the catchment contribute to the peak flow.

3.3.3 Storm Types

Storm events may be convective, frontal, cyclonic, or monsoonal in nature, depending on the region. In urban micro-drainage, short-duration intense storms are often critical because they produce rapid surface runoff. In larger peri-urban or rural catchments, longer-duration storms may be more important because they generate sustained flow and larger runoff volumes.

3.3.4 Design Storms and Frequency

For engineering purposes, rainfall data are commonly interpreted through design storms, return periods, and intensity-duration-frequency relationships. These concepts help the engineer select a rainfall event appropriate to the function, service level, and acceptable risk of the drainage system. The selected design storm should reflect both technical considerations and the importance of the infrastructure being protected.

3.4 Runoff Generation Processes

3.4.1 Basic Concept

Runoff is generated when rainfall reaching the ground exceeds the capacity of the land surface and soil to absorb, store, or delay water. The runoff response depends on rainfall characteristics, soil properties, surface condition, slope, land cover, and antecedent wetness.

3.4.2 Infiltration-excess Runoff

Infiltration-excess runoff occurs when rainfall intensity exceeds the infiltration capacity of the soil. This mechanism is common on paved surfaces, compacted ground, bare urban soil, and highly disturbed land. In such cases, runoff begins quickly and may contribute strongly to urban flood peaks.

3.4.3 Saturation-excess Runoff

Saturation-excess runoff occurs when the soil is already saturated and cannot accept additional water. Further rainfall therefore moves across the surface as runoff. This mechanism may be important in low-lying areas, fine-textured soils, depressions, and areas receiving prolonged rainfall.

3.4.4 Runoff in Developed Catchments

In settlements, roofs, road cambers, kerbs, paved courtyards, storm inlets, and drains increase hydraulic connectivity and speed up runoff transfer. Developed areas therefore produce faster and often larger runoff peaks than undeveloped land. Understanding this difference is fundamental to drainage design because inappropriate assumptions about runoff generation can lead to under design or uneconomical overdesign.

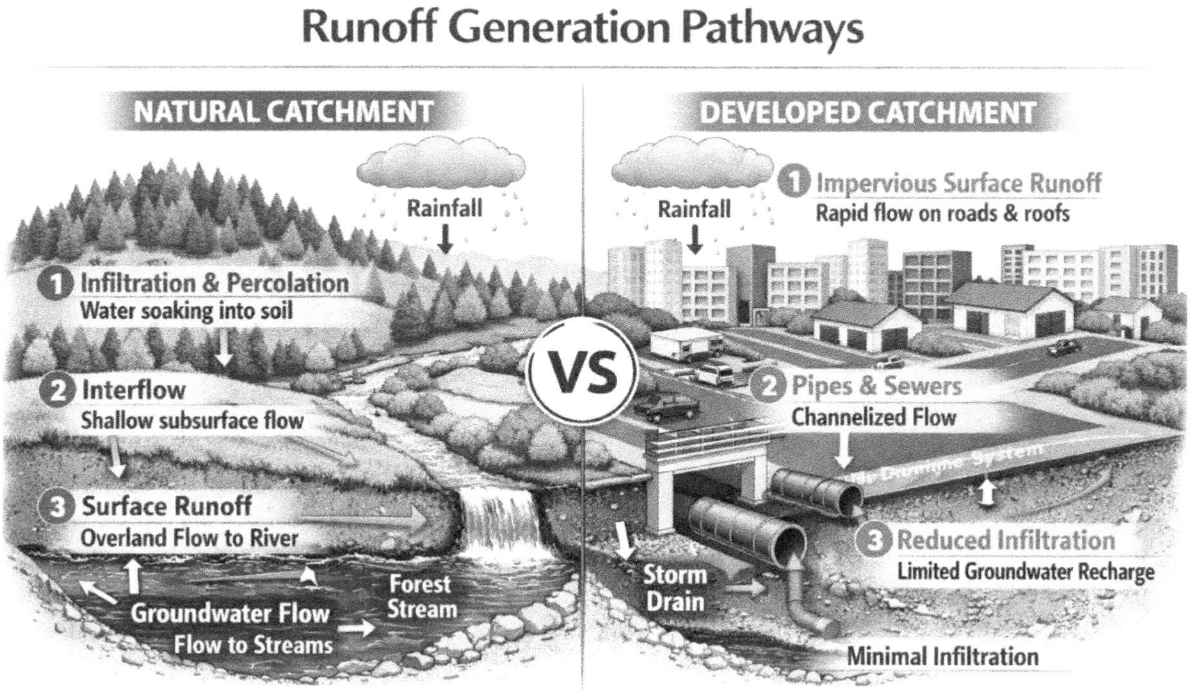

Figure 3.3 Runoff generation pathways in natural and developed catchments

3.5 Catchment Characteristics Affecting Runoff

3.5.1 Catchment Area and Shape

A catchment is the area that drains to a common point. Its hydrologic response depends first on area and shape. In general, a larger catchment produces greater runoff volume, although peak discharge does not always increase proportionally because flow travel times are longer. A compact catchment tends to concentrate runoff more rapidly than an elongated one and may therefore produce a sharper peak flow.

3.5.2 Slope and Relief

Slope affects runoff velocity and concentration time. Steeper catchments usually produce faster runoff movement and higher peak flow rates, while flatter terrain slows drainage but may increase ponding and prolonged water accumulation.

3.5.3 Land Use and Surface Condition

Land use strongly influences runoff behavior. Paved, roofed, and compacted surfaces generate rapid runoff, whereas vegetated and permeable surfaces promote infiltration and temporary storage. Surface roughness and depression storage further affect how quickly runoff moves across the catchment.

3.5.4 Soil type and Drainage Density

Permeable soils absorb more water than clayey, shallow, or compacted soils. Drainage density, defined by the extent of natural or artificial flow paths, also affects runoff response. A catchment with well-developed drainage pathways typically conveys water more rapidly than one with limited connectivity.

3.5.5 Modification by Development

Catchment behavior is often modified by development even without large changes in total area. Plot filling, road construction, boundary walls, and encroachment on natural flow paths can substantially alter runoff response. Drainage design must therefore consider both original and modified catchment characteristics.

Runoff generation within a catchment is influenced by a combination of meteorological factors (precipitation intensity, duration, distribution) and the physical characteristics of the catchment itself.

The physical properties of the drainage basin determine how rain is partitioned into interception, infiltration, evaporation, and surface flow. The following table summarizes key catchment characteristics and their typical influence on the volume and timing of runoff.

Table 3.1 Catchment characteristics and their influence on runoff generation

Catchment Characteristic	**Category / Parameter**	**Primary Influence on Hydrologic Processes**	**Typical Effect on Runoff Generation**
Size (Area)	Large Catchment	Greater storage capacity (in channel, soil, vegetation) and evaporation.	High total runoff volume, but typically lower peak flow *per unit area* due to attenuation.
	Small Catchment	Responds quickly to rainfall intensity; minimal attenuation.	Higher peak flow per unit area; faster response time.
Shape	Fern-shaped (Narrow, elongated)	Long water travel path from the remotest point; different arrival times.	Lower, delayed peak flow; longer baseflow duration.
	Fan-shaped (Circular, compact)	Water travel paths are more uniform; simultaneous arrival of flow.	Higher, sharper peak flow; faster response time.
Topography (Slope)	Steep Slope	High velocity of surface flow; reduced time for infiltration.	Greater runoff volume; faster peak flow arrival (shorter Time of Concentration, T_c).
	Gentle Slope / Flat	Low flow velocity; increased time for infiltration and storage.	Reduced runoff volume; delayed, lower peak flow.
Soil Type & Geology	Clay / Shale / Solid Rock	Low permeability and infiltration capacity; minimal subsurface storage.	High surface runoff volume and quick peaks.

(Contd.)

Catchment Characteristic	Category / Parameter	Primary Influence on Hydrologic Processes	Typical Effect on Runoff Generation
	Sand / Gravel / Fractured Rock	High permeability and infiltration capacity; significant storage.	Reduced surface runoff; significant baseflow contribution from groundwater.
Land Use / Land Cover	Urban (Impervious surfaces)	Negligible infiltration; smooth surfaces increase flow velocity.	Extremely high runoff volume and very fast, high peak flows.
	Forest / Densely Vegetated	High interception by canopy; leaf litter increases roughness and infiltration.	Significantly reduced runoff volume and delayed peak flow.
	Agriculture (Cropped)	Moderate infiltration; contouring can reduce flow velocity; bare soil increases runoff.	Variable runoff depending on soil management and crop type; higher than forested areas.
Drainage Density	High Density (Many streams)	Efficient water collection and transport system.	Fast response time; steeper hydrograph rising and falling limbs.
	Low Density (Few streams)	Water travels longer as overland flow.	Slower response time; flatter hydrograph shape.

3.6 Time of Concentration

3.6.1 Definition

Time of concentration is the time required for runoff from the hydraulically most distant point of a catchment to reach the outlet or design point. It is one of the most important parameters in drainage hydrology because it helps determine the storm duration that produces the critical peak discharge.

3.6.2 Engineering Significance

If rainfall duration is shorter than the time of concentration, not all parts of the catchment contribute simultaneously to peak flow at the outlet. If rainfall duration equals or exceeds the time of concentration, the entire catchment is assumed to contribute, and simplified design methods such as the rational method reach their critical condition.

3.6.3 Factors Affecting Time of Concentration

Time of concentration depends on flow path length, slope, roughness, ponding, and the nature of the flow path, whether it occurs as sheet flow, shallow concentrated flow, gutter flow, drain flow, or channel flow. Urban surfaces and storm drains often shorten travel time, while rougher surfaces and shallow depressions may lengthen it.

3.6.4 Importance of Accurate Estimation

Accurate estimation of time of concentration is important because it directly affects the rainfall intensity selected for design. If it is underestimated, the chosen intensity may be too high and the resulting design may become uneconomical. If it is overestimated, the selected intensity may be too low and the drainage system may be under designed.

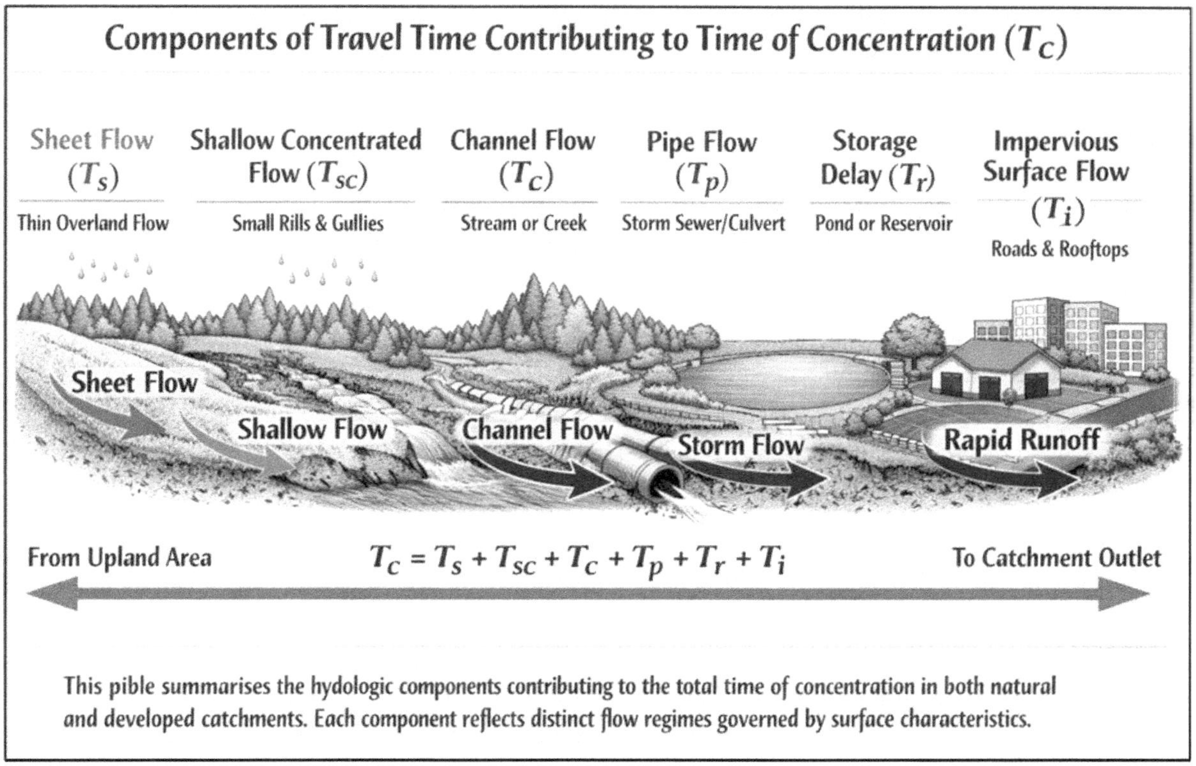

Figure 3.4 Components of travel time contributing to time of concentration

3.7 Estimation of Peak Discharge

3.7.1 Peak Discharge as Design Flow

Peak discharge is the maximum rate of runoff expected at a given location during a storm event. It is the key design quantity used to size drains, culverts, pipes, storm sewers, and outfall structures.

3.7.2 Rational Method

For small urban catchments, the rational method is one of the most widely used approaches. In general form, it relates peak runoff to runoff coefficient, rainfall intensity, and catchment area. The rainfall intensity is usually selected for a duration equal to the time of concentration and for a suitable return period.

The runoff coefficient represents the fraction of rainfall that becomes runoff and reflects land use, surface condition, and imperviousness. The rational method is especially useful for small, relatively homogeneous catchments such as road drainage areas, parking areas, and localized urban surface drainage systems.

3.7.3 Other Methods

For larger catchments or more complex conditions, peak discharge may be estimated using hydrograph methods, empirical regional equations, unit hydrograph techniques, or hydrologic simulation models. These methods account more explicitly for runoff timing, storage, and storm distribution.

3.7.4 Practical Caution

Peak discharge estimation must always reflect current and likely future catchment conditions. Urbanisation, land-use change, blocked flow paths, and reduced infiltration can significantly increase peak flow, so reliance on outdated assumptions may produce unsafe designs.

Table 3.2 Common peak discharge estimation approaches and typical applications

Approach	Basic idea	Typical applications	Advantages and limitations
Rational Method	Estimates peak discharge from rainfall intensity, drainage area, and runoff coefficient .	Small urban catchments, storm sewers, inlets, culverts, and other minor drainage structures.	**Advantages:** Simple, quick, widely used for small basins. **Limitations:** Best only for small, fairly uniform catchments; weak for large basins or complex routing.
Empirical formulae	Uses basin area and region-specific constants or exponents to estimate peak flow.	Quick estimates in regions with established local formulas, preliminary planning, and small basins.	**Advantages:** Fast and easy when local equations exist. **Limitations:** Site- and region-specific; may not transfer well elsewhere.
Flood frequency analysis	Fits statistical distributions to observed streamflow records to estimate discharges for selected return periods.	Bridge and culvert design, floodplain mapping, and gaged-river design studies.	**Advantages:** Based on measured flow data; useful for return-period estimates. **Limitations:** Needs long, reliable records; uncertain in short or altered records.
Unit hydrograph method	Converts rainfall excess into a runoff hydrograph to estimate peak flow and timing.	Watershed response studies, design storms, and medium to larger catchments where runoff timing matters.	**Advantages:** Captures runoff timing and hydrograph shape. **Limitations:** Requires more input data and assumptions; less suited to highly variable basins without calibration.

Approach	Basic idea	Typical applications	Advantages and limitations
NRCS/SCS methods	Uses curve number, runoff depth, and watershed lag/time of concentration to derive peak discharge.	Rural and mixed watersheds, ungaged basins, and hydrologic design using TR-55/TR-20 style procedures.	**Advantages:** Practical for ungaged basins; widely accepted in design. **Limitations:** Sensitive to curve number and land-use assumptions; can be less accurate in highly urbanized or unusual conditions.
Envelope curves	Uses regional envelope relationships that bound historical peak flows.	Screening-level flood estimation and regional hydrologic assessments where limited data are available.	**Advantages:** Useful where data are scarce; gives conservative upper-bound estimates. **Limitations:** Not a precise prediction method; mainly for broad regional screening.
Slope-area or hydraulic methods	Reconstructs flood discharge from channel geometry, slope, and flow marks after an event.	Post-flood investigations, historical flood estimation, and sites without direct gage data.	**Advantages:** Valuable when no direct measurements exist. **Limitations:** Sensitive to field survey accuracy and assumptions about flow conditions.
Modified rational / calibrated models	Adjusts simple methods with storage, calibration, or local observations.	Urban drainage networks, detention design, and projects where local performance data are available.	**Advantages:** Can better reflect local conditions than a basic formula. **Limitations:** Needs calibration data and may still be limited in transferability.

3.8 Fundamentals of Open Channel Flow

3.8.1 Meaning of Open Channel Flow

Open channel flow refers to the movement of water with a free surface exposed to atmospheric pressure. Most roadside drains, natural channels, lined ditches, and trapezoidal drains operate under open channel conditions.

3.8.2 Governing Factors

Open channel flow behavior depends on gravity, channel shape, slope, boundary roughness, and flow depth. It may be uniform or non-uniform, steady or unsteady. In routine drainage design, uniform steady flow is often used as an approximation for sizing channels, with the energy slope, bed slope, and water-surface slope treated as approximately equal over the design reach.

3.8.3 Hydraulic Properties

Important hydraulic properties include flow area, wetted perimeter, hydraulic radius, top width, and depth. These properties determine the section's capacity and velocity. Common open sections include rectangular, trapezoidal, triangular, and compound forms.

3.8.4 Engineering Significance

The choice of open drain section depends on space, expected discharge, maintenance needs, lining material, public safety, and settlement context. Open drains are often economical and easy to inspect, but they also require regular maintenance and may create nuisance or safety issues in dense urban areas.

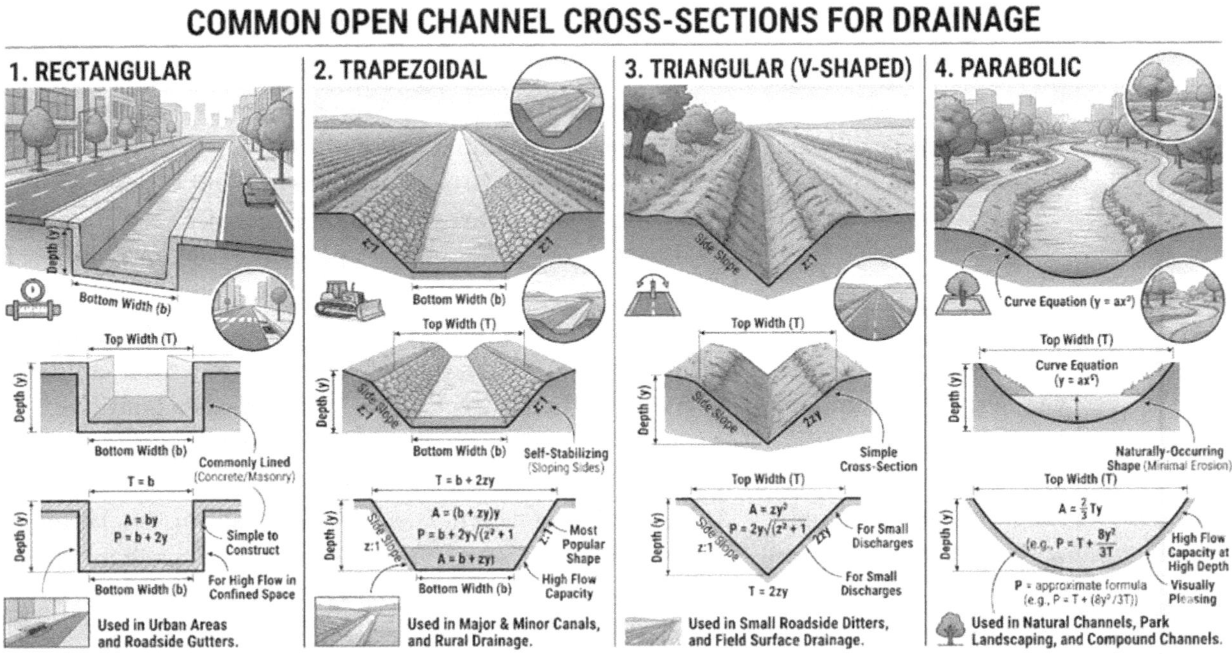

Figure 3.5 Common open channel cross-sections used in drainage systems

3.9 Flow in Closed Conduits

3.9.1 Meaning of Closed Conduit Flow

Closed conduit flow occurs when water moves through pipes, covered drains, and storm sewers. Such systems may flow partially full like open channels or full under pressure, depending on hydraulic conditions.

3.9.2 Gravity and Surcharged Conditions

In gravity drainage systems, design is usually based on partially full or full-flow conditions without sustained internal pressure. However, surcharge may occur during major storms when inflow exceeds normal carrying capacity. Under such conditions, head losses, junction behavior, and upstream water levels require careful consideration.

3.9.3 Use in Urban Settings

Closed conduits are commonly used in dense urban areas where open drains are impractical because of limited space, traffic demands, safety, or aesthetic concerns. They are also used at crossings, beneath roads, and where continuity under infrastructure is required.

3.9.4 Maintenance Considerations

Because closed systems are not directly visible, they require better access for inspection and cleaning. They can be hydraulically efficient where properly designed, but they are also more vulnerable to hidden blockage, surcharge, and maintenance neglect.

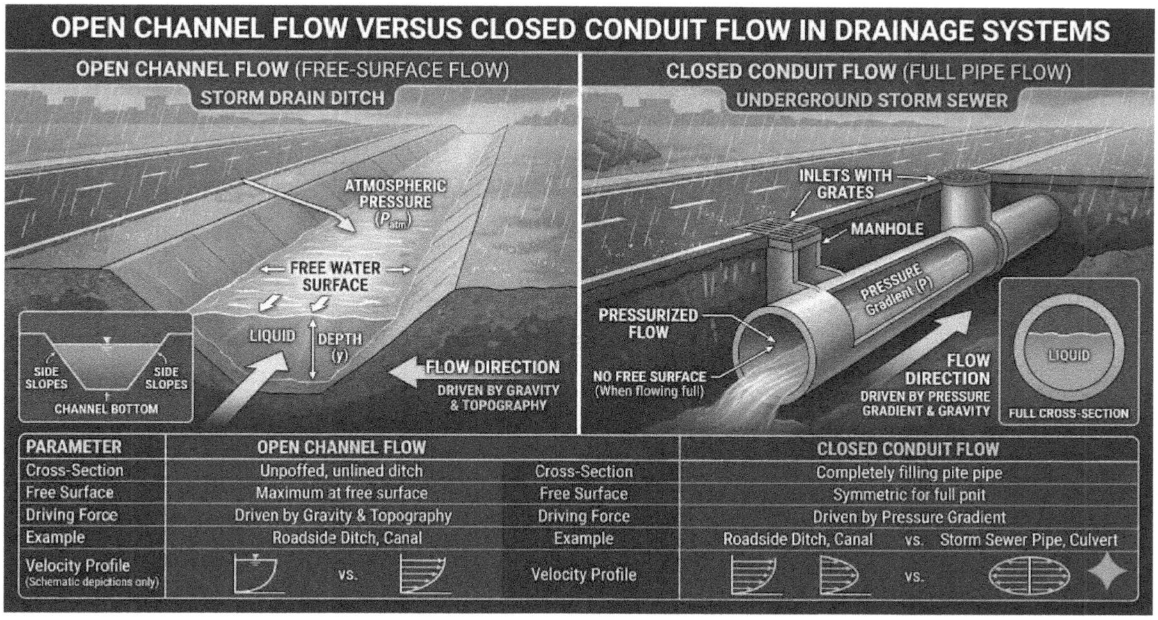

Figure 3.6 Open channel flow versus closed conduit flow in drainage systems

3.10 Flow Resistance, Slope, and Velocity

3.10.1 Flow Resistance

Water velocity in a drain or pipe is controlled by the slope of the energy grade line, hydraulic radius or diameter, and boundary resistance. Resistance arises from friction along the bed and walls, as well as from irregularities, vegetation, bends, and junctions.

3.10.2 Effect of Slope

A steeper slope generally increases flow velocity and carrying capacity. However, excessive velocity may produce erosion, scour, structural damage, and unsafe outlet conditions. Very low slope may result in sluggish flow, sedimentation, and stagnation.

3.10.3 Roughness and Hydraulic Efficiency

Rougher surfaces such as earth drains with vegetation have higher resistance and lower velocity for the same depth and slope, while smoother pipe materials provide lower resistance and greater carrying efficiency. Hydraulic resistance in open channels is often represented using roughness coefficients such as Manning's factor.

3.10.4 Balancing Velocity

In practical design, slope and velocity must always be considered together. Minimum velocities are often checked for self-cleansing, while maximum permissible velocities are checked to prevent erosion or structural damage. Good design therefore seeks hydraulic adequacy without creating instability.

Table 3.3 Influence of slope, roughness, and section properties on flow velocity

Parameter	Symbol	Effect on Flow Velocity	Physical Interpretation	Design Implications
Channel Slope	(S)	Velocity increases with slope; steeper slopes yield higher gravitational acceleration.	Represents energy gradient driving flow; higher slope → greater potential energy converted to kinetic energy.	Steeper slopes improve conveyance but risk erosion and structural instability; flatter slopes reduce velocity, requiring larger cross-sections to maintain capacity.
Surface Roughness	(n) (Manning's coefficient)	Velocity decreases with increasing roughness; rougher surfaces dissipate more energy.	Indicates resistance due to boundary friction; vegetation, debris, or rough lining increase (n).	Smooth linings (concrete, PVC) enhance velocity and reduce sediment deposition; rough natural channels slow flow, promote infiltration, and improve ecological function but reduce hydraulic efficiency.
Hydraulic Radius	(R = A/P)	Velocity increases with larger hydraulic radius; deeper or wider sections reduce relative friction.	Ratio of flow area to wetted perimeter; higher (R) means less boundary contact per unit area.	Larger hydraulic radius improves efficiency and reduces head losses; trapezoidal or circular sections are preferred for maximizing (R) and minimizing maintenance.
Flow Depth	(y)	Velocity generally increases with depth up to uniform flow conditions.	Greater depth increases hydraulic radius and reduces relative roughness effects.	Adequate depth ensures channel stability and capacity; excessive depth may cause overtopping, instability, or safety hazards, requiring structural reinforcement.

Parameter	Symbol	Effect on Flow Velocity	Physical Interpretation	Design Implications
Channel Shape	—	Velocity varies with geometry; circular and trapezoidal sections are more efficient than rectangular.	Determines distribution of shear stress and wetted perimeter.	Circular and trapezoidal sections provide self-cleansing velocities and minimize sediment deposition; rectangular sections are easier to construct but less hydraulically efficient.

3.11 Sediment Transport and Self-cleansing Conditions

3.11.1 Sediment in Drainage Systems

Drainage systems rarely carry only clear water. Runoff usually contains silt, sand, organic matter, grit, debris, and other particles from roads, exposed soil, construction areas, and unpaved surfaces. Whether this material is transported or deposited depends on velocity, turbulence, slope, particle size, and boundary condition.

3.11.2 Deposition and Transport

If flow velocity is too low, sediment settles and accumulates within the drain, reducing hydraulic capacity and increasing maintenance needs. If the velocity is high enough, deposited material may be re-entrained and carried downstream.

3.11.3 Self-cleansing Concept

The chapter source identifies self-cleansing velocity as a key idea in drainage design. It represents a minimum velocity intended to reduce the likelihood of persistent sediment deposition under expected operating conditions. This concept is important in both open drains and storm sewers because frequent deposition can make otherwise adequate systems ineffective.

3.11.4 Practical Limitations

Self-cleansing must be applied carefully. A velocity sufficient for fine sediment may still be too low for coarse particles, and a velocity that is safe in a lined drain may cause erosion in an unlined earth channel. In addition, storm drains often carry intermittent rather than continuous flow, so complete self-cleansing cannot always be achieved.

3.11.5 Design Implication

Practical drainage design should therefore aim to reduce sediment entry, provide suitable slopes and section geometry, and ensure access for periodic cleaning. Hydraulic design and maintenance planning must work together to manage sediment-related failure.

3.12 Indian Context Notes

3.12.1 Urban Drainage Relevance

In Indian cities, high-intensity monsoon rainfall, rapid surface sealing, mixed land use, and variable maintenance conditions make hydrologic and hydraulic understanding especially important. Urban catchments often experience shortened runoff response times because roofs, roads, and paved courtyards accelerate flow toward drains.

3.12.2 Peri-urban Complexity

In peri-urban environments, catchment behavior is often transitional. Agricultural land, open plots, roads, drains, and developing layouts may coexist within the same drainage area, making runoff estimation more complex than in fully urbanized terrain.

3.12.3 Village and Small-town Settings

In village and small-town drainage, rough surfaces, shallow roadside drains, local depressions, and discontinuous flow paths may lengthen travel time and alter runoff concentration patterns. The same design concept therefore requires different interpretation depending on settlement context.

3.13 Planning and Design Implications

1. The drainage design should never be separated into hydrology alone or hydraulics alone. Peak runoff estimation, time of concentration, channel capacity, pipe behaviour, slope selection, and sediment management are interdependent.
2. The simplified methods must be used with sound judgment. Small urban catchments may be suitable for rational-method-based estimates, but larger or more complex systems may require more detailed hydrologic treatment. Similarly, hydraulic design must account not only for section capacity but also for maintenance practicality and sediment behaviour.

3.14 Summary Points

1. Hydrology explains how rainfall becomes runoff, while hydraulics explains how runoff moves through drainage systems.
2. The hydrologic cycle is relevant to drainage because it governs runoff generation, timing, and storage.
3. Rainfall intensity, duration, distribution, and antecedent condition all influence drainage response.
4. Runoff is commonly generated through infiltration-excess or saturation-excess mechanisms.
5. Catchment characteristics such as area, shape, slope, soil type, land use, and drainage density strongly affect runoff volume and timing.

6. Time of concentration is a critical parameter for estimating peak discharge and selecting design rainfall intensity.
7. Peak discharge estimation methods include the rational method, empirical formulas, flood frequency analysis, unit hydrograph techniques, NRCS/SCS methods, envelope curves, hydraulic reconstruction, and modified/calibrated models.
8. Open channel flow depends on gravity, slope, roughness, and section geometry, while closed conduit flow occurs in pipes and storm sewers under gravity or surcharged conditions.
9. Flow resistance, slope, roughness, hydraulic radius, and depth all influence velocity and must be balanced to avoid erosion or sedimentation.
10. Sediment transport and self-cleansing velocity are essential considerations for maintaining hydraulic efficiency and reducing maintenance needs.
11. In the Indian context, urban, peri-urban, and village drainage systems require different interpretations of hydrologic and hydraulic principles due to varied land use and surface conditions.
12. Effective drainage planning integrates hydrology and hydraulics, applying simplified methods with judgment and ensuring designs account for both technical adequacy and maintenance practicality.

3.15 References

1. Butler, D., & Davies, J. W. (n.d.). Urban drainage. CRC Press.
2. Chow, V. T. (n.d.). Open-channel hydraulics. McGraw-Hill.
3. Subramanya, K. (n.d.). Engineering hydrology. McGraw-Hill.
4. Department of Transportation. (n.d.). Rational method guidance.
5. North Central Texas Council of Governments. (n.d.). Hydrology manuals for urban drainage applications in small impervious catchments.
6. International Water Association Publishing. (n.d.). Studies on time of concentration estimation in urban drainage systems.
7. American Society of Civil Engineers. (n.d.). Research on self-cleansing drainage design and sediment transport in storm sewers.
8. National Library of Medicine. (n.d.). Review literature on urban drainage planning under urbanization and changing rainfall conditions.

Chapter 4

Investigation and Diagnosis of Drainage Problems

Learning Objectives

By the end of this chapter, the learner should be able to:

1. Conduct a preliminary engineering assessment of a drainage system.
2. Identify key field indicators of hydraulic and structural problems.
3. Diagnose likely causes of recurring drainage failure.
4. Prioritise areas requiring detailed investigation or intervention.
5. Apply systematic field-based methods for mapping drainage paths, flood-prone zones, and problem structures.

4.1 Introduction

Effective drainage improvement begins with correct diagnosis. A drainage system may fail because of insufficient capacity, poor alignment, inadequate outfall, structural deterioration, blockage, encroachment, or weak maintenance, but the visible symptom alone rarely reveals the true cause. For this reason, investigation and diagnosis are among the most important stages in drainage engineering practice.

Drainage failures often present themselves through symptoms such as flooding, waterlogging, sediment accumulation, foul stagnant flow, or repeated overtopping at a particular location. However, the controlling cause may lie upstream, downstream, or within the broader infrastructure context rather than at the visible point of failure. A technically sound response, therefore, begins with systematic field observation, evidence gathering, and interpretation.

This chapter explains how drainage problems should be identified, examined, and interpreted in urban, semi-urban, and village settings. It emphasises field-based assessment, structured inspection, and root-cause diagnosis so that later planning, design, rehabilitation, and management decisions are technically justified and prioritised effectively.

4.2 Purpose of Drainage Investigation

4.2.1 Why is a Drainage Investigation Necessary

The purpose of drainage investigation is to understand how an existing drainage system performs, where it fails, why it fails, and what type of corrective action is required. Investigation provides the factual basis for repair, redesign, upgrading, rehabilitation, or complete redevelopment of drainage infrastructure.

A proper investigation does not focus only on the visible problem location. It studies the contributing catchment, the route of runoff, intermediate drains and structures, and the ultimate outfall. Flooding at one point may actually be caused by an obstruction far downstream, while stagnant flow may result from poor slope, encroachment, or wastewater intrusion rather than rainfall runoff alone.

4.2.2 Distinguishing Temporary and Recurring Problems

Drainage investigation also helps distinguish between temporary and recurring failures. Some problems appear only during unusually intense rainfall events, while others occur repeatedly after moderate storms because of chronic deficiencies. Understanding frequency, duration, severity, and trigger conditions is essential for selecting a realistic engineering response.

4.2.3 Role in Budgeting and Prioritization

Many settlements face multiple drainage problems but limited resources. A systematic investigation helps identify which interventions are urgent, which are preventive, and which can be phased over time. This makes investigation not only a technical activity but also a practical basis for infrastructure planning and budget allocation.

4.3 Reconnaissance Survey and Preliminary Assessment

4.3.1 Nature of Reconnaissance Survey

A reconnaissance survey is the first field-level examination of a drainage area. Its purpose is to develop a broad understanding of site conditions before detailed measurement, survey, or hydraulic analysis is undertaken. Although it is usually rapid, it is often highly informative and can define the direction of the entire investigation.

4.3.2 What to Observe During Reconnaissance

During reconnaissance, the engineer should observe topography, road levels, visible drains, ponding points, culvert locations, low-lying plots, outlet conditions, and visible signs of distress such as erosion, siltation, wall damage, foul flow, and accumulated debris. Local residents can also provide valuable information on flood depth, flood duration, seasonal patterns, previous interventions, and the history of recurring failures.

4.3.3 Preliminary Interpretation

A preliminary assessment should identify whether the issue appears primarily hydraulic, structural, operational, environmental, or planning-related. For example, ponding near a structurally intact drain filled with silt and waste may suggest maintenance deficiency, while a cracked and displaced drain may indicate structural instability. If no effective outlet is visible, the problem may be related to system layout or outfall inadequacy.

4.3.4 Defining the Scope of Further Investigation

Reconnaissance also helps define the limits of detailed investigation. It indicates where survey levels are needed, which structures require closer inspection, what data should be collected, and whether the problem extends beyond the immediate site. This prevents later analysis from being confined too narrowly to the visible symptom.

Table 4.1 Suggested reconnaissance checklist for drainage investigation

S. No.	Item / Parameter	Details to Observe / Record	Remarks
1	Area Identification	City name, ward/zone, street/locality, GPS coordinates	
2	Land Use Characteristics	Residential, commercial, industrial, mixed-use density	
3	Impervious Surface	% paved area (roads, rooftops, parking)	
4	Road Network	Road width, camber, condition affecting runoff	
5	Existing Drain Type	Open drains, closed conduits, stormwater pipes	
6	Drain Dimensions	Width, depth, slope, material (RCC, brick, etc.)	
7	Drain Connectivity	Continuity of network, missing links, dead ends	
8	Flow Direction	Confirm actual flow vs. designed flow direction	
9	Inlets/Catch Basins	Spacing, condition, blockage status	
10	Manholes	Location, spacing, accessibility, condition	
11	Waterlogging Spots	Low-lying areas, duration after rainfall	
12	Encroachments	Buildings, vendors, illegal covers over drains	
13	Solid Waste Accumulation	Garbage, plastics, debris inside drains	
14	Silt Deposition	Sediment level reducing carrying capacity	
15	Outfall Condition	Discharge into river/nala/lake, backflow risk	
16	Pumping Stations	Capacity, operational condition, power backup	
17	Rainfall Intensity	Local rainfall pattern (short intense storms)	
18	Flood History	Past urban flooding events and affected areas	
19	Utility Interference	Water supply lines, sewer lines, cables crossing drains	
20	Sewer Interaction	Cross-connections or mixing of sewage & stormwater	
21	Green Infrastructure	Parks, recharge pits, permeable surfaces	
22	Maintenance Practices	Cleaning frequency, responsible authority	
23	Public Complaints	Frequent issues reported by residents	
24	Traffic Impact	Flooding impact on traffic movement	
25	Safety Concerns	Open drains, damaged covers, accident risks	
26	Climate Consideration	Urban heat, extreme rainfall trends	
27	Photographic Documentation	Photos of critical points and problem areas	
28	Immediate Issues	Urgent problems needing quick action	
29	Improvement Scope	Desilting, widening, redesign, smart drainage	
30	Remarks/Recommendations	Short-term and long-term measures	

4.4 Field Identification of Drainage Paths

4.4.1 Meaning of Drainage Path

A drainage path is the route followed by runoff from the point of generation to the final outfall. Identifying drainage paths is one of the most critical tasks in diagnosing drainage failure because water follows the easiest available route, whether or not that route matches the intended engineering layout.

4.4.2 Designed Versus Actual Drainage Path

The designed path is the route the drainage system is supposed to use; the actual path is the route runoff follows in real storm conditions. Differences between the two often reveal the source of failure.

In semi-urban and village settings especially, the actual path may deviate because of road humps, plot filling, blocked drains, wall construction, farm bunds, or natural depressions. Water may bypass formal drains and move through lanes, courtyards, vacant land, or adjoining fields.

4.4.3 Field Indicators of Runoff Routes

Field indicators that help identify drainage paths include sediment trails, flow marks, erosion lines, water stains on walls, vegetation patterns, deposited debris, and resident observations. In many cases, these indicators reveal routes that are not apparent from plan drawings alone.

4.4.4 Supporting Methods

Drainage-path identification is often strengthened by contour analysis, road profile examination, and local flood mapping. In more advanced studies, field observations may also be combined with GIS layers or elevation models to verify runoff routes and locate hydraulic inconsistencies.

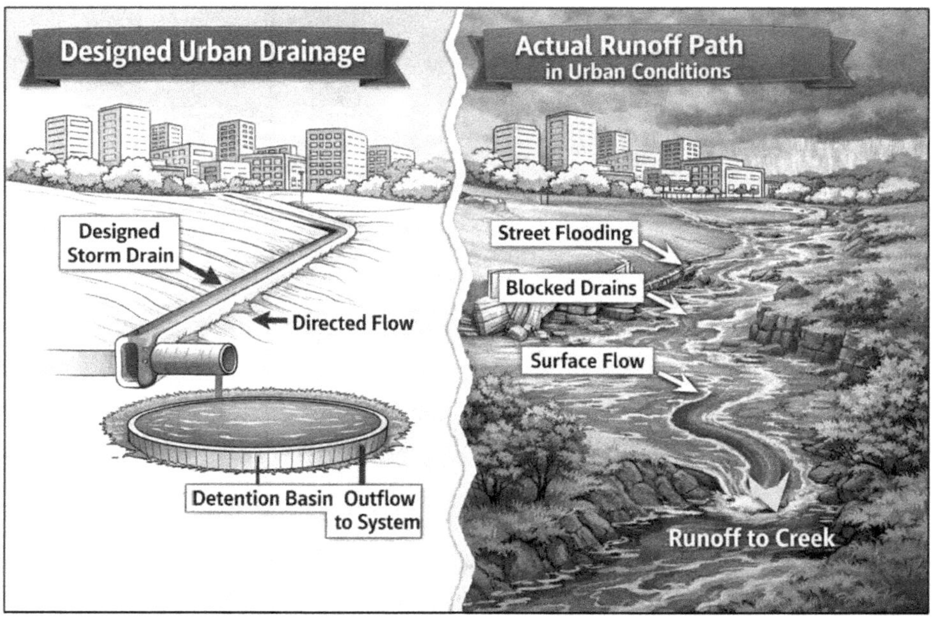

Figure 4.1 Designed drainage path versus actual runoff path under field conditions

4.5 Mapping of Low-Lying and Flood-Prone Zones

4.5.1 Importance of Flood-prone Mapping

Low-lying and flood-prone areas are locations where runoff tends to accumulate because the ground is lower than surrounding land or because discharge is hydraulically restricted. Mapping such areas is essential for understanding the spatial extent of drainage problems and planning effective interventions.

4.5.2 Causes of Low-lying and Flood-prone Conditions

Such zones may arise naturally because of terrain, but they may also be created artificially through development. Common causes include excavation, differential settlement, repeated road raising, plot filling around older properties, obstruction of natural channels, and poor site grading.

4.5.3 What the Map Should Record

Flood-prone mapping should record location, areal extent, depth, duration, and probable frequency of flooding. It should also identify critical receptors such as houses, schools, hospitals, roads, markets, and utility corridors. This helps distinguish nuisance ponding from serious infrastructure risk.

4.5.4 Mapping Methods

Flood-prone areas may be mapped using field sketches, survey levels, GIS layers, community reporting, or historical flood records. Even a simple map showing low points, drainage directions, blocked structures, and outfall limitations can provide major insight during diagnosis. More advanced flood-inundation mapping may support prioritization and emergency planning where resources permit.

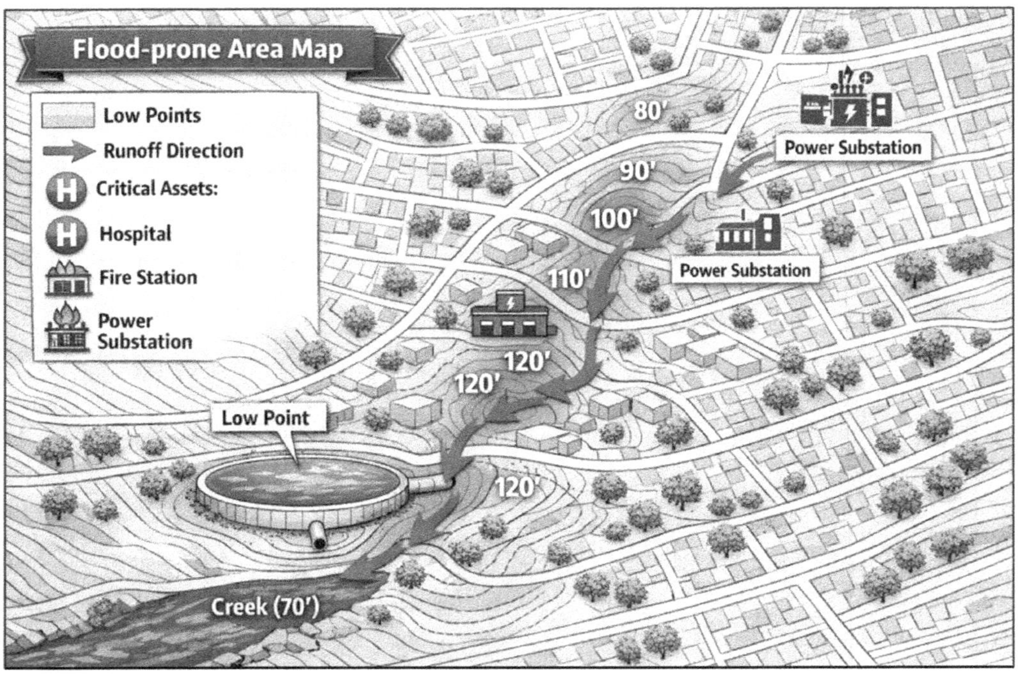

Figure 4.2 Example flood-prone area map with low points, runoff direction, and critical assets

4.6 Inspection of Drains, Culverts, and Outfalls

4.6.1 Scope of Inspection

Drainage inspection involves direct examination of the physical condition and hydraulic function of system components. The most common elements inspected are open drains, covered drains, stormwater pipes, culverts, chambers, inlets, and outfalls.

4.6.2 Inspection of Open Drains and Covered Drains

For open drains, inspection should record section shape, dimensions, lining type, slope continuity, bed condition, sediment depth, vegetation growth, damage to walls, illegal crossings, and visible obstructions. For covered drains and pipes, access chambers, manholes, and inspection openings become essential because internal defects are not directly visible.

Signs such as surcharge, overflow marks, odor, repeated blockage, backwater staining, or local settlement may indicate hidden internal problems. These observations are important in dense urban areas where the system is largely concealed.

4.6.3 Inspection of Culverts

Culverts require special attention because they often act as control points. Inspection should include inlet blockage, barrel condition, alignment, headwall stability, scour at the outlet, sediment deposition, and adequacy of waterway opening. Even minor blockage or local damage at a culvert may significantly affect upstream drainage performance.

4.6.4 Inspection of Outfalls

Outfalls should be checked for submergence, erosion, backflow, structural damage, sediment deposition, and whether they discharge into an effective receiving system. A well-maintained upstream drain may still perform badly if the outfall is submerged or hydraulically restricted.

4.6.5 Dry-weather and Wet-weather Inspection

The chapter source recommends that inspection in both dry-weather and wet-weather conditions is recommended. Some defects are visible only when water is flowing, such as surcharge, reverse flow, backwater effects, or outlet submergence. A drain that appears satisfactory in dry conditions may fail completely during rainfall.

Table 4.2 Inspection items for drains, culverts, chambers, and outfalls

S. No.	Component	Inspection Item	Details to Check / Observe	Condition Rating (Good/ Fair/Poor/Critical)
1	Drains	Physical Condition	Cracks, damage, deformation of lining	
2	Drains	Blockage	Presence of debris, silt, garbage	
3	Drains	Flow Condition	Free flow / stagnant water / reverse flow	
4	Drains	Capacity	Adequacy during rainfall, overtopping signs	
5	Drains	Slope/Gradient	Proper slope for self-cleansing velocity	
6	Drains	Encroachment	Structures obstructing drain alignment	
7	Drains	Vegetation Growth	Weeds, roots affecting flow	
8	Drains	Lining Condition	Erosion, missing lining, leakage	
9	Culverts	Structural Integrity	Cracks, settlement, joint failure	
10	Culverts	Inlet Condition	Blockage, erosion, debris accumulation	
11	Culverts	Outlet Condition	Scour, erosion, sediment deposition	
12	Culverts	Barrel Condition	Obstruction, sedimentation inside barrel	
13	Culverts	Headwalls/Wing Walls	Damage, displacement, cracks	
14	Culverts	Hydraulic Performance	Adequacy during peak flow	
15	Chambers	Structural Condition	Cracks, leakage, corrosion	
16	Chambers	Cover Condition	Missing, damaged, improperly seated covers	
17	Chambers	Accessibility	Ease of access for maintenance	
18	Chambers	Silt Level	Accumulated sediment inside chamber	
19	Chambers	Flow Condition	Smooth entry/exit of flow	
20	Chambers	Odour/Contamination	Signs of sewage mixing or foul smell	
21	Outfalls	Discharge Condition	Free discharge or submerged outlet	
22	Outfalls	Backflow Risk	Tidal or floodwater backflow possibility	

S. No.	Component	Inspection Item	Details to Check / Observe	Condition Rating (Good/ Fair/Poor/Critical)
23	Outfalls	Erosion/Scour	Soil erosion near outlet point	
24	Outfalls	Sediment Deposition	Silt accumulation at outlet	
25	Outfalls	Protection Works	Condition of apron, riprap, retaining structures	
26	Outfalls	Environmental Impact	Pollution, contamination of receiving water body	
27	Outfalls	Obstructions	Waste, vegetation blocking discharge	
28	All Components	Safety	Open hazards, risk to public/ traffic	
29	All Components	Maintenance Status	Cleaning frequency, last maintenance date	
30	All Components	Photographic Record	Photos for documentation and reporting	

4.7 Assessment of Structural Condition

4.7.1 Objective of Structural Assessment

Structural condition assessment evaluates the integrity, stability, and serviceability of drainage assets. Its purpose is to determine whether the structures are capable of performing safely under expected hydraulic and loading conditions.

4.7.2 Common Structural Defects

Typical structural defects include cracked masonry, broken concrete, displaced slabs, spalled surfaces, joint failure, pipe deformation, wall bulging, settlement, undermining, corrosion, and edge collapse. In culverts and chambers, distress may also appear as headwall cracking, apron failure, ring displacement, and slab sagging.

4.7.3 Linking Defect, Cause, and Consequence

Structural assessment should consider not only the visible defect but also its cause and consequence. A crack may be superficial or may indicate foundation movement. A displaced slab may be local or may reflect broader settlement of the chamber structure. Erosion at the toe of a drain may eventually cause collapse if left untreated.

4.7.4 Condition Rating

A useful field approach is to classify condition into categories such as good, fair, poor, and critical. This supports prioritization of repairs and helps convert field observations into decision-ready information.

However, structural assessment should always be interpreted alongside hydraulic assessment because a structurally damaged drain may still convey flow, while an intact drain may fail hydraulically.

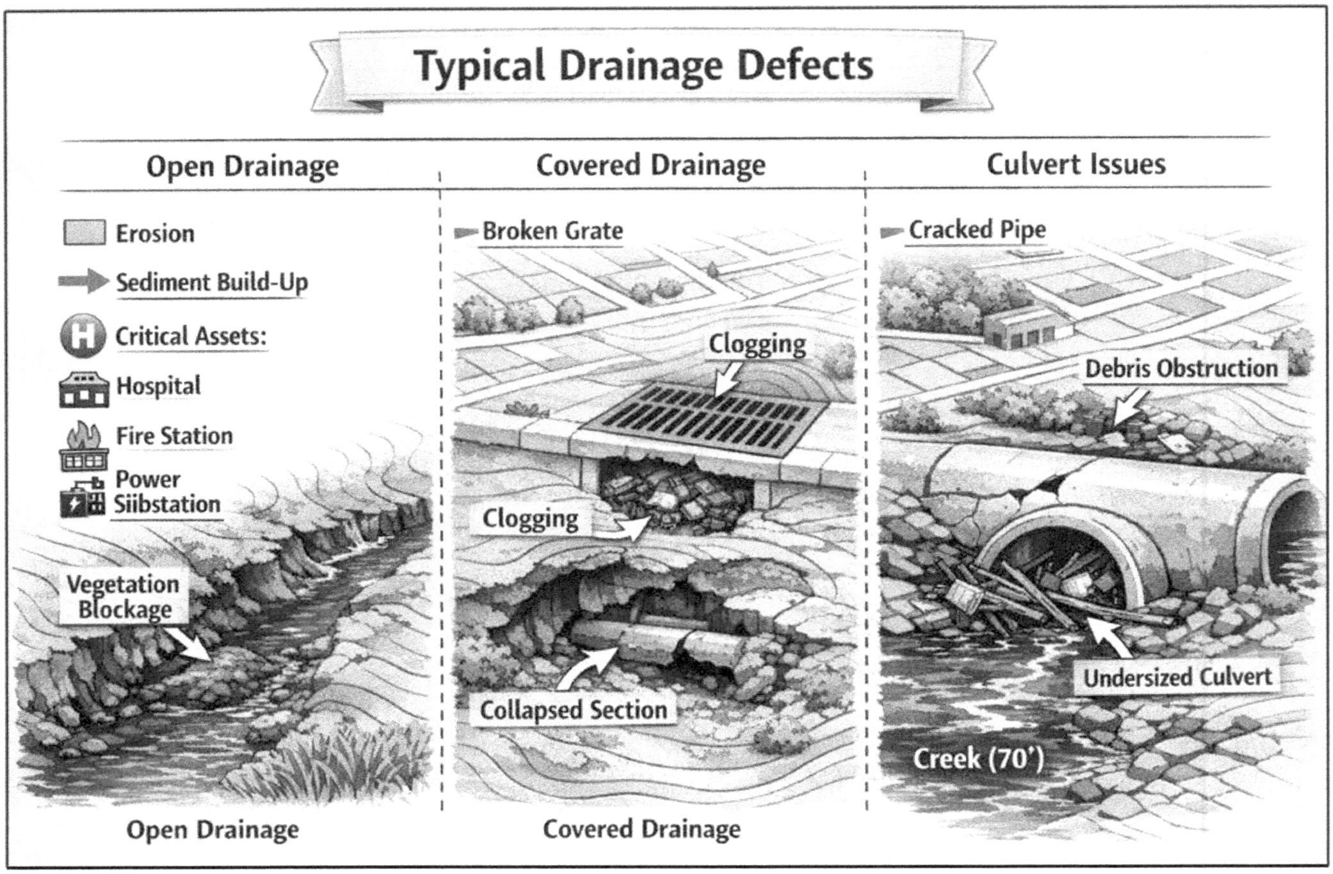

Figure 4.3 Typical structural defects in open drains, covered drains, and culverts

1. Open Drains: Erosion of sidewalls, sediment deposition reducing capacity, and vegetation blockage restricting flow.
2. Covered Drains: Broken or missing grates, clogging from debris, and collapsed sections due to subgrade failure.
3. Culverts: Cracked concrete, debris obstruction, and undersized culverts causing overtopping and backflow.

4.8 Assessment of Hydraulic Condition

4.8.1 Purpose of Hydraulic Assessment

Hydraulic condition assessment examines whether the drainage system conveys flow as intended. It is concerned with capacity, continuity, slope, depth, velocity, inlet efficiency, surcharge behavior, ponding, and outfall performance.

4.8.2 Typical Field Symptoms

Hydraulic problems often appear as overtopping, sluggish flow, prolonged ponding, sediment accumulation, reverse flow, repeated flooding, and poor inlet capture. During assessment, the engineer should compare actual field conditions with likely hydraulic requirements and observed catchment response.

4.8.3 Key Questions to Ask

Practical questions for hydraulic assessment include: is the drain large enough for the contributing catchment, is the longitudinal slope sufficient, are there local restrictions at bends or crossings, are inlets properly spaced and functioning, and does the outlet remain free during storm conditions. These questions provide a useful diagnostic structure for field investigation.

4.8.4 Minor Deficiencies as Combined Causes

In many systems, hydraulic deficiency results not from one major error but from a series of small ones. Slight reverse slopes, discontinuous invert levels, undersized connecting pipes, narrow crossings, and submerged outlets may each appear minor in isolation, but together they create repeated failure.

4.8.5 Supporting Measurements

Hydraulic assessment is strengthened by measuring dimensions, levels, water marks, sediment thickness, and the extent of temporary ponding. Where possible, these observations should later be supported by simple hydraulic checks or more detailed analysis.

4.9 Identification of Encroachments and Obstructions

4.9.1 Meaning of Encroachment and Obstruction

Encroachments are unauthorized or poorly planned intrusions into drainage corridors, channels, or outfall paths. Obstructions are physical barriers that reduce flow area or interrupt water movement. Both are major causes of drainage failure in built environments.

4.9.2 Common Forms in Practice

Typical encroachments include building extensions over drains, filling of natural channels, narrowing of drains for access, construction of boundary walls across runoff paths, unplanned road crossings, and conversion of drain margins into storage or parking space. In village areas, informal crossings, blocked culverts, and field bunds may obstruct continuity.

Obstructions may consist of solid waste, vegetation, sediment deposits, fallen debris, collapsed masonry, temporary earth bunds, or utility works. Even a partial obstruction at a critical location can greatly reduce discharge capacity.

4.9.3 Technical and Administrative Implications

The diagnosis of encroachment-related problems requires attention to both hydraulic and administrative aspects. Some encroachments are physically small but hydraulically severe. Others have become normalized over time even though they disrupt the intended drainage network. Their location, extent, and impact should therefore be recorded clearly for both engineering action and institutional follow-up.

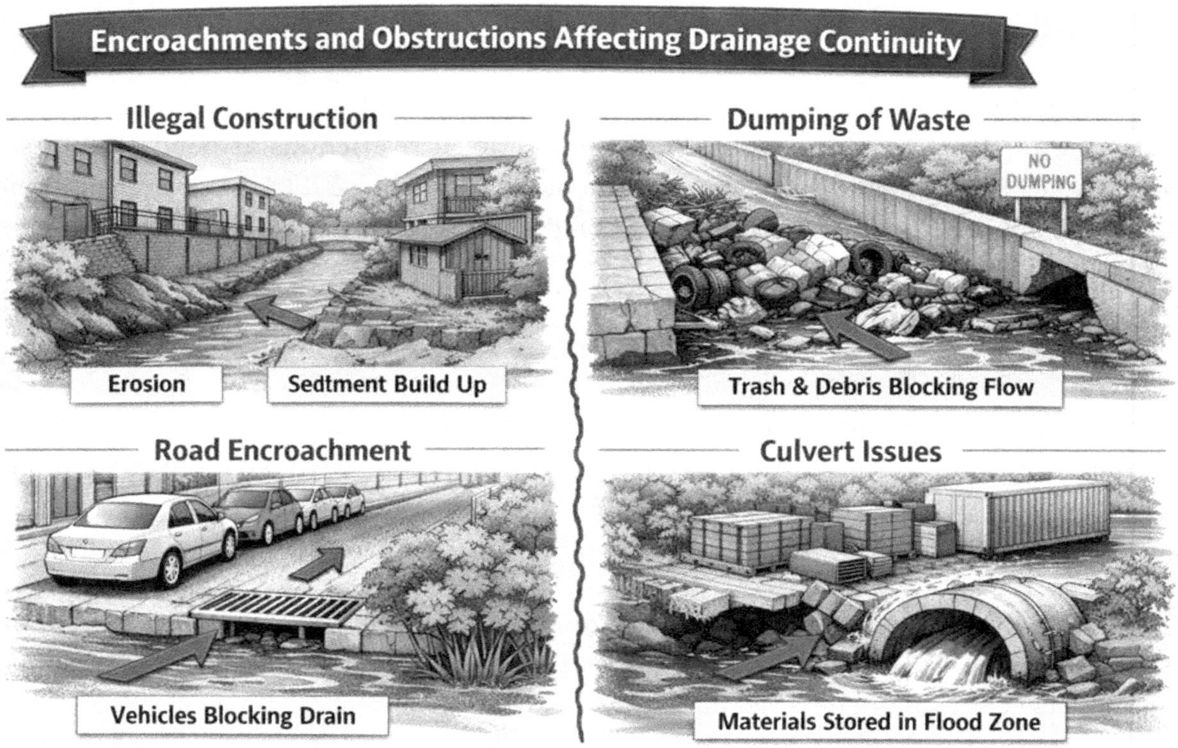

Figure 4.4 Examples of encroachments and obstructions affecting drainage continuity

It highlights four common scenarios:

- **Illegal Construction:** Buildings encroaching on canal margins, reducing hydraulic capacity.
- **Dumping of Waste:** Trash and debris blocking flow paths.
- **Road Encroachment:** Vehicles parked over storm drains, obstructing inlet access.
- **Storage in Floodplain:** Materials and containers placed in flood zones, impeding flow and increasing flood risk.

4.10 Root Cause Analysis of Drainage Failure

4.10.1 Meaning of Root Cause Analysis

Root cause analysis is the process of identifying the fundamental reason a drainage problem occurs rather than merely describing its visible symptoms. A flooded lane is a symptom; the root cause may be an undersized culvert, a submerged outfall, poor road grading, encroachment on a natural channel, or lack of maintenance.

4.10.2 Primary Causes and Secondary Effects

A good diagnosis must distinguish between primary causes and secondary effects. If a drain is full of sediment, the primary cause may be upstream erosion and low velocity rather than cleaning neglect alone. If a culvert overtops, the underlying cause may be increased runoff due to land-use change rather than the culvert barrel itself.

4.10.3 Sequence of Failure

An example of failure sequence: land use change or construction alters runoff, existing drain capacity becomes insufficient, flow slows and sediment accumulates, maintenance becomes more difficult, and recurrent flooding then appears even during moderate rainfall. Understanding this sequence helps the engineer recommend durable solutions rather than repeated temporary fixes.

4.10.4 Importance in Multi-agency Contexts

Root cause analysis is especially important where multiple agencies are involved, because visible symptoms are often treated without addressing the real hydraulic, planning, or maintenance deficiency. A technically sound drainage program therefore depends on evidence-based cause identification.

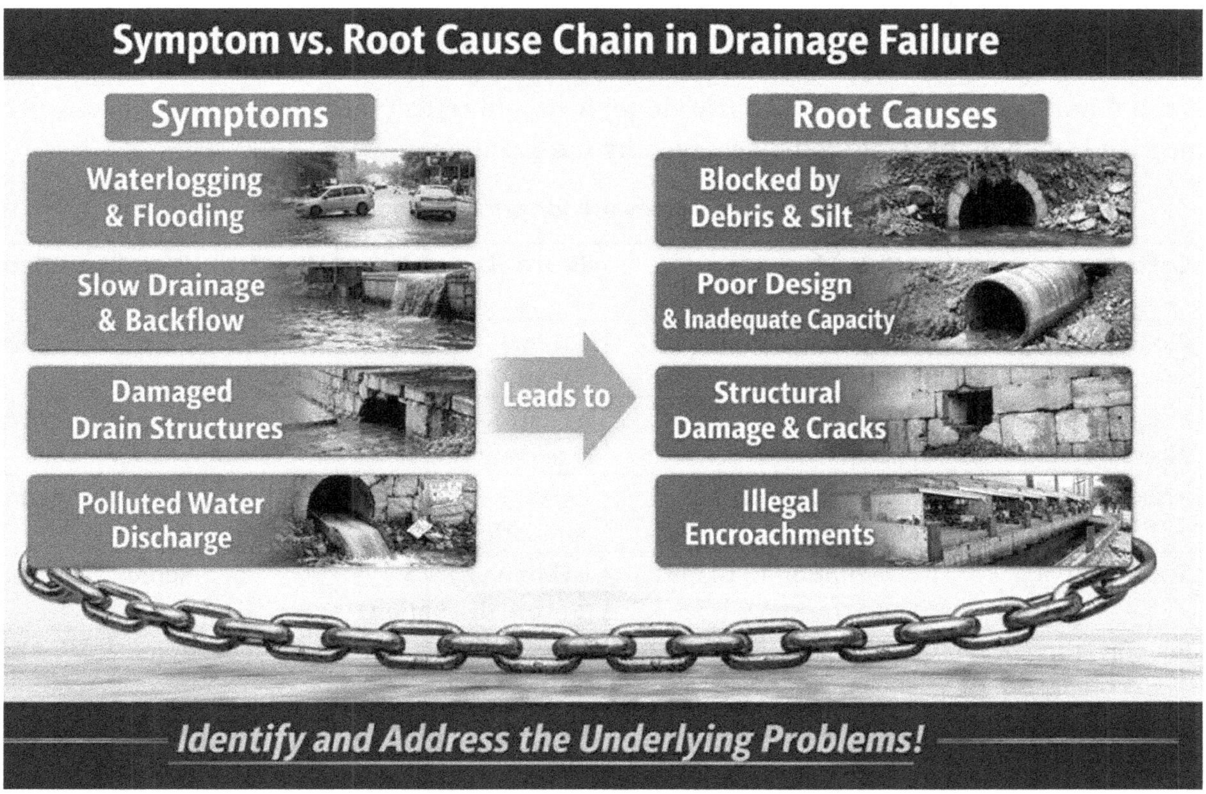

Figure 4.5 Symptom versus root cause chain in drainage failure

4.11 Prioritization of Intervention Areas

4.11.1 Need for Prioritization

Not all drainage problems can be corrected at once, especially in large settlements with limited resources. Prioritization is therefore essential to identify which areas need immediate action, which require medium-term improvement, and which can be addressed through routine maintenance or phased upgrading.

4.11.2 Priority Criteria

Priority may be based on severity of flooding, risk to life and property, importance of the affected road or facility, frequency of failure, extent of structural deterioration, effects on sanitation and environmental quality, and cost and feasibility of intervention. These criteria provide a balanced basis for technical decision-making.

4.11.3 Example of Priority Logic

A blocked drain near a hospital access road may deserve higher priority than a similar defect near an unused open plot. Likewise, a culvert that repeatedly cuts off village access during monsoon conditions may be more urgent than a short-duration nuisance ponding point in a low-use area. Prioritization must therefore combine hydraulic severity with social consequence.

4.11.4 Ranking Approaches

Structured prioritization may use scoring or ranking systems, but engineering judgment remains essential. Priority should not be determined by visibility alone; it should reflect technical urgency, exposure, public importance, and the benefit likely to be achieved by intervention.

Table 4.3 (A) Example scoring framework for prioritising drainage interventions

S. No.	Criteria	Description	Score Range	Weight (%)	Weighted Score
1	Flood Severity	Depth and extent of waterlogging	1 (Low) – 5 (Severe)	20	Score × Weight
2	Frequency of Flooding	How often flooding occurs	1 (Rare) – 5 (Frequent)	15	Score × Weight
3	Affected Population	Number of people impacted	1 (Few) – 5 (Large population)	15	Score × Weight
4	Traffic Impact	Disruption to roads and transport	1 (Minimal) – 5 (Severe disruption)	10	Score × Weight
5	Economic Impact	Damage to property/ business	1 (Low) – 5 (High loss)	10	Score × Weight
6	Condition of Infrastructure	Physical state of drains/culverts	1 (Good) – 5 (Critical)	10	Score × Weight

S. No.	Criteria	Description	Score Range	Weight (%)	Weighted Score
7	Blockage/Siltation Level	Degree of obstruction in system	1 (Clear) – 5 (Fully blocked)	5	Score × Weight
8	Environmental Impact	Pollution, contamination risk	1 (Low) – 5 (Severe)	5	Score × Weight
9	Safety Risk	Risk to public health and safety	1 (Low) – 5 (High risk)	5	Score × Weight
10	Ease of Intervention	Feasibility of fixing the issue	1 (Difficult) – 5 (Easy)	5	Score × Weight

Table 4.3 (B) Priority Classification (Example)

Total Score Range	Priority Level	Action Required
80 – 100	Critical	Immediate intervention
60 – 79	High	Urgent action
40 – 59	Medium	Planned improvement
< 40	Low	Routine maintenance

4.12 Indian Field Context Notes

4.12.1 Dense Urban Settings

In Indian metropolitan areas, drainage investigation often involves dense road networks, covered drains, mixed utility corridors, informal encroachments, and repeated local modifications of road and plot levels. Field diagnosis in such areas must pay particular attention to inlet performance, concealed conduit condition, downstream continuity, and outfall restriction.

4.12.2 Peri-urban and Transitional Zones

In peri-urban areas, the main investigative challenge is often identifying the actual drainage path in landscapes where former rural runoff routes coexist with new roads, raised layouts, and incomplete drainage networks. Such zones require careful mapping of both designed and actual drainage behavior.

4.12.3 Village Settings

In village environments, drainage diagnosis often depends heavily on field observation, local knowledge, and simple level relationships rather than on formal network drawings. Informal crossings, shallow roadside drains, blocked local outfalls, and poor continuity are common issues that must be traced practically on site.

4.13 Planning Implications

1. The effective drainage improvement begins with investigation rather than design calculation alone. Field evidence, drainage-path tracing, structure inspection, flood-prone mapping, and root-cause analysis are essential to avoid misdirected interventions.
2. The investigation should be system-based, not symptom-based. The visible flood point is often not the true controlling point. Integrated diagnosis therefore improves both technical accuracy and investment efficiency, especially where multiple drainage problems compete for limited funds.

4.14 Summary Points

1. Effective drainage improvement begins with correct diagnosis; symptoms alone rarely reveal the true cause.
2. Investigation must consider the entire catchment, runoff routes, intermediate structures, and outfalls, not just the visible failure point.
3. Reconnaissance surveys provide rapid but valuable insight into site conditions, guiding detailed investigation.
4. Field identification of drainage paths distinguishes between designed and actual runoff routes, often revealing hidden causes of failure.
5. Mapping low-lying and flood-prone zones helps identify areas of recurrent waterlogging and prioritize interventions.
6. Inspection of drains, culverts, chambers, and outfalls should cover physical condition, hydraulic function, blockage, erosion, and safety.
7. Structural assessment identifies defects such as cracks, displacement, settlement, and corrosion, linking them to causes and consequences.
8. Hydraulic assessment evaluates capacity, slope, continuity, inlet efficiency, surcharge, and outfall performance, supported by field measurements.
9. Encroachments and obstructions—including illegal construction, waste dumping, and blocked channels—are major contributors to drainage failure.
10. Root cause analysis distinguishes between primary causes and secondary effects, ensuring durable solutions rather than temporary fixes.
11. Prioritization of interventions balances hydraulic severity with social consequence, using scoring frameworks and engineering judgment.
12. A systematic approach to investigation, diagnosis, and prioritization ensures that drainage planning and rehabilitation are technically justified and resource-efficient.

4.15 References

1. CIRIA. (n.d.). *Culvert, screen and outfall manual.*

2. CIRIA. (n.d.). *Culvert condition assessment manuals and storm drain inspection manuals.*

3. Various authors. (n.d.). *Technical references on flood-prone mapping, drainage diagnosis, and hydraulic risk assessment.*

4. Various authors. (n.d.). *Guidance documents on drainage investigation, culvert inspection, and outfall management.*

5. Various authors. (n.d.). *Review and planning literature on resilient urban drainage and risk-based prioritisation.*

6. Various authors. (n.d.). *Textbooks and field manuals on drainage inspection, municipal infrastructure assessment, and rehabilitation planning.*

Chapter 5

Surveying, Data Collection, and Geospatial Tools

Learning Objectives

By the end of this chapter, the learner should be able to:

1. List the data required for drainage investigation and design.
2. Select appropriate survey methods for different drainage settings.
3. Use GIS and remote sensing concepts in drainage analysis.
4. Organise field and spatial data into a usable engineering database.
5. Explain the role of digital terrain data, GNSS, total station surveys, and drone imaging in drainage studies.

5.1 Introduction

Reliable drainage planning and design depend on reliable data. Even a technically sound method can produce poor results if the underlying information on topography, rainfall, land use, drain alignment, structure condition, or outfall performance is incomplete or inaccurate. Surveying and data collection therefore form a critical bridge between field diagnosis and engineering design.

This chapter explains the main categories of data required for settlement drainage studies and the methods used to collect, organize, and interpret them. It also introduces the growing role of geospatial technologies such as GIS, digital elevation models, remote sensing, and drone surveys in urban, semi-urban, and village drainage engineering.

The chapter emphasizes a practical principle: advanced tools can significantly improve drainage analysis, but they do not replace careful field observation, level control, and engineering judgment. The best results come from integrating field survey, hydrologic records, terrain analysis, and digital data management into a coherent engineering database.

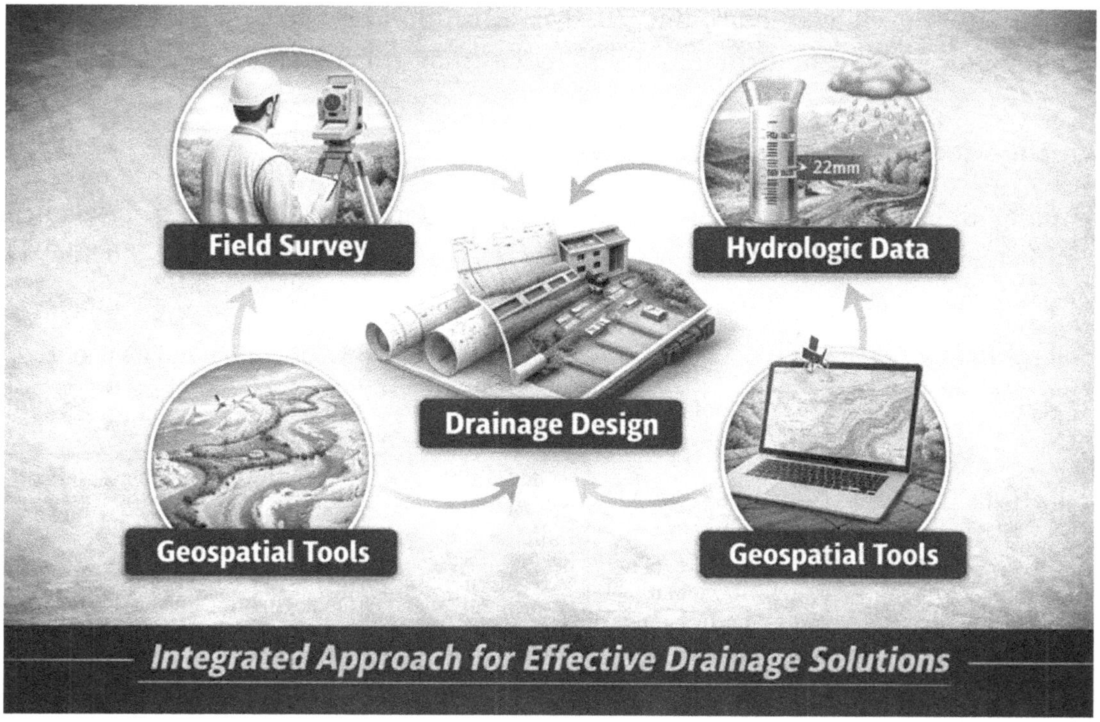

Figure 5.1 Relationship between field survey, hydrologic data, geospatial tools, and drainage design

5.2 Data Requirements for Drainage Studies

5.2.1 Nature of Drainage Data

Drainage studies require a combination of physical, hydrologic, hydraulic, environmental, infrastructural, and social data. The exact requirement depends on the scale and purpose of the study, but most projects need enough information to understand catchment behavior, drainage pathways, existing infrastructure, and receiving conditions.

5.2.2 Core Data Categories

At a minimum, the following basic requirements are identified: topographic data to define slopes and flow direction, rainfall data for runoff estimation, land use and surface condition data for assessing runoff generation, drain and culvert inventory for evaluating hydraulic capacity, outfall data for discharge feasibility, and field-condition data on blockage, erosion, and structural defects.

These categories are fundamental because they link the landscape, the rainfall input, the drainage network, and the observed performance of the system. Without this integration, drainage design may be technically neat but practically misleading.

5.2.3 Additional Data for Complex Situations

Where the system is more complex, additional information may be needed on soil type, groundwater condition, utility location, road levels, historical flood records, wastewater intrusion, and maintenance history. In peri-urban and village environments, it is also important to understand informal drainage routes, seasonal ponding areas, and community practices that affect water movement.

5.2.4 Data Quality Considerations

Drainage planning depends not only on the quantity of data but also on its relevance, accuracy, consistency, and timeliness. Good engineering judgment is therefore required to decide what level of data is necessary for the task at hand.

Table 5.1 Major data categories required for drainage investigation and design

S. No.	Data Category	Type of Data Required	Purpose in Drainage Design	Typical Data Sources
1	Topographic Data	Contours, elevations, slope, DEM	To determine flow direction and drainage patterns	Survey of India maps, Total Station survey, LiDAR surveys, GIS portals
2	Hydrological Data	Rainfall intensity, duration, frequency (IDF), runoff	To estimate peak discharge	India Meteorological Department, Central Water Commission (CWC), rainfall stations
3	Meteorological Data	Temperature, evaporation, climate trends	To assess climate impact	India Meteorological Department, state meteorological departments
4	Land Use / Land Cover	Urban area, vegetation, imperviousness	To calculate runoff coefficients	Satellite imagery (NRSC/ISRO), master plans, GIS datasets
5	Soil Data	Soil type, permeability, infiltration	To estimate infiltration	National Bureau of Soil Survey and Land Use Planning, soil maps, field tests
6	Geological Data	Subsurface strata, rock type	To assess structural stability	Geological Survey of India (GSI), borehole data
7	Existing Drainage Network	Layout, size, condition of drains	To evaluate system performance	Municipal Corporation records, field surveys
8	Hydraulic Data	Flow velocity, water levels, discharge	To design drainage sections	Field measurements, CWC, hydraulic studies
9	Flood History Data	Flood levels, frequency, duration	To identify risk zones	Municipal records, disaster management reports, local community
10	Population Data	Density, growth trends	To estimate future demand	Census of India, urban local bodies
11	Infrastructure Data	Roads, utilities, buildings	To avoid conflicts	Municipal GIS, PWD, utility departments
12	Environmental Data	Water quality, pollution sources	To reduce environmental impact	Pollution Control Boards, environmental agencies
13	Groundwater Data	Water table, seasonal variation	To assess sub-surface drainage	Central Ground Water Board, borewell data
14	Remote Sensing / GIS Data	Satellite images, spatial layers	Watershed delineation and mapping	Indian Space Research Organisation, NRSC, GIS portals

5.3 Topographic and Surveys

5.3.1 Topographic Data Matter

Topographic and levelling surveys are the foundation of drainage engineering because water movement is controlled by elevation difference and surface form. Without accurate level information, it is not possible to determine flow direction, slope adequacy, invert continuity, ponding zones, or outfall feasibility.

5.3.2 Scope of Topographic Survey

A topographic survey records ground levels, contours, break lines, road edges, drain alignments, culvert positions, channel sections, building edges, and other physical features relevant to flow behavior. In drainage work, the survey should capture both broad terrain form and small but hydraulically important local level differences such as road crowns, kerb heights, inlet levels, and shallow depressions at junctions.

5.3.3 Importance of Level Survey

Level surveys are particularly important for longitudinal drain profiles. They help identify whether a drain has adequate slope, whether reverse gradients exist, and whether culvert inverts and outfalls align properly. In urban drainage, even small vertical errors may misrepresent hydraulic continuity.

5.3.4 Need for Survey Control

Topographic and level data should be tied to a consistent datum and checked for closure and accuracy. Poor survey control can produce incorrect diagnosis of drainage failure and flawed design recommendations. This is especially important in flat areas where drainage gradients are small and highly sensitive to minor elevation differences.

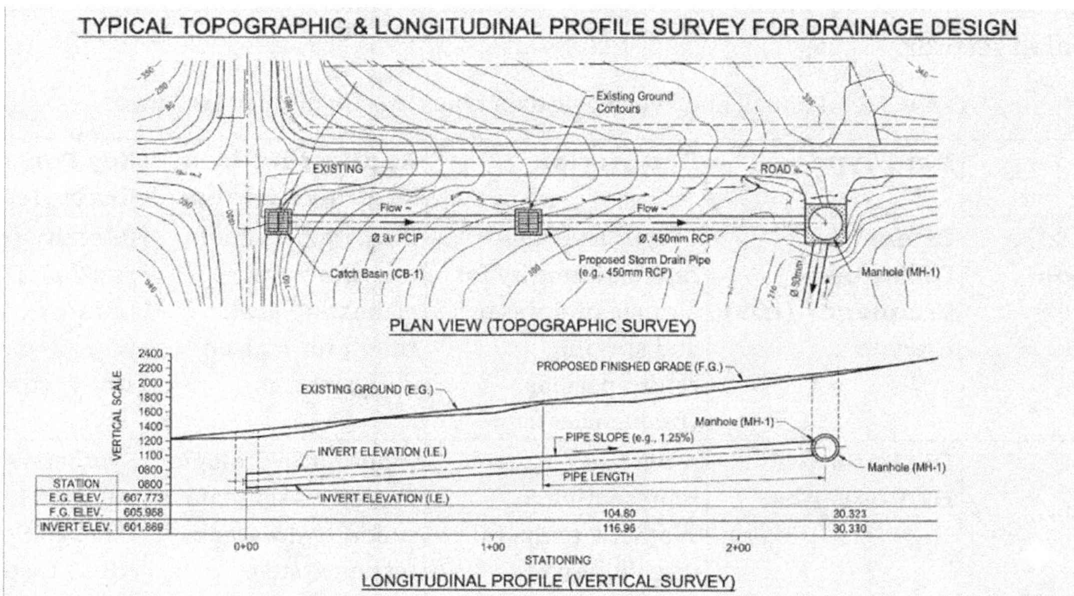

Figure 5.2 Typical topographic and longitudinal profile survey information for drainage design

5.4 Collection of Rainfall and Hydrologic Data

5.4.1 Rainfall as Hydrologic Input

Rainfall data are required to estimate runoff and determine design drainage capacity. The most important rainfall variables are depth, intensity, duration, frequency, and seasonal pattern. Depending on the scale of the project, hydrologic data may also include runoff observations, ponding duration, flood depth, and local water-level records.

5.4.2 Data Sources

Rainfall information is typically obtained from meteorological agencies, hydrologic departments, municipal records, research institutions, and project-specific monitoring. For local drainage systems, short-duration rainfall data are especially important because micro-drainage often responds to intense storms over minutes or a few hours rather than to daily totals alone.

5.4.3 Supplementary Hydrologic Evidence

In areas with limited formal records, supplementary evidence may be collected from community observations, historical flood reports, photographs, disaster records, and temporary rain gauges. Although such evidence may be less precise, it can still help identify practical design thresholds and recurring failure conditions.

5.4.4 Interpreting Hydrologic Data Over Time

Rainfall data should not be interpreted in isolation. Urbanization, land development, and climate variability may alter runoff response even when long-term rainfall statistics remain similar. Therefore, hydrologic analysis should be linked to evolving catchment conditions rather than relying only on historical rainfall records.

Table 5.2 Rainfall and hydrologic data types used in drainage studies

Category	Data Type	Description	Application in Drainage Design	Key Parameters & Examples
I. Rainfall & Precipitation	**Intensity-Duration-Frequency (IDF) Curves**	Graphs showing rainfall intensity for a range of durations and specific return periods (frequencies).	Sizing pipes, inlets, and structures; estimating peak runoff for a given design storm.	Intensity (in/hr or mm/hr), Duration (min/hr), Frequency (e.g., 2-yr, 10-yr, 100-yr storms).
	Design Storm Hyetographs	A time-series graph representing a synthetic temporal distribution of rainfall depth for a specific event.	Input for hydrologic models to simulate runoff hydrographs; testing system performance.	Synthetic storm patterns (e.g., Chicago Storm, SCS/NRCS Distributions), Total Depth, Peak Intensity, Duration.

Category	Data Type	Description	Application in Drainage Design	Key Parameters & Examples
	Historical Rainfall Records (Rain Gauge)	Long-term measured precipitation depths at specific point locations.	Analyzing storm statistics; validating design methods; performing continuous hydrologic simulation.	Daily, hourly, or sub-hourly depths (inches or mm); Station ID, Duration of record.
	Areal Averaged Rainfall	Calculated or estimated average precipitation depth over a defined watershed or basin.	Required input for basin-wide hydrologic models where point data is insufficient.	Thiessen Polygons, Isohyetal Method, Radar Rainfall estimates (spatially distributed).
II. Catchment & Hydrologic Characteristics	**Watershed Delineation & Topography**	Definition of the contributing drainage area boundary and surface slope from elevation data.	Determining the area that produces runoff for a point of interest; defining sub-basins.	Area (acres, hectares), Slopes (%), Flow Paths, Digital Elevation Models (DEMs).
	Land Use & Land Cover (LULC)	Classification of the physical land surface (e.g., developed, forest, concrete).	Fundamental for estimating runoff volume and timing; directly affects interception and infiltration.	Percent Impervious Cover, Vegetation Type, Developed Areas (e.g., Residential, Commercial).
	Soil Classification	Information on subsurface soil types and their infiltration characteristics.	Determining water infiltration and storage capacity; key input for runoff volume methods.	NRCS Hydrologic Soil Groups (HSG A, B, C, D); Infiltration Rate, Hydraulic Conductivity.
	Runoff Coefficient (C) / Curve Number (CN)	Empirical values that represent the portion of rainfall that becomes surface runoff.	Estimating peak flow (Rational Method, C) or runoff volume (NRCS Curve Number, CN).	C values for different land uses; CN for combination of Soil Type, Land Use, and Moisture condition.
	Time of Concentration (T_c)	The travel time for water to flow from the hydraulically most remote point of the watershed to the outlet.	Determines the critical storm duration for calculating peak runoff rates.	Flow Path Lengths, Slopes, Velocity; calculated using Manning's equation, TR-55 methods.
	Groundwater & Infiltration Data	Measured or estimated rates of water loss to the ground.	Crucial for modeling long-term water balance, infiltration losses, and interflow.	Infiltration rates (Horton, Green-Ampt), Watertable depth, Baseflow estimates.

(*Contd.*)

5.5 Land Use, Soil, and Infrastructure Data

5.5.1 Land Use Data

Runoff generation depends strongly on land use, surface condition, and built form. Land use data should identify built-up areas, roads, open spaces, industrial plots, commercial zones, agricultural land, vacant plots, and water bodies. Surface properties such as paving, compaction, roof density, and vegetation cover are important because they affect infiltration, runoff coefficient, and drainage connectivity.

5.5.2 Soil Data

Soil information is useful for evaluating infiltration potential, erosion susceptibility, side-slope stability, and the suitability of sustainable drainage measures. Fine-grained soils often generate more runoff and support waterlogging, whereas coarse soils may promote infiltration but behave differently under erosion or instability conditions. In village settings, local soil behavior may determine whether open drains remain stable or require lining.

5.5.3 Infrastructure Data

Infrastructure data include roads, drains, culverts, bridges, outfalls, ponds, canals, embankments, utilities, and sanitation features. These features shape drainage pathways and often create both constraints and opportunities for improvement. The drainage engineer must therefore treat infrastructure mapping as part of catchment analysis rather than as a separate exercise.

5.5.4 Rapidly Changing Land Conditions

Rapidly changing land use is especially important in semi-urban and peri-urban catchments, where drainage systems may lag behind development. In such areas, current land-use mapping is essential because outdated information may not reflect actual runoff behavior

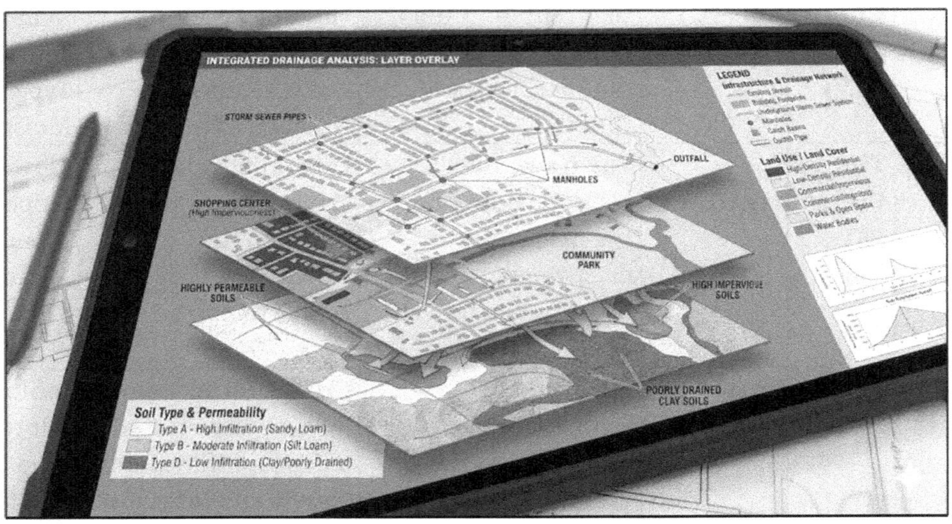

Figure 5.3 Overlay of land use, soil, and infrastructure layers in drainage analysis

5.6 Use of Global Navigation Satellite System (GNSS) and Total Station Surveys

5.6.1 GNSS Applications

Modern drainage surveys often use GNSS for accurate positioning and for integration with GIS platforms. GNSS is useful for locating drainage assets, mapping survey control points, and establishing spatial reference for field observations. It is especially effective in open areas where satellite visibility is good.

5.6.2 Limitations of GNSS

GNSS may be less reliable in dense urban corridors, tree-covered areas, or places with signal obstruction. This limitation is important because many urban drainage investigations occur in environments with buildings, overhead structures, and narrow corridors.

5.6.3 Total Station Applications

A total station is widely used for detailed topographic and engineering survey because it measures angles, distances, and elevations accurately. It is well suited to drain alignment surveys, road profiles, cross-sections, culvert positioning, and congested built environments where vertical control is critical.

5.6.4 Choosing the Survey Method

The choice between GNSS, total station, conventional leveling, or a combination of methods depends on accuracy needs, terrain condition, project purpose, and available resources. For drainage design, vertical accuracy is often more important than horizontal mapping alone. As a result, GNSS and total station methods are often combined with benchmark-based leveling for reliable elevation control.

Table 5.3 Comparison of GNSS, total station, and conventional survey methods in drainage work

S. No.	Parameter	GNSS (GPS Survey)	Total Station	Conventional Methods (Level & Tape)
1	Principle	Satellite-based positioning	Electronic distance and angle measurement	Manual measurement using level & tape
2	Accuracy	Medium (±10–30 mm with RTK)	High (±2–5 mm)	Moderate (±5–10 mm)
3	Speed of Survey	Very fast for large areas	Fast	Slow
4	Suitability for Drainage	Good for large catchments & mapping	Best for detailed design & profiles	Suitable for small/simple works
5	Terrain Requirement	Requires open sky view	Works in most conditions	Works in all conditions
6	Obstruction Issues	Affected by buildings/trees	Minimal obstruction issues	No major issues
7	Data Output	Direct digital coordinates (GIS-ready)	High-precision digital data	Manual records, needs digitization
8	Manpower Requirement	Low (1–2 persons)	Moderate (2–3 persons)	High (3–5 persons)

(Contd.)

S. No.	Parameter	GNSS (GPS Survey)	Total Station	Conventional Methods (Level & Tape)
9	Cost of Equipment	High initial cost	High cost	Low cost
10	Skill Requirement	Moderate (GNSS operation & processing)	High (skilled operator needed)	Basic skills sufficient
11	Use in Longitudinal Profile	Moderate accuracy	Excellent accuracy	Good but time-consuming
12	Use in Cross Sections	Good for wide areas	Very accurate	Moderate
13	Real-time Capability	Yes (RTK GNSS)	Limited	No
14	Dependency on Weather	Affected by signal conditions	Slightly affected	Minimal impact
15	Best Use Case in Drainage	Catchment mapping, GIS integration	Detailed drain alignment & design	Small drains, verification work

Table 5.4 Decision Matrix for Selecting Survey Method in Drainage Projects

S. No.	**Project Condition / Requirement**	**Recommended Method**	**Reason**
1	Large area / catchment mapping	GNSS (RTK)	Fast data collection over wide areas, GIS-ready output
2	Detailed drain alignment design	Total Station	High accuracy needed for slope and alignment
3	Longitudinal profile survey	Total Station	Precise elevation control is critical
4	Cross-section survey (urban streets)	Total Station	Accurate and reliable in built-up areas
5	Open rural terrain with clear sky	GNSS	Efficient and quick with minimal obstruction
6	Dense urban area (buildings, trees)	Total Station	GNSS signal obstruction issues
7	Small project / short drain	Conventional Method	Cost-effective and sufficient accuracy
8	Budget constraints (low-cost survey)	Conventional Method	Minimal equipment cost
9	High precision requirement (design stage)	Total Station	Millimeter-level accuracy
10	Preliminary reconnaissance survey	GNSS	Quick overview and mapping
11	Integration with GIS / mapping systems	GNSS	Direct coordinate-based data
12	Verification / checking levels	Conventional Method	Simple and reliable for checks
13	Real-time positioning required	GNSS (RTK)	Instant coordinate output
14	Complex utility corridor survey	Total Station	Better control and detail in congested areas
15	Remote / difficult terrain	GNSS	Reduces need for line-of-sight setup

5.7 GIS for Drainage Mapping

5.7.1 GIS as an Integration Platform

Geographic Information Systems provide a framework for storing, analyzing, and visualizing drainage-related spatial data. GIS is one of the most useful tools in modern drainage engineering because it allows terrain, land use, infrastructure, flood-prone zones, catchment boundaries, and field observations to be integrated into a single decision-support environment.

5.7.2 Mapping and Analysis Functions

GIS supports mapping of existing drains, culverts, inlets, manholes, outfalls, and low-lying zones. It also supports analyses such as flood-hotspot mapping, drainage density assessment, network tracing, land-use overlay, and prioritization of problem areas. When linked with survey and remote sensing data, GIS can become the core platform for drainage inventory and planning.

5.7.3 Use in Different Settlement Settings

In urban areas, GIS is particularly useful in handling dense infrastructure networks and overlapping datasets. In semi-urban regions, it helps reveal how new development interacts with old natural drainage patterns. In village settings, GIS can assist in documenting relatively simple but often poorly mapped drainage systems, local depressions, ponds, and agricultural runoff connections.

5.7.4 GIS for Long-Term Management

The value of GIS lies not only in mapping but also in data organization and updating. A GIS-based drainage inventory can support both one-time planning studies and long-term asset management.

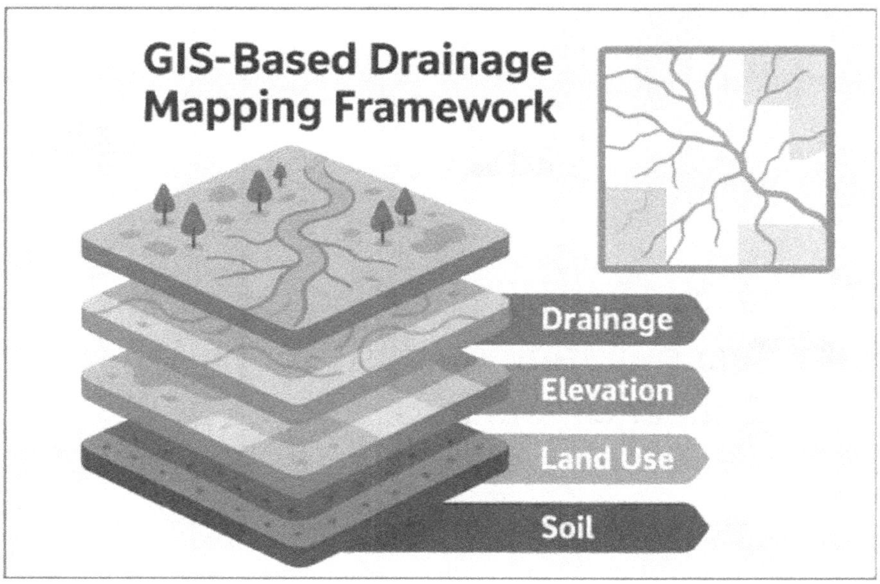

Figure 5.4 GIS-based drainage mapping framework and thematic layers

5.8 Digital Elevation Models and Terrain Analysis

5.8.1 Meaning of a DEM

A Digital Elevation Model, or DEM, is a digital representation of terrain height. DEMs are highly valuable in drainage engineering because they help identify slopes, flow directions, watershed boundaries, depressions, and natural drainage paths over large areas.

5.8.2 Terrain Analysis Applications

Terrain analysis based on DEMs supports catchment delineation, low-lying area detection, flow accumulation mapping, and preliminary identification of drainage corridors. In peri-urban and rural areas, DEM-based analysis is especially useful because terrain still strongly controls water movement.

5.8.3 Limits in Built-Up Areas

Coarse DEMs may be less useful in dense urban areas because buildings, roads, embankments, and small local level differences strongly affect actual runoff movement. This is an important limitation because terrain models may suggest theoretical flow routes that do not match field reality in built-up settings.

5.8.4 Complement, not Replacement

For detailed engineering design, DEMs should complement rather than replace field survey. They are valuable for regional understanding and preliminary planning, but local invert levels, road cambers, and structure elevations still require direct measurement.

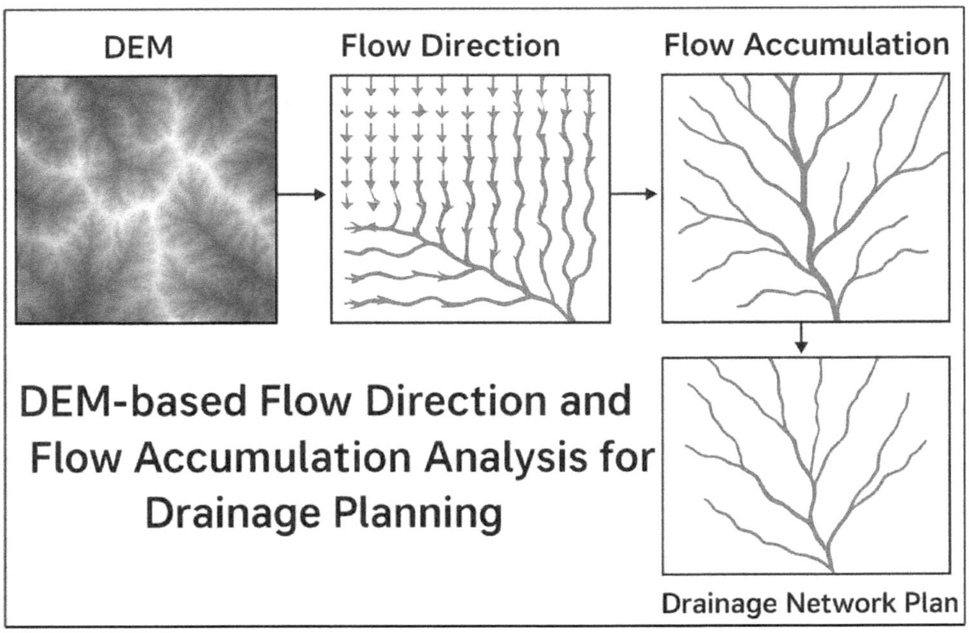

Figure 5.5 DEM-based flow direction and flow accumulation analysis for drainage planning

5.9 Remote Sensing for Drainage Assessment

5.9.1 Role of Remote Sensing

Remote sensing involves obtaining information about the ground surface from satellite or aerial imagery. In drainage engineering, it is useful for identifying land use change, drainage encroachment, expansion of built-up area, surface-water spread, vegetation cover, and historical flood extent.

5.9.2 Major Strengths

One major advantage of remote sensing is that it provides a synoptic view over a large area. This is especially valuable when drainage problems extend beyond a single site or where field access is limited. Multi-date imagery can also reveal how drainage-related features have changed over time, such as pond filling, channel narrowing, wetland loss, or expansion of impervious surfaces.

5.9.3 Application by Settlement Type

In urban and peri-urban studies, remote sensing can help detect altered drainage patterns caused by rapid development. In village settings, it may help map tanks, natural depressions, agricultural drainage links, and seasonal water accumulation zones.

5.9.4 Need for Ground Verification

Remote sensing should always be supported by ground verification. Surface patterns seen in imagery cannot always be interpreted correctly without field evidence. Remote sensing is therefore most effective when combined with survey data and GIS analysis.

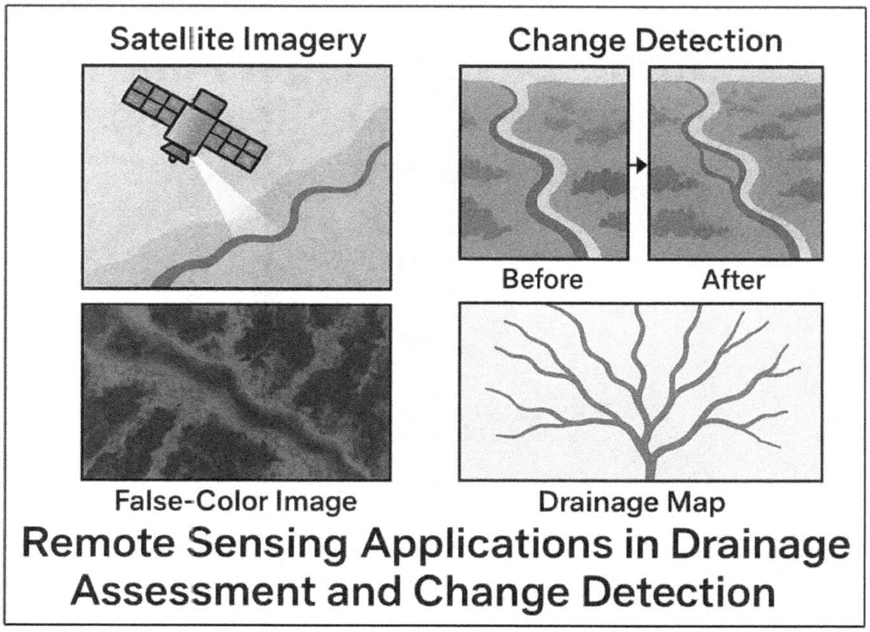

Figure 5.6 Remote sensing applications in drainage assessment and change detection

5.10 Drone Surveys and Imaging Applications

5.10.1 Why Drones Are Useful

Drone surveys provide high-resolution aerial imagery and, in some cases, detailed terrain information for drainage assessment. They are especially useful where conventional survey is difficult, where rapid coverage is needed, or where visual documentation of drainage problems is important.

5.10.2 Typical Drainage Applications

Drones can capture the condition of open drains, encroachments, blocked channels, flood-prone streets, culvert approaches, erosion sites, and outfall areas. Their high spatial resolution is particularly valuable in built-up environments where small features strongly affect drainage performance.

5.10.3 Advanced Outputs

Drone-based photogrammetry can support the preparation of detailed surface models and orthomosaic maps. These products can assist route tracing, cross-drainage inventory, and communication of drainage problems to planners, authorities, and communities. Drones also improve inspection safety and efficiency in infrastructure environments where access is difficult or hazardous.

5.10.4 Practical Cautions

Drone use requires proper regulatory permission, safe flight conditions, suitable weather, and appropriate ground control for technically reliable outputs. Drone data therefore enhance drainage assessment but still require careful processing and engineering interpretation.

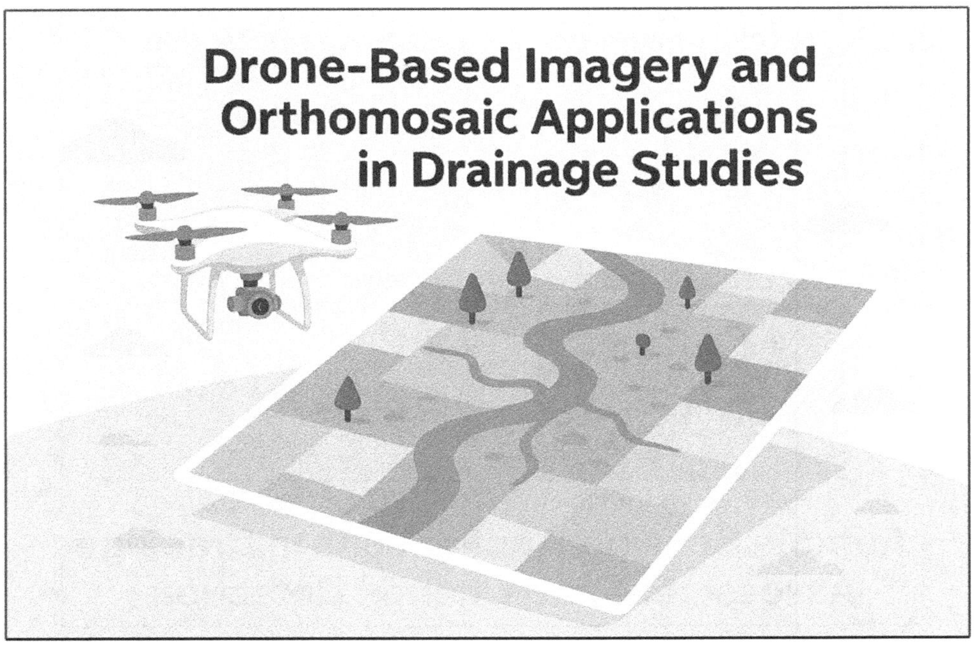

Figure 5.7 Drone-based imagery and orthomosaic applications in drainage studies

5.11 Preparation of Digital Drainage Databases

5.11.1 Purpose of a Drainage Database

A digital drainage database is an organized system for storing information on drainage assets, field observations, survey levels, hydraulic structures, maintenance history, and problem locations. It converts scattered observations into a structured engineering resource that can support planning, design, operation, and rehabilitation.

5.11.2 Typical Database Contents

A drainage database typically includes:

- Drainage network layout and attributes (open drains, covered drains, pipes, culverts, chambers, outfalls).
- Survey levels, invert elevations, slopes, and profiles.
- Hydraulic data such as flow capacity, velocity, and surcharge points.
- Structural condition records including cracks, displacement, erosion, and lining status.
- Maintenance logs, cleaning frequency, and repair history.
- Flood-prone zones, waterlogging points, and encroachment records.
- Photographic documentation and geotagged field observations.
- Integration with GIS layers for spatial visualization.

5.11.3 Benefits of Digital Databases

Digital databases improve consistency, accessibility, and long-term management. They allow engineers to track recurring problems, compare interventions, and prioritize resources. They also support transparency and accountability by providing evidence-based records for municipal authorities and communities.

5.11.4 Core Principle

The chapter makes an important practical point: data should not remain buried in reports alone. Once collected, drainage information should be organized so that it remains accessible, updateable, and useful throughout the lifecycle of the system. UBCH-5.docx

Table 5.5 Suggested fields for a digital drainage asset database

A. Asset Identification & Classification

Field Name	**Data Type**	**Description**
Asset_ID	Text / Integer	Unique stormwater drain identifier
Drain_Name	Text	Road name / locality-based drain name
Drain_Type	Coded Text	Open drain / Closed conduit / Box drain
Drain_Function	Coded Text	Stormwater only / Combined (storm + sewer)
Drain_Order	Coded Text	Primary / Secondary / Tertiary
Ownership_Agency	Text	Municipal Corporation, Development Authority
Jurisdiction	Text	Ward / Zone / Circle

B. Spatial & GIS Attributes

Field Name	**Data Type**	**Description**
Geometry	Polyline	GIS geometry of the drain
Start_Node_ID	Text	Upstream node / inlet ID
End_Node_ID	Text	Downstream outfall / junction ID
Latitude	Decimal	Midpoint latitude (WGS 84)
Longitude	Decimal	Midpoint longitude (WGS 84)
Road_Association	Text	Adjacent road or ROW
Catchment_Area_ha	Float	Contributing urban catchment area
Land_Use_Dominance	Coded Text	Residential / Commercial / Industrial / Mixed

C. Structural & Physical Characteristics

Field Name	**Data Type**	**Description**
Drain_Length_m	Float	Length of drain segment
Internal_Width_m	Float	Clear width
Internal_Depth_m	Float	Clear depth
Shape	Coded Text	Rectangular / Trapezoidal / Circular
Construction_Type	Coded Text	RCC / Brick masonry / Precast
Bed_Level_m	Float	Invert or bed level (m)
Cover_Level_m	Float	Top slab or ground level (m)
Year_of_Construction	Integer	Construction year

D. Hydraulic & Design Parameters (Urban-Specific)

Field Name	**Data Type**	**Description**
Design_Return_Period_yr	Integer	Design rainfall return period
Design_Discharge_cumecs	Float	Designed peak discharge
Existing_Capacity_cumecs	Float	Existing carrying capacity
Slope_percent	Float	Longitudinal slope
Roughness_n	Float	Manning's roughness coefficient
Freeboard_Available_m	Float	Freeboard under design flow
Backflow_Risk	Yes/No	Backflow from downstream system

E. Drainage Appurtenances (Critical for Urban Systems)

Field Name	Data Type	Description
Inlet_Type	Coded Text	Grating / Curb inlet / Gully
No_of_Inlets	Integer	Count of connected inlets
Manhole_Presence	Yes/No	Manhole availability
Culvert_Connection	Yes/No	Connection with road culvert
Outfall_Type	Text	River / Nallah / Lake / STP
Flap_Gate_Presence	Yes/No	Backflow prevention gate

F. Condition Assessment & Encroachment

Field Name	Data Type	Description
Structural_Condition	Coded Text	Good / Fair / Poor
Silt_Depth_cm	Float	Average silt thickness
Siltation_Percentage	Float	% reduction in hydraulic capacity
Encroachment_Type	Coded Text	Building / Wall / Utility / None
Utility_Conflict	Yes/No	Clash with water, sewer, OFC
Illegal_Connections	Yes/No	Unauthorized sewer inflow

G. Flood Risk & Performance Indicators

Field Name	Data Type	Description
Waterlogging_Prone	Coded Text	Low / Medium / High
Flood_Depth_cm	Float	Typical flood depth
Flood_Duration_hr	Float	Duration of waterlogging
Critical_Location	Yes/No	Hospital, subway, junction
Complaints_Frequency	Integer	Public complaints per year

H. Environmental & Public Health Aspects

Field Name	Data Type	Description
Sewage_Admixture	Yes/No	Sewage mixing in storm drain
Odour_Issue	Yes/No	Odour nuisance
Solid_Waste_Load	Coded Text	Low / Medium / High
Vector_Risk	Yes/No	Mosquito breeding risk

I. Maintenance & Operations (O&M)

Field Name	Data Type	Description
Last_Desilting_Date	Date	Last cleaning date
Desilting_Method	Text	Manual / Jetting / Mechanical
Desilting_Frequency	Text	Monthly / Pre-monsoon
Maintenance_Agency	Text	O&M contractor / ULB
Priority_Class	Coded Text	Emergency / Routine
Estimated_O&M_Cost	Float	Annual maintenance cost

J. Data Source & Digital Metadata

Field Name	Data Type	Description
Survey_Method	Text	Drone / Field / GPS / Records
Orthomosaic_Year	Integer	Drone survey year
DEM_Resolution_cm	Float	DEM/Grid resolution
Attribute_Accuracy	Coded Text	High / Medium / Low
Last_Updated	Date	Last database update
Remarks	Text	Field notes

5.12 Integration of Survey and Spatial Data

5.12.1 Need for Integration

Drainage studies often generate diverse datasets: survey levels, rainfall records, land use maps, soil data, infrastructure inventories, and remote sensing imagery. Integration ensures that these datasets can be analysed together rather than in isolation.

5.12.2 GIS as a Common Platform

GIS provides the most effective platform for integration. Survey data can be georeferenced, rainfall stations can be mapped, land use can be overlaid, and hydraulic structures can be linked to condition records. This allows engineers to visualize drainage problems spatially and to analyze relationships between terrain, infrastructure, and hydrologic response.

5.12.3 Practical Outcomes

Integrated data supports:

- Catchment delineation and runoff estimation.
- Identification of mismatches between designed and actual drainage paths.
- Prioritization of intervention areas based on combined hydraulic, structural, and social criteria.
- Long-term asset management and monitoring.

5.13 Role of Modern Geospatial Tools

5.13.1 Expanding Capabilities

Modern geospatial tools such as LiDAR, high-resolution DEMs, drone photogrammetry, and advanced GIS software have expanded the ability to analyze drainage systems. They provide detailed terrain models, accurate positioning, and powerful visualization.

5.13.2 Application in Drainage Engineering

These tools are applied in:

- Mapping catchments and sub-catchments.
- Detecting low-lying and flood-prone zones.
- Tracing flow accumulation and drainage density.
- Documenting encroachments and obstructions.
- Communicating findings to stakeholders through maps and 3D models.

5.13.3 Limitations

Despite their strengths, geospatial tools cannot replace field verification. Terrain models may misrepresent built-up areas, and imagery may not capture underground conditions. Engineering judgment and ground truth remain essential.

5.14 Use of Field Photography and Documentation

5.14.1 Importance of Photographic Records

Photographs provide visual evidence of drainage conditions, defects, and problem areas. They help communicate issues to stakeholders, support diagnosis, and create a permanent record for comparison over time.

5.14.2 What to Capture

Field photography should cover:

- Drain alignment and section condition.
- Culverts, chambers, and outfalls.
- Waterlogging points and flood-prone zones.
- Encroachments, obstructions, and illegal crossings.
- Structural defects such as cracks, erosion, or collapse.
- Maintenance practices and cleaning operations.

5.14.3 Integration with Databases

Photographs should be geotagged and linked to survey data, inspection records, and GIS layers. This integration ensures that visual evidence supports technical analysis and asset management.

5.15 Community-Based Data Collection

5.15.1 Role of Local Knowledge

Community observations often provide valuable information about flood depth, duration, seasonal patterns, and historical interventions. Residents may recall events not captured in formal records.

5.15.2 Methods of Collection

Community-based data can be gathered through interviews, participatory mapping, focus group discussions, and household surveys.

5.15.3 Benefits

Such data helps identify recurring problems, validate technical findings, and prioritize interventions based on social impact. It also fosters community ownership of drainage improvement programs.

5.16 Data Validation and Quality Control

5.16.1 Need for Validation

Drainage studies depend on accurate data. Errors in survey levels, rainfall records, or land use mapping can lead to flawed designs. Validation ensures reliability.

5.16.2 Methods of Validation

- Cross-checking survey data with benchmarks.
- Comparing rainfall records from multiple sources.
- Ground-truthing GIS and remote sensing outputs.
- Reconciling community observations with technical measurements.

5.16.3 Quality Control Practices

Data should be checked for consistency, completeness, and logical accuracy. Outliers should be investigated, and metadata should record source, date, and method of collection.

5.17 Summary Points

1. Reliable drainage design depends on accurate and relevant data.
2. Core data categories include topography, rainfall, land use, soil, infrastructure, and outfall conditions.
3. Topographic and evelling surveys provide the foundation for slope and profile analysis.
4. Rainfall and hydrologic data must be interpreted in relation to evolving catchment conditions.

5. Land use, soil, and infrastructure mapping are essential for understanding runoff generation and drainage pathways.
6. GNSS and total station surveys provide complementary methods for accurate positioning and elevation control.
7. GIS integrates diverse datasets into a single platform for mapping, analysis, and management.
8. DEMs support terrain analysis but must be complemented by field surveys in built-up areas.
9. Remote sensing offers synoptic views and change detection, but requires ground verification.
10. Drone surveys provide high-resolution imagery and terrain models, enhancing inspection and communication.
11. Digital drainage databases organize survey, inspection, and maintenance data for long-term management.
12. Community-based data collection adds social insight and validates technical findings.
13. Data validation and quality control are essential to ensure reliability and prevent flawed designs.

5.18 References

1. Various authors. (n.d.). Relevant textbooks on engineering surveying and geospatial analysis for civil and water resources applications.
2. Various authors. (n.d.). Drainage mapping and municipal asset inventory manuals.
3. Various authors. (n.d.). Publications on GIS and remote sensing applications in drainage and flood assessment.
4. Tarboton, D. G. (n.d.). Terrain analysis using digital elevation models in hydrology. Proceedings.
5. Various authors. (n.d.). Studies and practical guidance on drone-based infrastructure inspection and imaging.
6. Various authors. (n.d.). Technical literature on DEM-based drainage analysis, flow accumulation mapping, and watershed analysis.

Chapter 6

Monitoring and Performance Evaluation

Learning Objectives

By the end of this chapter, the learner should be able to:

1. Explain the role of monitoring in drainage system management.
2. Identify instruments and techniques used for system performance assessment.
3. Evaluate drainage performance using technical indicators.
4. Propose a basic monitoring strategy for settlement drainage.
5. Distinguish between serviceability and reliability in drainage performance evaluation.

6.1 Introduction

A drainage system cannot be managed effectively unless its performance is observed over time. Design drawings and capacity calculations describe intended behavior, but only monitoring can show how the system actually performs under real field conditions, seasonal variation, maintenance cycles, and storm events. For this reason, monitoring and performance evaluation are essential components of modern drainage engineering.

This chapter explains how drainage systems are monitored, what parameters are measured, and how performance is assessed in urban, semi-urban, and village contexts. It also shows how routine observation, instrumentation, and evaluation indicators can support maintenance planning, rehabilitation, and long-term improvement of drainage assets.

The core idea of this chapter is that drainage management should be preventive rather than purely reactive. A system that is monitored regularly is more likely to reveal deterioration early, allowing corrective action before visible flooding, structural failure, or service breakdown occurs.

6.2 Need for Monitoring in Drainage Systems

6.2.1 Meaning of Monitoring

Monitoring is the systematic observation and recording of drainage-system behavior over time. Its purpose is to determine whether the system is functioning as intended, detect emerging problems early, and provide evidence for maintenance, upgrading, and operational decisions.

6.2.2 Why Post-Construction Monitoring Matters

Many drainage interventions fail not because the original design was entirely wrong, but because no mechanism exists to observe system behavior after construction. A drain may gradually lose capacity due to sedimentation, a culvert may become partially blocked, or a pumping arrangement may operate inefficiently without attracting attention until a storm produces visible flooding. Monitoring helps avoid this reactive mode of management.

6.2.3 Changing Performance Over Time

Drainage performance changes with time. Land-use change, road widening, utility installation, encroachments, and climate-related rainfall shifts may alter the hydraulic demand placed on the network. A system that once performed adequately may become insufficient years later even if no obvious structural failure is present.

6.2.4 Institutional Value

From an institutional perspective, monitoring supports accountability. It helps agencies verify whether maintenance is effective, whether service levels are being met, and whether capital investment priorities are justified by observed system behavior rather than by isolated complaints alone.

6.3 Rainfall and Water Level Monitoring

6.3.1 Core Monitored Variables

Rainfall and water level are among the most fundamental variables in drainage monitoring. Rainfall represents the input to the system, while water level shows the response of drains, culverts, channels, detention structures, ponds, and outfalls.

6.3.2 Rainfall Monitoring Methods

Rainfall may be monitored using standard rain gauges, tipping-bucket gauges, automatic weather stations, or data from nearby meteorological networks. For drainage systems, short-duration rainfall records are especially important because local flooding often results from intense storms over short periods rather than from daily totals alone. Continuous records also help link observed failures to actual storm characteristics.

6.3.3 Water Level Monitoring Methods

Water-level monitoring is useful for assessing surcharge, ponding, backwater effects, culvert submergence, and outfall behavior. In open drains, staff gauges and marked reference points may provide simple but useful field information. In more advanced systems, automated level sensors and telemetry-based devices may be installed at critical nodes.

6.3.4 Interpreting Rainfall and Water Level Together

Combined interpretation of rainfall and water-level data is especially powerful. A modest rainfall event producing unusually high water levels may indicate blockage or loss of capacity, while a normal water-level response during intense rainfall may confirm satisfactory performance. Monitoring both variables together therefore helps distinguish hydraulic deficiency from mere storm severity.

Table 6.1 Rainfall and water level monitoring parameters in drainage systems

A. Rainfall Monitoring Parameters

Category	**Parameter**	**Unit**	**Description / Purpose**
Basic Rainfall	Rainfall Amount	mm	Total rainfall recorded over a time period
	Rainfall Intensity	mm/hr	Rate of rainfall, critical for urban runoff
	Rainfall Duration	minutes / hours	Length of rainfall event
	Rainfall Frequency	events/year	Number of rainfall events
Temporal Characteristics	Time of Onset	hh:mm	Start time of rainfall
	Time to Peak Intensity	minutes	Time taken to reach peak rainfall
	Antecedent Dry Period	days	Gap between rainfall events
Spatial Attributes	Rain Gauge Location	Coordinates	Geo-location of rain gauge
	Areal Rainfall	mm	Average rainfall over catchment
	Catchment Coverage	%	Area influenced by rainfall station
Design & Analysis	Return Period	years	Design storm reference (e.g., 2yr, 25yr)
	IDF Curve Reference	Text	Intensity–Duration–Frequency basis
	Design Storm Type	Coded Text	Short burst / Continuous / Cloudburst
Data Quality	Recording Interval	min	Time resolution of data
	Missing Data Flag	Yes/No	Identifies data gaps
	Data Source	Text	AWS, IMD, Manual Gauge, Radar

B. Water Level Monitoring Parameters (Drains, Channels & Outfalls)

Category	**Parameter**	**Unit**	**Description/Purpose**
Water Level Basics	Water Level (Stage)	m	Depth of water above datum
	Normal Water Level	m	Typical dry-weather level
	Maximum Water Level	m	Observed peak during rainfall
	Freeboard Available	m	Vertical clearance before overflow
Hydraulic Response	Rate of Rise	cm/hr	Speed of water level increase
	Time to Peak Level	minutes	Lag between rainfall and peak stage
	Recession Time	hr	Time to return to normal level

Category	Parameter	Unit	Description/Purpose
	Flow Condition	Coded Text	Subcritical / Supercritical
Flood & Overflow Indicators	Bankfull Level	m	Threshold before overflow
	Overflow Occurrence	Yes/No	Indicates flooding
	Overflow Duration	hr	Length of surcharging
	Flood Depth	cm	Water depth during flood
Monitoring Location	Monitoring Point Type	Text	Manhole, drain section, outfall
	Sensor Location	Coordinates	Exact sensor position
	Critical Node ID	Text	Linked junction or asset ID
Instrumentation	Sensor Type	Text	Pressure, Ultrasonic, Radar
	Recording Interval	min	Sampling frequency
	Telemetry Status	Yes/No	Real-time data transmission
Data Reliability	Sensor Accuracy	± cm	Measurement precision
	Calibration Date	Date	Last sensor calibration
	Data Gaps Flag	Yes/No	Reliability check

C. Integrated Rainfall–Water Level Performance Indicators

Indicator	Unit	Purpose
Rainfall–Runoff Lag Time	minutes	Drain response efficiency
Peak Rainfall vs Peak Water Level Ratio	–	System sensitivity
Drain Capacity Utilization	%	Stress on drainage network
Frequency of Critical Level Exceedance	events/ year	Flood risk indicator
Drainage Recovery Time	hr	Post-event performance
Waterlogging Threshold Exceedance	Yes/No	Service level assessment

6.4 Flow measurement in Drains and Channels

6.4.1 Purpose of Flow Measurement

Flow measurement helps determine the actual discharge passing through a drainage element and provides direct evidence of hydraulic performance. It is especially useful in trunk drains, major channels, culverts, and outfalls where flow capacity and distribution are important for planning, diagnosis, and rehabilitation.

6.4.2 Direct and Indirect Methods

Flow may be measured directly or estimated indirectly. In open drains, discharge may be derived from measured depth and velocity, from section properties and rating relationships, or from temporary field methods such as float-based velocity approximation in simple conditions. In pipes and culverts, flow assessment may depend on water level, flow area, hydraulic condition, and hydraulic equations.

6.4.3 Where to Measure

Continuous discharge measurement across an entire drainage network is often impractical. Monitoring should therefore focus on representative or critical points such as flood-prone reaches, major trunk drains, culvert bottlenecks, detention outlets, and receiving channels.

6.4.4 Diagnostic Value

Flow measurement is useful not only for validating design assumptions but also for diagnosis. If flow at a downstream point is lower than expected, this may indicate upstream diversion, blockage, leakage, sedimentation, or poor hydraulic continuity. Repeated measurement over time can reveal whether performance is stable, improving, or deteriorating.

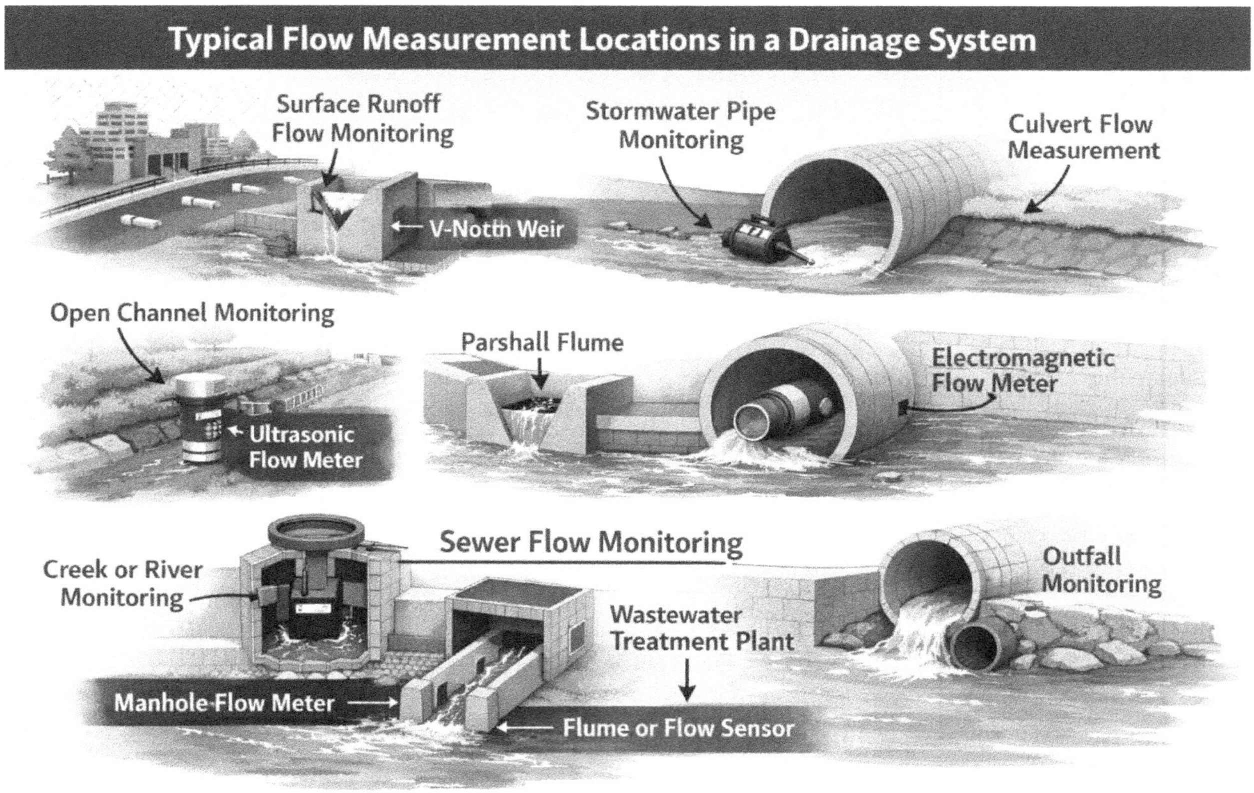

Figure 6.1 Typical flow measurement locations in a drainage system

6.5 Inspection of Underground Drains

6.5.1 Why Underground Drains Require Inspection

Underground drainage systems are difficult to evaluate because their internal condition is not visible from the ground surface. Yet they are highly susceptible to hidden problems such as sediment accumulation, joint displacement, infiltration, root intrusion, deformation, and partial or total collapse.

6.5.2 Basic Inspection Approach

Basic underground inspection begins at manholes, junction chambers, and inlet structures. Engineers may observe flow condition, sediment build-up, odor, surcharge marks, invert alignment, and evidence of structural distress. These observations are especially useful during dry-weather and post-rainfall inspection.

6.5.3 Internal Examination

In larger or more critical systems, internal examination may involve visual tools, probes, or camera-based inspection. Such work helps locate local defects, identify illegal connections, detect partial blockage, and understand why flooding occurs even when the surface system appears intact.

6.5.4 Need for Periodic Inspection

Inspection of underground drains should be periodic rather than purely complaint-driven. Many failures develop gradually, and routine inspection can reveal deterioration before it causes major service breakdown.

Table 6.2 Inspection observations for underground drainage systems

A. General Asset & Inspection Details

Field Name	Data Type	Description / Observation
Asset_ID	Text / Integer	Unique underground drain/manhole ID
Inspection_Date	Date	Date of inspection
Inspection_Type	Coded Text	Manual / CCTV / Drone (confined use)
Inspector_Name	Text	Inspection engineer/agency
Location_Reference	Text	Road name, chainage, landmark
Weather_Condition	Text	Dry / Wet / Rainy
Flow_Condition	Coded Text	No flow / Low / Moderate / High flow

B. Access Structure (Manhole / Inlet) Condition

Observation Parameter	Recorded Value	Remarks
Manhole Cover Condition	Good / Cracked / Missing	Safety and access issue
Frame Condition	Intact / Damaged / Tilted	Load bearing status
Manhole Depth	m	Vertical depth from ground
Step Irons / Ladder	Good / Corroded / Missing	Accessibility
Signs of Surcharge	Yes / No	Evidence of past overflow
Inflow at Manhole	Clean / Sewage / Mixed	Cross-connection indication

C. Internal Pipe / Drain Condition (CCTV / Visual)

Observation Parameter	Severity (Low/Med/High)	Description
Cracks / Fractures	Low / Medium / High	Longitudinal or circumferential
Joint Displacement	Low / Medium / High	Open or offset joints
Deformation	%	Ovality / structural distortion
Corrosion / Surface Erosion	Low / Medium / High	Chemical or abrasive
Pipe Collapse	Yes / No	Structural failure
Root Intrusion	Low / Medium / High	Tree root ingress

D. Hydraulic Performance Observations

Observation Parameter	Recorded Value	Remarks
Standing Water Depth	cm	Indicates low slope or blockage
Silt Deposition Depth	cm	Loss of effective section
Grease / Sludge Buildup	Low / Medium / High	Common in mixed drains
Obstruction Type	Solid waste / Debris / Tree roots	Cause of blockage
Flow Restriction	%	Estimated reduction

E. Structural & External Influences

Observation Parameter	Status	Details
Surrounding Soil Ingress	Yes / No	Through cracks/joints
Infiltration Signs	Yes / No	Groundwater seepage
Exfiltration Signs	Yes / No	Leakage to surrounding soil
Utility Crossings	Yes / No	Water, sewer, OFC conflict
Nearby Construction Impact	Yes / No	Induced settlement/damage

F. Environmental & Safety Observations

Observation Parameter	Status	Remarks
Odour Presence	Yes / No	Anaerobic/sewage mixing
Gas Hazard Indication	Yes / No	H_2S, methane risk
Vector Breeding Evidence	Yes / No	Stagnant water
Solid Waste Accumulation	Low / Medium / High	Public health impact

G. Defect Classification & Condition Rating

Parameter	Classification
Defect Class	Minor / Moderate / Severe / Critical
Structural Condition Grade	1 (Excellent) – 5 (Failed)
Hydraulic Condition Grade	1 (Free flow) – 5 (Blocked)
Overall Asset Condition	Good / Fair / Poor
Failure Risk Level	Low / Medium / High

H. Recommended Actions & Priority

Field Name	Description
Recommended_Action	Cleaning / Repair / Relining / Replacement
Method_Suggested	Jetting / Mechanical / Trenchless
Urgency_Level	Immediate / Short-term / Routine
Estimated_Cost_Class	Low / Medium / High
Followup_Inspection_Date	Target date for review
Remarks	Additional engineering notes

6.6 CCTV and Sensor-Based Diagnostics

6.6.1 CCTV Inspection

Advanced drainage monitoring increasingly uses camera systems and digital diagnostics to examine inaccessible structures. CCTV inspection allows the internal condition of pipes and covered drains to be assessed without excavation. It can reveal cracks, root intrusion, joint offset, deposits, illegal connections, pipe deformation, and collapse more reliably than surface observation alone.

6.6.2 Sensor-based Diagnostics

Sensor-based diagnostics may include level sensors, flow sensors, pump-operation sensors, gate-position sensors, sediment monitoring devices, and telemetry-linked alarms. At critical drainage nodes, such instrumentation can detect surcharge, pumping failure, abnormal rise in water level, or blockage during storms.

6.6.3 Value in Smart Drainage Systems

In more advanced drainage networks, sensors support real-time event tracking and operational response. Such systems can improve detection time, support decision-making, and strengthen preventive maintenance strategies.

6.6.4 Limits of Advanced Tools

Despite their advantages, advanced diagnostics should not replace engineering judgment. Their usefulness depends on proper installation, calibration, interpretation, and maintenance. In many practical settings, the best approach combines simple field inspection, community observation, and selective use of advanced tools.

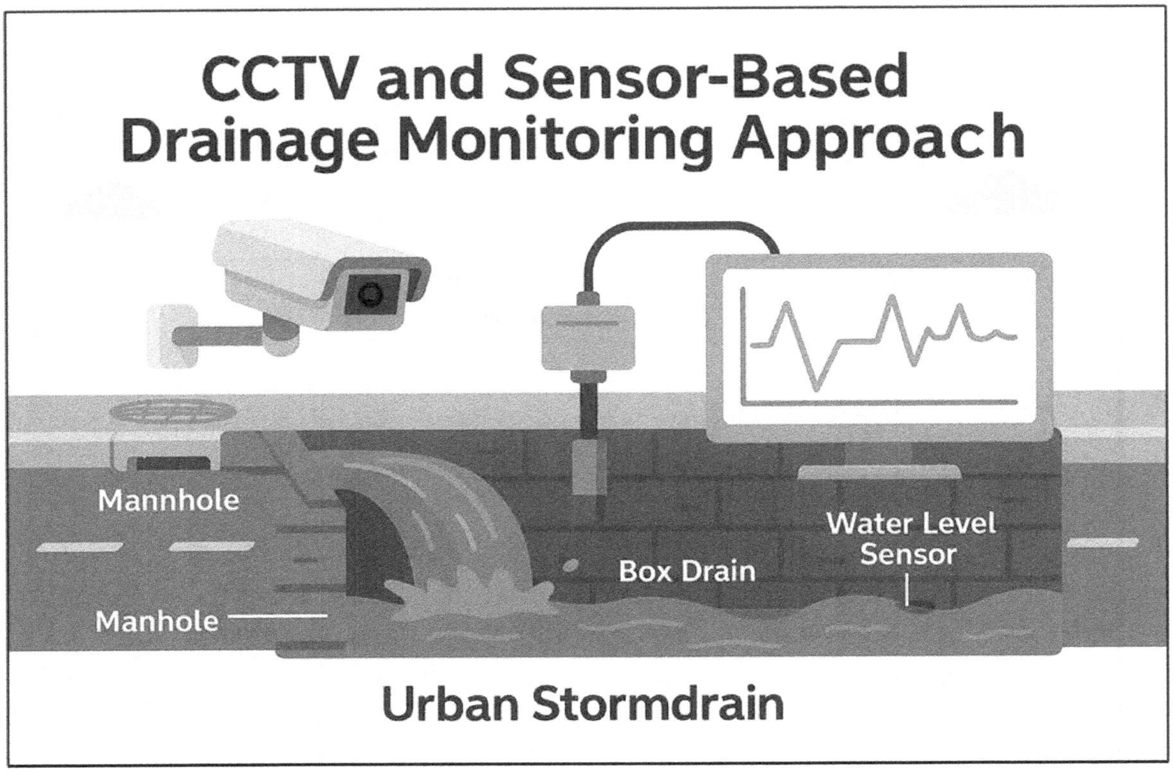

Figure 6.2 CCTV and sensor-based drainage monitoring approach

6.7 Indicators of Drainage Performance

6.7.1 Meaning of Performance Indicators

Performance indicators are measurable signs that show whether a drainage system is working effectively. They convert field observation into engineering evaluation and help compare the performance of different system components or management strategies.

6.7.2 Common Indicators

Common indicators include frequency of flooding, duration of waterlogging, maximum observed water depth, sediment accumulation rate, blockage frequency, maintenance interval, condition of structures, pump availability, and number of complaints from affected users.

6.7.3 Hydraulic, Operational, and Environmental Indicators

- Hydraulic indicators relate to conveyance efficiency, surcharge behavior, inlet capture, and outfall performance.
- Operational indicators relate to ease of maintenance, cleaning frequency, and operational reliability.
- Environmental indicators may include stagnant-water persistence, visible pollution, and the condition of receiving waters.

6.7.4 Need for quantification

Indicators should be practical and comparable over time. Describing a drain as "often blocked" is less useful than recording the number of cleaning events required during a monsoon season. Quantified indicators improve objectivity and support better technical decision-making.

Table 6.3 Technical, operational, and environmental indicators of drainage performance

A. Technical Performance Indicators

Indicator	Unit / Scale	Description / Performance Aspect
Design Capacity Utilization	%	Ratio of peak flow to design capacity
Hydraulic Efficiency	Qualitative	Ability to convey runoff without surcharge
Freeboard Availability	m	Vertical clearance before overflow
Structural Condition Index	1–5	Physical integrity of drains/pipes
Slope Adequacy	Yes / No	Compliance with minimum design slope
Drain Connectivity	Complete / Partial	System continuity and absence of dead ends
Frequency of Structural Failure	events/year	Incidence of cracks, collapses, or joint failure
Response Time to Rainfall	minutes	Lag between rainfall peak and flow/stage peak

B. Operational Performance Indicators

Indicator	Unit / Scale	Description / Performance Aspect
Overflow / Surcharge Events	events/year	Operational stress on the system
Waterlogging Incidents	events/year	Surface flooding due to drainage failure
Flood Duration	hours	Time taken to drain floodwater
Maintenance Frequency Compliance	%	Planned vs actual maintenance cycles
Siltation Level	% or cm	Reduction in effective flow area
Blockage Occurrence	events/year	Operational reliability
Drainage Recovery Time	hours	Time to restore normal conditions post-rainfall
Complaint Frequency	complaints/year	Citizen-reported service issues

C. Environmental Performance Indicators

Indicator	Unit / Scale	Description / Environmental Impact
Sewage Admixture	Yes / No	Presence of wastewater in storm drains
Water Quality Status	Clean / Polluted	Visual or sampled quality
Solid Waste Accumulation	Low / Medium / High	Environmental degradation risk
Odour Nuisance	Yes / No	Public health and comfort indicator
Vector Breeding Risk	Yes / No	Mosquito and pest generation
Groundwater Infiltration Risk	Low / Medium / High	Exfiltration / infiltration impacts
Impact on Receiving Water Body	Low / Medium / High	Effect on lakes, rivers, or wetlands
Environmental Compliance Status	Compliant / Noncompliant	Adherence to regulations and guidelines

D. Integrated Performance Indicators (Optional)

Indicator	**Purpose**
Drainage Performance Index (DPI)	Composite overall performance score
Flood Risk Index	Identifies high-risk urban zones
Service Level Benchmark Achievement	Compliance with urban SLB norms
Climate Resilience Indicator	Performance under extreme rainfall

6.8 Serviceability and reliability assessment

6.8.1 Serviceability

A drainage system may exist physically but still provide poor service. Serviceability refers to whether the system performs its intended function under normal or expected conditions. A serviceable system removes runoff within acceptable time, prevents harmful ponding, and maintains access and sanitary conditions.

If nuisance flooding occurs during ordinary storms, serviceability is low even where no major structural damage is visible. Serviceability is therefore closely related to the user's practical experience of drainage performance.

6.8.2 Reliability

Reliability refers to the consistency of performance over time and across varying conditions. A reliable drainage system continues to function despite seasonal variability, moderate blockage risk, and normal operational stress. A system that works only when recently cleaned, or fails during every moderately intense rainfall event, has low reliability.

6.8.3 Technical and User Perspectives

Both technical and user perspectives should be included in serviceability and reliability assessment. Engineers may evaluate water levels, hydraulic adequacy, structural integrity, and maintenance history, while residents may provide information on flooding frequency, access disruption, odor, and seasonal failure patterns. Combining both perspectives gives a more realistic picture of actual performance.

6.8.4 Importance for Decision-making

Serviceability and reliability are useful because they shift attention from mere infrastructure presence to actual infrastructure function. A physically complete drainage network may still be unsatisfactory if it performs inconsistently or fails under common conditions.

6.9 Monitoring Plans for Different Settlement Types

6.9.1 Urban Areas

Monitoring strategies should reflect settlement type, asset complexity, available resources, and risk level. In urban areas, monitoring plans may include rainfall gauges, water-level sensors at critical nodes, underground-drain inspection, pump-station records, flood-hotspot mapping, complaint tracking, and structured maintenance logs. Dense infrastructure and higher asset value justify more systematic and often technology-supported monitoring.

6.9.2 Semi-urban and Peri-urban Areas

In semi-urban and peri-urban areas, monitoring should focus on drainage continuity, culvert crossings, interaction with natural channels, new development impacts, and seasonal flood spread. Because such areas change rapidly, repeated field observation and periodic map updating may be more valuable than sophisticated but infrequent monitoring.

6.9.3 Village Settings

In village settings, monitoring can often be based on practical field inspection, rainfall observation, culvert and outfall checks, seasonal ponding records, and community reporting. Since resources may be limited, the monitoring approach should be simple, understandable, and directly linked to maintenance action. Community participation is especially useful where residents observe system behavior continuously.

6.9.4 Essential Elements of A Monitoring Plan

A good monitoring plan should clearly state what is to be monitored, where, how often, by whom, and for what purpose. It should also ensure that the information collected is actually used for maintenance and improvement rather than being stored without follow-up.

Table 6.4 Monitoring plans for urban, semi-urban, and village drainage systems

A. System Context & Objectives

Parameter	Urban Drainage	Semi-Urban Drainage	Village Drainage
Typical Drain Type	Closed storm sewers, box drains	Mix of open & closed drains	Open earth / lined drains
Rainfall Characteristics	High intensity, short duration	Moderate intensity	Low–moderate, longer duration
Primary Objective	Flood prevention & service continuity	Flood mitigation & asset protection	Quick runoff disposal & hygiene
Key Risks	Waterlogging, surcharge, asset failure	Local flooding, siltation	Stagnation, erosion, health issues

B. Monitoring Instruments & Methods

Monitoring Component	**Urban**	**SemiUrban**	**Village**
Rainfall Measurement	Automatic Weather Stations (AWS)	Tipping bucket / manual gauges	Manual rain gauge
Water Level Monitoring	Fixed sensors (ultrasonic/ pressure)	Staff gauge / limited sensors	Visual observation
Flow Monitoring	Modelled & sensorderived	Crosssectional estimation	Not normally monitored
Structural Inspection	CCTV + manual	Manual with limited CCTV	Visual inspection
Spatial Monitoring	GIS + drone/orthomosaic	GIS (limited)	Sketch maps
Underground Drain Inspection	CCTV / periodic manhole checks	Manhole inspection / occasional CCTV	Manhole and culvert checks
Pumping Station Monitoring	Automated logs / SCADA	Manual logs	Not applicable (usually absent)
Flood Hotspot Mapping	GIS-based mapping	Field sketch / periodic update	Community reporting
Complaint Tracking	Structured database / helpline	Local registers / municipal records	Informal reporting to village council
Maintenance Records	Digital logs / asset management system	Paper-based / semi-digital	Manual registers / notebooks

C. Monitoring Frequency

Parameter	Urban Drainage	Semi-Urban Drainage	Village Drainage
Rainfall Monitoring Frequency	Continuous / automated	Daily / event-based	Manual observation during rainfall
Water Level Monitoring Frequency	Continuous / hourly	Event-based / weekly	Seasonal / event-based
Underground Drain Inspection	Quarterly / annual	Annual / complaint-driven	Seasonal / after monsoon
Responsibility	Municipal engineering department	Local authority / development body	Village council / community volunteers

D. Use of Monitoring Information

Parameter	**Urban Drainage**	**Semi-Urban Drainage**	**Village Drainage**
Maintenance Planning	Preventive, scheduled	Preventive + reactive	Reactive, resource-based
Rehabilitation Decisions	Based on performance indicators	Based on field observation	Based on visible failure
Community Involvement	Limited (formal complaint channels)	Moderate (local feedback)	High (direct observation and reporting)

E. Key Monitoring Indicators

E(1). Technical & Hydraulic Indicators

Indicator	**Urban**	**SemiUrban**	**Village**
Capacity Utilization	✓	✓	✗
Freeboard Availability	✓	✓	✗
Surcharge Occurrence	✓	✓	✗
Flow Obstruction	✓	✓	✓
Structural Deterioration	✓	✓	✓

E(2). Operational & Service Indicators

Indicator	**Urban**	**SemiUrban**	**Village**
Waterlogging Incidents	✓	✓	✓
Flood Duration	✓	✓	✓
Maintenance Compliance	✓	✓	✓
Citizen Complaints	✓	Limited	Informal
Response Time	✓	✓	✗

E(3). Environmental & Public Health Indicators

Indicator	**Urban**	**SemiUrban**	**Village**
Sewage Admixture	✓	✓	✓
Solid Waste Accumulation	✓	✓	✓
Odour & Nuisance	✓	✓	✓
Vector Breeding Risk	✓	✓	✓
Soil Erosion / Scouring	✗	✓	✓

F. Data Management & Responsibility

Aspect	**Urban**	**SemiUrban**	**Village**
Data Platform	GISbased dashboard	Spreadsheet / GIS	Register / logbook
Responsible Agency	ULB / Smart City SPV	Nagar Panchayat / Block	Gram Panchayat
Skill Requirement	Engineers + IT staff	Junior engineers	Trained local staff
Use of Data	Realtime control & planning	Planning & prioritization	Maintenance & hygiene

G. Suggested Outputs

Output	**Urban**	**SemiUrban**	**Village**
Drainage Performance Index	✓	✓	Simplified
Flood Hotspot Maps	✓	✓	✗
Maintenance Priority List	✓	✓	✓
Annual Monitoring Report	✓	✓	Seasonal note

Monitoring and performance evaluation are indispensable for effective drainage management. They transform static infrastructure into dynamic systems that can be observed, assessed, and improved over time. By combining rainfall and water-level monitoring, flow measurement, underground inspection,

CCTV and sensor-based diagnostics, and performance indicators, engineers and communities can ensure that drainage systems remain serviceable and reliable.

The chapter emphasizes that monitoring must be tailored to settlement type—urban, semi-urban, or village—while always maintaining clarity of objectives, frequency, responsibility, and practical use of collected data. Ultimately, preventive monitoring strengthens resilience, reduces failures, and supports sustainable drainage management.

6.10 Data Management in Drainage Monitoring

Effective monitoring requires not only field observation but also systematic data management. Rainfall, water-level, flow, and inspection records must be stored in a structured format that allows retrieval, comparison, and analysis.

- Digital databases are preferable to paper records, as they support long-term archiving and integration with GIS.
- Metadata such as location, date, instrument type, and calibration status should accompany each dataset to ensure reliability.
- Data visualization tools (graphs, charts, maps) help interpret trends and communicate findings to stakeholders.

6.11 Indian Practice Notes

Monitoring and performance evaluation are indispensable for effective drainage management. They transform static infrastructure into dynamic systems that can be observed, assessed, and improved over time. By combining rainfall and water-level monitoring, flow measurement, underground inspection, CCTV and sensor-based diagnostics, and performance indicators, engineers and communities can ensure that drainage systems remain serviceable and reliable.

The chapter emphasizes that monitoring must be tailored to settlement type—urban, semi-urban, or village—while always maintaining clarity of objectives, frequency, responsibility, and practical use of collected data. Ultimately, preventive monitoring strengthens resilience, reduces failures, and supports sustainable drainage management.

6.11.1 Relevance to Indian Drainage Management

- In Indian drainage practice, monitoring is often stronger during crisis response than during routine asset management. This creates a tendency toward complaint-driven action rather than evidence-based preventive maintenance.

6.11.2 Need for Practical Monitoring Systems

- Given the wide variation in municipal capacity, many Indian systems benefit from simple but structured monitoring frameworks built around rainfall logs, water-level marks, flood hotspot records, desilting schedules, culvert checks, and periodic CCTV use where feasible. Such approaches are often more realistic than highly complex monitoring systems that cannot be maintained.

6.11.3 Scope for Smart Monitoring

- At the same time, recent work on smart drainage monitoring suggests that sensor-based systems can improve real-time awareness, reduce response delays, and support better event management in urban areas. Their long-term value, however, depends on calibration, maintenance, communications reliability, and institutional integration.

6.12 Role of Community Participation

Community involvement strengthens monitoring, especially in resource-limited settings. Residents often observe drainage behavior continuously and can provide valuable information on:

- Frequency and duration of waterlogging.
- Locations of recurring blockages.
- Seasonal changes in performance.
- Public health impacts such as odor or mosquito breeding.

Structured reporting channels, such as mobile apps or village registers, can convert informal observations into actionable monitoring data.

6.13 Integration with Maintenance Planning

Monitoring is most effective when directly linked to maintenance planning.

- **Routine cleaning schedules** should be adjusted based on observed siltation or blockage frequency.
- **Repair priorities** should reflect structural condition ratings and failure risk levels.
- **Budget allocation** should be justified by performance indicators rather than anecdotal complaints.

This integration ensures that monitoring results are not just archived but actively used to improve drainage serviceability and reliability.

6.14 Challenges in Monitoring And Evaluation

Drainage monitoring faces several challenges:

- **Resource constraints**: Limited funds for instruments, staff, and data systems.
- **Technical limitations**: Inaccurate sensors, poor calibration, or incomplete coverage.
- **Institutional gaps**: Weak accountability, fragmented responsibilities, or lack of follow-up.
- **Environmental variability**: Unpredictable rainfall patterns and climate change impacts.

Addressing these challenges requires a balance between advanced tools and practical, low-cost methods suited to local conditions.

6.15 Future Directions in Drainage Monitoring

Emerging approaches include:

- **Smart sensors** with real-time telemetry and automated alerts.
- **Remote sensing** for flood mapping and catchment-scale monitoring.
- **Machine learning models** to predict performance based on rainfall and land-use data.
- **Citizen science platforms** enabling residents to contribute observations digitally.

These innovations can enhance resilience and efficiency, but they must be adapted to local contexts and supported by institutional capacity.

6.16 Summary Points

Monitoring and performance evaluation are essential for effective drainage management.

1. They provide evidence of system behavior under real conditions.
2. Rainfall, water-level, and flow monitoring form the technical foundation.
3. Underground inspection, CCTV, and sensor-based diagnostics reveal hidden defects.
4. Performance indicators quantify hydraulic, operational, and environmental aspects.
5. Serviceability and reliability assessments combine technical and user perspectives.
6. Monitoring plans must be tailored to urban, semi-urban, and village contexts.
7. Data management, community participation, and integration with maintenance planning ensure practical impact.

Ultimately, preventive monitoring strengthens accountability, improves service levels, and supports sustainable drainage infrastructure.

6.17 References

1. Various authors. (n.d.). *Manuals and guidance on stormwater drainage management and operations.*
2. Various authors. (n.d.). *Storm sewer CCTV inspection manuals and related inspection specifications.*
3. Various authors. (n.d.). *Guidance documents on water-level sensing and low-cost monitoring in drainage applications.*
4. Various authors. (n.d.). *Technical publications on sensor-based and smart drainage monitoring systems.*
5. Various authors. (n.d.). *Research on reliability assessment and performance evaluation of drainage systems.*
6. Various authors. (n.d.). *Textbooks and field manuals on infrastructure monitoring, stormwater maintenance, and drainage asset management.*

Chapter 7

Planning Principles and Design Standards

Learning Objectives

By the end of this chapter, the learner should be able to:

1. Define service levels and planning objectives for drainage systems.
2. Select basic design criteria appropriate to settlement type.
3. Explain the importance of standards and regulatory compliance.
4. Integrate drainage planning with broader settlement infrastructure planning.
5. Explain how return periods, risk levels, and phased development influence drainage planning decisions.

7.1 Introduction

Drainage planning is the process of deciding how runoff will be managed across a settlement in a safe, efficient, and sustainable manner. It links field diagnosis, hydrologic understanding, engineering design, land-development control, and long-term infrastructure management. Without sound planning, even well-designed individual drains may fail because they are disconnected from the larger catchment system or from the realities of future growth.

This chapter introduces the planning principles and design standards that guide drainage development in urban, semi-urban, and village settings. It explains how service goals are defined, how planning units are selected, how return periods and risk levels are chosen, and how drainage should be integrated with roads, sanitation, land development, and resource limitations.

The central message is that drainage planning is broader than hydraulic sizing alone. It involves deciding what level of protection is required, what degree of risk is acceptable, what standards should apply, and how the drainage system should fit within the broader physical and institutional structure of the settlement.

7.2 Planning Objectives for Drainage Systems

7.2.1 Need to Define Objectives

The first step in drainage planning is to define the objectives of the system clearly. A drainage network is not built merely to convey water away; it is planned to provide a level of protection and service appropriate to settlement type, land use, public safety requirements, and environmental setting.

7.2.2 Typical Objectives

Typical planning objectives include preventing frequent flooding of roads, homes, and public facilities; reducing the duration and depth of waterlogging; protecting building foundations and transport corridors; maintaining hygienic conditions by avoiding stagnant water; preserving natural drainage paths where feasible; reducing downstream adverse impacts; and ensuring maintainability and long-term functionality.

These objectives show that drainage planning is both hydraulic and social in character. A successful system is expected not only to carry runoff but also to protect access, support health, preserve infrastructure, and avoid transfer of problems to downstream areas.

7.2.3 Context-specific Objectives

Objectives should be realistic and measurable. A roadside drain in a small settlement may be intended to prevent persistent ponding during routine seasonal storms, while a trunk drain in a major city may be expected to carry larger design flows associated with more severe rainfall. Likewise, a village drainage system may prioritize rapid removal of water from habitation areas and access roads rather than the standard of a full urban storm sewer.

7.2.4 Role in Planning And Evaluation

Clear planning objectives help define what level of investigation, design, and investment is necessary. They also provide the basis for later performance evaluation, asset management, and prioritization of improvements.

7.3 Service Levels and Performance Targets

7.3.1 Meaning of Service Level

A service level describes the expected standard of drainage performance for a given area or type of infrastructure. It translates broad planning objectives into practical engineering targets. Without defined service levels, systems are often judged only after failure occurs rather than against agreed performance expectations.

7.3.2 Examples of Service Levels

Examples include:

- No prolonged ponding on major roads during routine design storms
- Limited allowable water depth in residential streets
- Rapid drainage of public spaces within a specified time after rainfall
- Protection of critical facilities such as hospitals and emergency routes
- Prevention of recurring waterlogging around houses in village settlements

These examples show that service levels should relate to the function and importance of the area. A major urban corridor usually requires a higher level of service than a low-use open space, and critical public infrastructure generally justifies stronger drainage performance targets.

7.3.3 Performance Targets

Performance targets make service levels operational. They may refer to maximum water depth, acceptable duration of ponding, allowable flooding frequency, or overtopping limits at selected structures. In larger systems, service targets may distinguish between minor drainage for frequent events and major drainage for rarer severe events.

7.3.4 Balancing Social Need and Feasibility

Service levels should reflect both engineering feasibility and social need. High-value, densely built, or safety-critical zones typically require higher service levels than low-risk areas, but resource constraints must also be recognized so that targets remain achievable and maintainable.

Table 7.1 Illustrative service levels and performance targets for different settlement contexts

A. Context-Specific Drainage Challenges

Context	**Dominant Challenges**
Coastal Settlements	High groundwater table, tidal backflow, storm surge, intense rainfall
Hill Settlements	Steep slopes, high runoff velocity, erosion, slope instability
Floodplain Settlements	River backwater effect, prolonged inundation, flat terrain

B. Hydraulic & Flood-Related Performance Targets

Indicator	**Coastal Areas**	**Hill Areas**	**Floodplain Areas**
Design Storm Return Period	10–25 years	10–15 years	5–10 years
Maximum Waterlogging Depth	≤ 15 cm	≤ 10 cm	≤ 30–50 cm
Maximum Waterlogging Duration	≤ 2 hours	≤ 1 hour	≤ 6–12 hours
Acceptable Flood Frequency	≤ 1 event in 2 years	Rare / exceptional	Seasonal flooding acceptable
Peak Drain Capacity Utilization	≤ 90%	≤ 85%	≤ 95%

C. Structural & Asset Safety Targets

Indicator	**Coastal**	**Hill**	**Floodplain**
Structural Condition Rating	≥ 4 (Good)	≥ 4 (Good)	≥ 3 (Fair)
Siltation / Sediment Tolerance	≤ 20%	≤ 15%	≤ 30%
Erosion / Scour Occurrence	Minimal	Zero tolerance	Controlled
Backflow Protection	Mandatory	Not normally required	Mandatory at outfalls
Asset Failure Acceptance	Nil	Nil	Very limited

D. Operational Performance Targets

Indicator	**Coastal**	**Hill**	**Floodplain**
Overflow / Surcharge Events	≤ 1 per year	Ideally zero	≤ 2 per year
Drainage Recovery Time	≤ 3 hours	≤ 2 hours	≤ 12–24 hours
Maintenance Frequency	Pre & post-monsoon + monthly	Frequent desilting (pre-monsoon)	Seasonal / pre-flood
Emergency Response Time	< 4 hours	< 2 hours	< 6 hours

E. Environmental & Public Health Targets

Indicator	**Coastal**	**Hill**	**Floodplain**
Sewage Mixing in Storm Drains	Zero tolerance	Zero tolerance	Zero tolerance
Saline Ingress Impact	Controlled	Not applicable	Not applicable
Soil / Slope Stability Risk	Low	Very Low (critical)	Moderate
Vector & Disease Risk	Low	Low	Medium (controlled)
Impact on Natural Systems	Minimal (mangroves, creeks)	Minimal (streams)	Managed seasonal impact

F. Monitoring & Preparedness Service Levels

Aspect	**Coastal**	**Hill**	**Floodplain**
Rainfall Monitoring	Dense AWS network	High-resolution gauges	Basin-level gauges
Water Level Monitoring	Drains + tidal gauges	Key drains only	Drains + river gauges
Early Warning System	Mandatory (rain + tide)	Event-based alerts	Flood forecasting linkage
Inspection Frequency	Monthly	Frequent, event-based	Seasonal
Data Review Cycle	Seasonal & annual	Annual	Pre & post-flood

G. Context-Sensitive Serviceability Interpretation

Aspect	**Coastal**	**Hill**	**Floodplain**
Serviceability Focus	Tidal resilience & quick drainage	Safe conveyance without erosion	Managed flooding
Reliability Focus	Performance during cyclone / surge	Performance during cloudburst	Performance during river floods
Failure Consequence	High (urban disruption)	Very high (landslide risk)	High but partly predictable

Key Planning Implications

- Coastal systems must prioritize backflow control, pumping, and redundancy.
- Hill systems demand structural robustness, slope safety, and rapid conveyance.
- Floodplain systems focus on adaptive service levels rather than flood elimination.

7.4 Watershed-based Planning Approach

7.4.1 Why Watershed Thinking Matters

Drainage planning is most effective when based on the natural movement of water across the landscape rather than on administrative boundaries alone. A watershed-based approach recognizes that runoff follows catchment limits, topography, and natural channels regardless of plot lines or jurisdictional divisions.

7.4.2 Planning From Upstream to Downstream

Under this approach, the drainage system is planned from upstream to downstream, considering runoff generation, concentration, conveyance, storage, and final disposal. This enables engineers to identify major drainage corridors, contributing sub-catchments, bottlenecks, and downstream limitations. It also helps prevent the common mistake of solving a local problem by transferring flooding elsewhere.

7.4.3 Relevance Across Settlement Types

Watershed-based planning is especially important in peri-urban areas where rapid development disrupts older rural drainage routes. It is also relevant in urban areas where multiple local networks ultimately discharge to a common receiving system. In villages, this approach helps ensure that settlement drainage does not adversely affect adjacent fields, ponds, or downstream habitation zones.

7.4.4 Integrated Planning Benefit

A watershed-based approach supports integrated planning of local, intermediate, and trunk drainage. It also strengthens coordination between land development, flood management, and environmental protection.

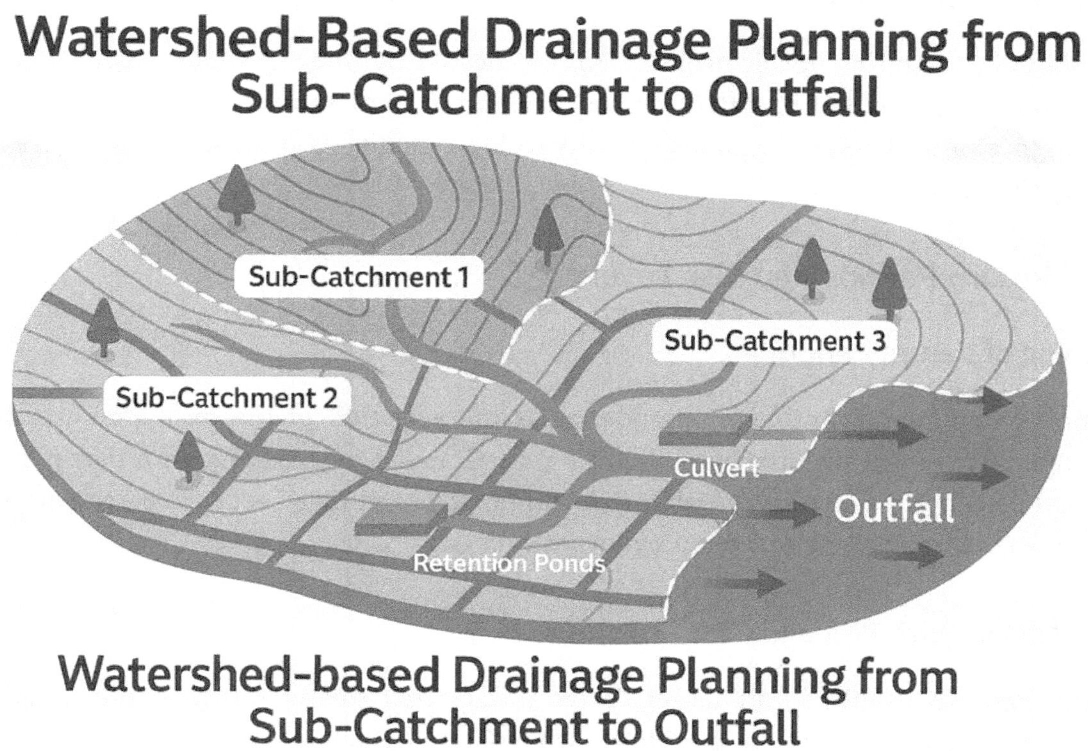

Figure 7.1 Watershed-based drainage planning from sub-catchment to outfall

7.5 Administrative Versus Hydrologic Planning Units

7.5.1 Two Different Planning Logics

Although water moves according to topography, infrastructure projects are usually planned and financed within administrative areas such as wards, municipalities, development blocks, or village jurisdictions. This creates an important distinction between hydrologic planning units and administrative planning units.

7.5.2 Hydrologic Planning Unit

A hydrologic planning unit is based on catchment behavior, drainage boundaries, and flow connectivity. It reflects how runoff is actually generated and conveyed through the landscape.

7.5.3 Administrative Planning Unit

An administrative unit is based on governance, budgeting, program implementation, and institutional responsibility. In practice, drainage planning must take both systems into account. A flooding problem may originate in one administrative area but cause damage in another, while corrective action may depend on who holds financial and legal authority.

7.5.4 Need for Coordination

Effective planning requires local projects to be nested within a wider hydrologic framework. Neighbourhood drains should be related to trunk drains and final outfalls, not treated as isolated works. Failure to align administrative decisions with hydrologic reality is a major cause of fragmented and ineffective drainage infrastructure.

7.6 Design Return Periods And Risk Levels

7.6.1 Meaning of Return Period

A key planning decision in drainage engineering is the selection of the design return period, which represents the average interval associated with a rainfall event or discharge of a particular magnitude. Return period selection reflects acceptable risk rather than certainty, because a system designed for a given return period can still be exceeded by larger events.

7.6.2 Relationship With Risk and Cost

Higher return periods usually imply larger design discharges, greater safety margins, and higher cost. The selected return period should therefore depend on the importance of the protected area, the consequences of flooding, and the economic practicality of providing additional capacity. Major roads, commercial districts, and critical public facilities often justify higher standards than local lanes or low-value open areas.

7.6.3 Different Standards for Different Components

Different parts of a drainage system may be designed for different return periods. Local roadside drains, branch drains, trunk drains, culverts, and detention elements do not necessarily require the same design standard. It is common to distinguish between minor systems for frequent storms and major systems or overland flow routes for rarer extreme events.

7.6.4 Planning Significance

Return period selection is not only a hydraulic issue but also a policy decision involving risk tolerance, land use, service expectation, and affordability. It should therefore be stated clearly and applied consistently across the drainage plan. Indian guidance has historically discussed return periods such as 1 in 5 years and 1 in 10 years for different urban settings, illustrating the importance of matching design level to settlement context.

Table 7.2 Illustrative return periods and risk levels for drainage components

A. Local Collection Components

Drainage Component	Typical Function	Illustrative Design Return Period	Risk Level if Exceeded	Planning Interpretation
Roof drains & downpipes	Building-scale runoff collection	2–5 years	Low	Temporary surcharge acceptable
Curb inlets / gully pits	Roadside runoff interception	2–5 years	Low	Short-term ponding tolerable
Local street drains	Convey runoff from local streets	5 years	Low–Moderate	Minor waterlogging permissible
Roadside open drains	Neighborhood drainage	5–10 years	Moderate	Service disruption limited in duration

B. Secondary Conveyance Components

Drainage Component	Typical Function	Illustrative Design Return Period	Risk Level if Exceeded	Planning Interpretation
Secondary storm drains	Collect flow from multiple streets	10 years	Moderate	Controlled surcharge tolerated
Cross-drain culverts	Drainage across roads/ embankments	10–25 years	Moderate–High	Traffic interruption risk
Junction chambers / manholes	Flow convergence points	10–25 years	Moderate	Intermediate flooding risk
Minor natural channels	Carry combined urban runoff	10–25 years	Moderate–High	Local flooding impacts possible

C. Primary Trunk & Outfall Components

Drainage Component	Typical Function	Illustrative Design Return Period	Risk Level if Exceeded	Planning Interpretation
Trunk storm drains	Convey bulk watershed runoff	25–50 years	High	City-scale flooding risk
Major box drains	High-capacity urban conveyance	25–50 years	High	Critical infrastructure protection
Pumping stations (if any)	Overcome flat terrain/backflow	25–50 years	High	System-wide disruption
Outfalls to river/sea	Final discharge point	50–100 years	Very High	Failure causes widespread inundation

D. Storage and Mitigation Components

Drainage Component	**Typical Function**	**Illustrative Design Return Period**	**Risk Level if Exceeded**	**Planning Interpretation**
Detention ponds	Peak flow attenuation	25–50 years	Moderate–High	Controlled overtopping acceptable
Retention ponds	Permanent storage & delay	25–50 years	High	Must avoid downstream surge
Floodplains (planned)	Excess flow accommodation	50–100 years	Managed risk	Controlled inundation strategy
Emergency overflow paths	Safe flood routing	Extreme events	Low systemic risk	Damage minimization approach

E. Context-Sensitive Risk Interpretation

Settlement / Terrain Context	**Emphasis on Return Periods**	**Risk Management Approach**
Dense urban core	Higher return periods for trunk systems	Flood prevention
Peri-urban / mixed land use	Balanced return periods	Flood mitigation
Coastal areas	High return period + backflow safety	Compound risk (rain + tide)
Hill settlements	Moderate return period, high velocity control	Structural safety
Floodplain settlements	Moderate return period with adaptive flooding	Risk accommodation

Table 7.3 Return Periods and Risk Levels from Sub-Catchment to Outfall

A. Watershed Hierarchy and Planning Logic

Watershed Level	Hydrological Role	Planning Focus
Micro / Local Catchment	Rainfall capture and initial runoff	Serviceability
Sub-Catchment	Flow concentration and routing	Reliability under design storms
Catchment / Sub-Basin	Peak attenuation and conveyance	Flood mitigation
Basin Outfall	Discharge control and system safety	Flood risk management

B. Integration of Drainage Components, Return Periods, and Risk

Watershed Planning Level	**Drainage Component Examples**	**Illustrative Design Return Period**	**Risk Level if Exceeded**	**Performance Interpretation**
Micro-Catchment (Plot / Street)	Roof drains, curb inlets, local street drains	2–5 years	Low	Short-term ponding acceptable

Watershed Planning Level	Drainage Component Examples	Illustrative Design Return Period	Risk Level if Exceeded	Performance Interpretation
Sub-Catchment	Roadside drains, secondary storm sewers	5–10 years	Low–Moderate	Minor localized flooding tolerable
Sub-Catchment Control Points	Junctions, minor culverts	10–25 years	Moderate	Intermediate service disruption
Catchment / Trunk System	Trunk drains, major box drains	25–50 years	High	City-level flooding risk
Catchment Storage Nodes	Retention / detention ponds	25–50 years	High	Controlled overtopping allowed
Basin Outfall	River / lake / sea outfall structures	50–100 years	Very High	Widespread inundation risk
Extreme Event Pathways	Floodplains, emergency overflow routes	> design event	Managed risk	Damage minimization

C. Risk Accumulation Across the Watershed

Planning Aspect	Upstream (Local)	Midstream (Sub-Catchment)	Downstream (Outfall)
Flow Volume	Low	Moderate	Very High
Failure Consequence	Local nuisance	Neighborhood disruption	City-scale flooding
Acceptable Overtopping	Yes	Limited	No
Design Conservatism	Low	Medium	High

D. Linking Return Periods with Serviceability and Reliability

Watershed Level	Serviceability Objective	Reliability Objective
Micro / Local	Maintain usability under frequent storms	Not critical
Sub-Catchment	Prevent repeated waterlogging	Maintain function during design peaks
Catchment / Trunk	Minimize flood extent	Operate under severe storms
Outfall	Safely pass extreme flows	Prevent catastrophic failure

E. Practical Use in Watershed-Based Planning

Planning Step	How the Table Is Used
Catchment Delineation	Assign drainage components to watershed level
Design Standard Selection	Select return period by watershed hierarchy
Risk Assessment	Identify components with unacceptable risk
Storage Planning	Locate detention where risk escalates
Performance Evaluation	Check serviceability vs reliability compliance
Climate Adaptation	Increase return periods at critical nodes

7.7 Codes, Standards, And Guidelines

7.7.1 Role of Standards

Drainage planning and design should follow recognized codes, standards, manuals, and technical guidelines. These documents provide accepted criteria for hydrologic analysis, hydraulic sizing, structural details, materials, slopes, freeboard, access, and safety requirements.

7.7.2 Why Standards Matter

Standards promote consistency, minimum quality, and regulatory compliance. They also reduce uncertainty by offering tested design ranges and commonly accepted performance criteria. In urban drainage, such standards may address pipe sizing, inlet spacing, culvert design, road drainage, and stormwater management.

7.7.3 Limits of Standardization

Standards should not be applied mechanically. Engineering judgment remains essential because local terrain, land use, maintenance capacity, and climate conditions may differ from the assumptions behind generic manuals. In some cases, national or state guidance may need adaptation to fit local settlement conditions.

7.7.4 Baseline and Adaptation

A sound drainage plan therefore combines compliance with codes and manuals with informed interpretation of site reality. Standards provide the baseline, but good planning may require going beyond minimum rules where local risk or future growth justifies it.

7.8 Integration With Roads, Sanitation, and Land Development

7.8.1 Why Integration is Necessary

Drainage cannot be planned in isolation because it interacts directly with roads, sanitation systems, plot grading, utilities, buildings, and public space. When these elements are planned separately, the result is often conflict, redundancy, inefficiency, or outright failure.

7.8.2 Road Interaction

Road design strongly influences drainage behavior. Road crowns, embankments, kerbs, shoulders, side slopes, and culvert crossings determine how runoff is intercepted and where it flows. A road project can therefore improve or worsen drainage depending on whether it is coordinated with the catchment system.

7.8.3 Sanitation Interaction

Sanitation infrastructure may affect drainage through greywater discharge, sewer crossings, wastewater intrusion, and contamination of open drains. This interaction is important because a drainage system carrying mixed stormwater and wastewater may experience blockage, odor, pollution, and reduced hydraulic efficiency.

7.8.4 Land Development Interaction

Land development also has a major impact. Plot filling, basement construction, surface paving, boundary walls, and encroachment on natural drainage channels all alter runoff generation and flow paths. If drainage planning does not keep pace with development control, frequent flooding and waterlogging become likely.

7.8.5 Meaning of Integrated Planning

Integrated planning means that drainage is considered from the beginning in road layouts, subdivision plans, utility coordination, building approvals, and expansion proposals. In villages, this also includes local roads, ponds, access crossings, and agricultural drainage connections.

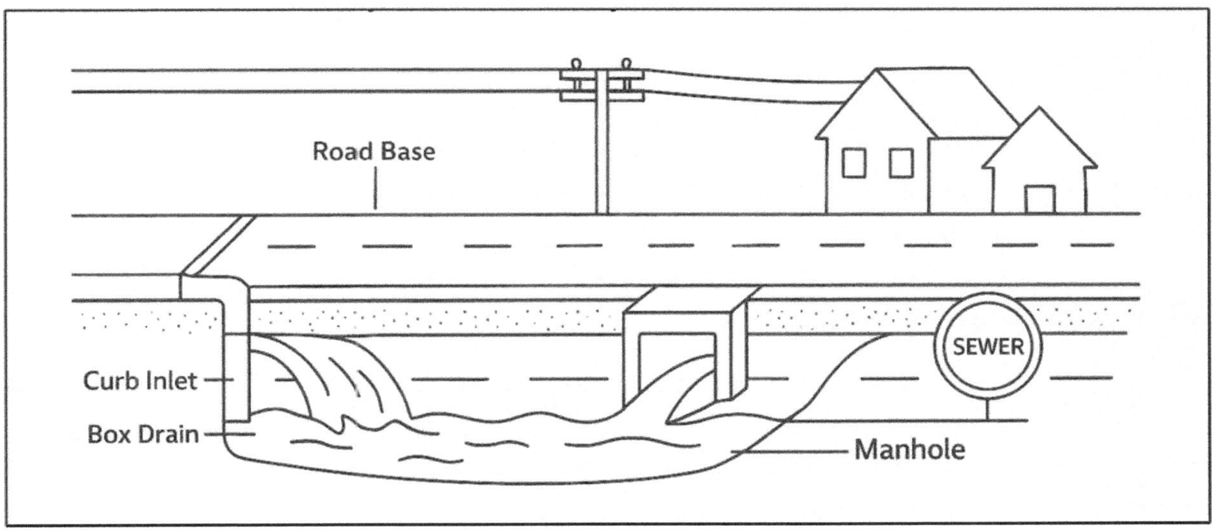

Figure 7.2 Integration of drainage with roads, sanitation, utilities, and land development

7.9 Planning Under Resource Constraints

7.9.1 Nature of Constraints

Many settlements, especially small towns, peri-urban areas, and villages, face financial, technical, institutional, land, and maintenance constraints. Drainage planning in such settings must therefore be realistic as well as technically sound.

7.9.2 Practical Planning Under Limits

Planning under resource constraints does not mean abandoning engineering principles. It means selecting practical priorities, maintainable layouts, staged improvements, context-appropriate standards, and materials that can be sustained locally.

7.9.3 Example of Appropriate Choice

A simple, well-graded open drain with proper outfall and regular maintenance may perform better than an expensive covered drain that cannot be cleaned properly. This captures an important design-planning principle that sustainability includes operational realism.

7.9.4 Space and Institutional Limitations

Limited space is a major issue in dense urban corridors and informal settlements. In such cases, planning may rely on localized widening, selective covering, improved inlets, culvert replacement, and protection of overland flow routes rather than complete reconstruction. Institutional and operational capacity must also be considered, because a technically advanced system may fail if it cannot be maintained.

7.10 Phased Drainage Development Plans

7.10.1 Why Phasing is Important

Drainage systems are often developed progressively rather than all at once. A phased drainage development plan sets out how improvements will be implemented over time while remaining consistent with the long-term vision of the network.

7.10.2 Relevance in Growing Settlements

Phasing is especially important in expanding urban and peri-urban areas where future development will increase runoff and alter drainage demand. It is also useful where funding is limited and interventions must be sequenced over several years. A phased plan helps avoid piecemeal construction that later proves incompatible with trunk systems or outfall capacity.

7.10.3 Typical Phases

Typical phases include:

- **Immediate works** for critical bottlenecks and flooding hotspots
- **Medium-term strengthening** through branch and trunk upgrades
- **Long-term expansion** through storage measures, sustainable drainage features, and climate-resilient improvements

7.10.4 Technical Compatibility

Phased development must ensure that each stage is technically compatible with the eventual full system. Interim measures should not obstruct future trunk alignments, reduce outfall capacity, or create maintenance burdens. Proper sequencing allows settlements to build resilience progressively while avoiding wasteful or conflicting investments.

7.11 Indian Practice Notes

7.11.1 Standards and Manuals

In India, drainage planning is influenced by manuals and guidance prepared for urban stormwater design and municipal engineering practice. These documents provide a planning and design framework but still require adaptation to local terrain, settlement pattern, and maintenance realities.

7.11.2 Return Period Considerations

Indian discussions on drainage design have highlighted the importance of selecting return periods appropriate to land use and urban density, including recommendations such as 1 in 5 years for some residential areas and 1 in 10 years for denser conditions. This reinforces the need to match design risk to settlement context rather than applying a single standard

7.11.3 Need for Integrated Planning

Rapid urbanization, informal growth, road expansion, and encroachment on natural drainage have made integrated drainage planning increasingly important in Indian cities and fringe areas. Planning that combines hydrology, land use, roads, and outfall management is better suited to these changing conditions than isolated component design.

7.12 Planning Implications

The transformed chapter makes clear that drainage planning begins before detailed design and continues beyond it. Objectives, service levels, planning units, risk tolerance, design standards, integration needs, and phasing strategy all shape the eventual success of the system.

The chapter also shows that a technically good drainage plan must be hydrologically coherent and institutionally workable. A system planned only by administrative convenience may fail hydrologically, while a system planned only by catchment logic may fail institutionally if responsibility and funding are unclear.

Finally, the chapter highlights that good drainage planning must be adaptive. Urbanisation, changing runoff behaviour, and infrastructure expansion require planning approaches that combine standards with flexibility, watershed logic with governance practicality, and technical ambition with operational realism.

7.13 Summary Points

1. Drainage planning links hydrology, design, land development, infrastructure coordination, and long-term management.
2. Planning objectives should be clear, realistic, and measurable.
3. Service levels translate planning goals into performance expectations.
4. Performance targets make service levels operational and assessable.
5. Watershed-based planning aligns drainage design with natural runoff movement.
6. Administrative and hydrologic planning units must be coordinated.
7. Return period selection reflects acceptable risk, consequence, and affordability.
8. Different drainage components may require different design standards.
9. Codes and manuals provide essential guidance but must be applied with judgment.
10. Drainage must be integrated with roads, sanitation, land development, and utilities.
11. Resource-constrained planning should still remain technically sound and maintainable.
12. Phased plans help settlements improve drainage systematically over time.

7.14 References

1. *Various authors. (n.d.). Manuals and design guidance for urban stormwater drainage planning and engineering.*
2. *Various authors. (n.d.). Stormwater drainage manuals addressing design standards and return periods.*
3. *Various authors. (n.d.). Review literature on urban drainage planning under urbanization and changing climate conditions.*
4. *Various authors. (n.d.). Publications on integrated land-use and drainage planning.*
5. *Various authors. (n.d.). Guidance and discussions related to return-period selection for drainage design in Indian conditions.*
6. *Various authors. (n.d.). Research on sustainable and multicriteria drainage infrastructure planning.*

Chapter 8

Hydrologic Analysis for Drainage Design

Learning Objectives

By the end of this chapter, the learner should be able to:

1. Delineate catchments for drainage design.
2. Estimate runoff using appropriate hydrologic methods.
3. Interpret rainfall data for engineering design purposes.
4. Evaluate how land development changes design discharge.
5. Explain the role of design storms, runoff coefficients, IDF relationships, and climate variability in drainage hydrology.

8.1 Introduction

Hydrologic analysis converts rainfall and catchment information into design runoff estimates. It is the stage of drainage engineering in which field data, rainfall records, catchment characteristics, and planning assumptions are translated into discharge values for sizing drains, culverts, storage systems, and outfalls. Without sound hydrologic analysis, hydraulic design becomes unreliable, no matter how carefully the physical structures are detailed.

This chapter explains the core steps involved in hydrologic analysis for drainage design in urban, semi-urban, and village settings. It covers catchment delineation, rainfall analysis, runoff estimation, design storm selection, and the influence of land use change and climate variability on design discharge.

The chapter serves as the direct transition from planning principles to quantitative drainage design. Its purpose is to help the learner understand not only the methods used to estimate runoff, but also the assumptions, limitations, and contextual factors that influence hydrologic design decisions.

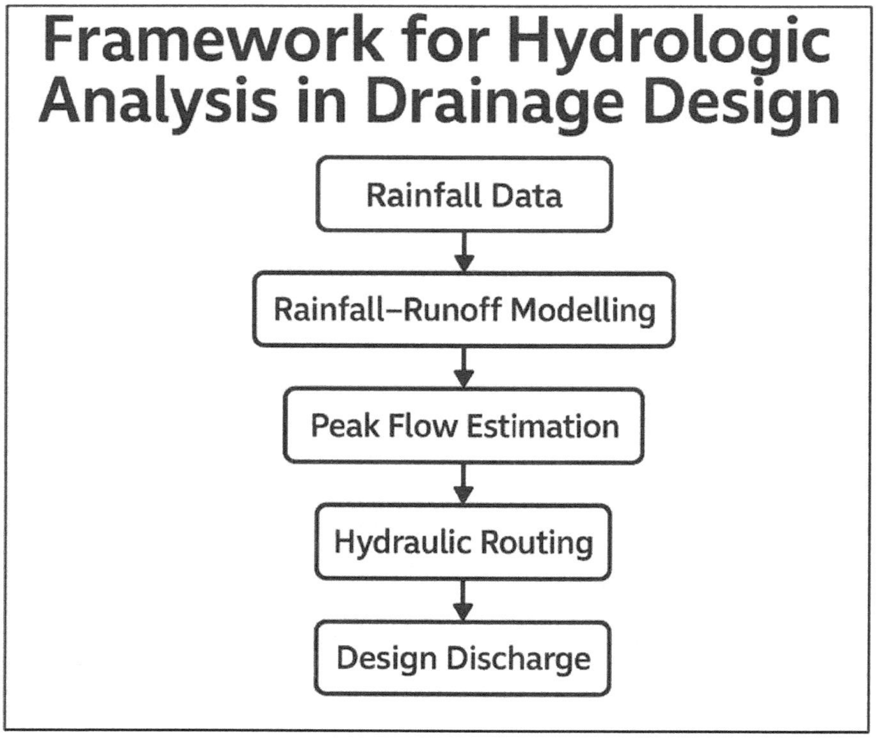

Figure 8.1 Framework for hydrologic analysis in drainage design

8.2 Catchment Delineation and Subdivision

8.2.1 Importance of Catchment Delineation

Hydrologic analysis begins with identification of the catchment, which is the area contributing runoff to a selected outlet or design point. In settlement drainage, delineation is essential because design discharge depends directly on the size, shape, slope, and surface condition of the contributing area.

8.2.2 Natural and Modified Catchments

Catchment boundaries are determined by topography, surface grading, road levels, built structures, and natural or artificial drainage paths. In undeveloped terrain, water divides are usually controlled by natural ridges and slopes. In urban and semi-urban areas, however, the effective catchment may be altered significantly by roads, kerbs, walls, embankments, drains, and plot development. As a result, the hydrologic catchment in built environments may differ from the natural topographic catchment.

8.2.3 Need for Subdivision

For design purposes, large catchments are often subdivided into smaller sub-catchments. This allows runoff generation to be analyzed more accurately by accounting for local differences in land use, imperviousness, slope, and drainage routing. Subdivision is especially important in complex urban systems where runoff enters the network at multiple points and does not behave as a single uniform flow.

8.2.4 Design Significance

Proper delineation and subdivision support phased design, network analysis, and identification of local control points. Errors at this stage affect every later step in hydrologic computation, so catchment definition must be treated as a primary engineering task rather than a preliminary formality.

8.3 Rainfall Data Analysis

8.3.1 Purpose of Rainfall Analysis

Rainfall data analysis is carried out to understand the magnitude, duration, frequency, and pattern of rainfall events relevant to drainage design. The objective is not simply to describe past rainfall, but to derive design values consistent with the selected level of drainage risk.

8.3.2 Data Sources and Types

Rainfall records may be obtained from meteorological stations, local gauges, engineering departments, and regional hydrologic agencies. These records may include daily totals, hourly depths, and shorter-duration intensities where available. For local drainage systems, short-duration rainfall is often most critical because peak runoff depends strongly on intense storms over relatively short periods.

8.3.3 Data Screening and Selection

Rainfall analysis may involve checking record length, continuity, missing data, consistency between stations, and suitability for the project location. In some cases, nearby stations must be compared to identify the most representative dataset. For drainage design, the engineer is especially interested in rainfall events capable of generating critical runoff in the catchment under study.

8.3.4 Supplementary Information

Where formal records are limited, supplementary evidence such as community recollection of major flood events, local damage records, and temporary rainfall measurements may help interpretation. Such information should be used cautiously and preferably alongside measured data.

Table 8.1 Rainfall Data Requirements for Urban Drainage Design

Parameter	Requirement for Urban Areas	Typical Values / Practice	Application in Urban Drainage Design
Design Rainfall Intensity	Short-duration, high-intensity storms	50–200 mm/hr (city & return period dependent)	Determines peak runoff in storm sewers
Rainfall Duration	Equal to or greater than time of concentration	5, 10, 15, 30, 60 minutes	Used in Rational Method (Q = CIA)
Return Period	As per importance of area	2–5 years (local streets), 10–25 years (arterial roads, trunk drains)	Defines design safety and service level
IDF Curves	Mandatory, city-specific	IMD-based or regional curves	Primary input for rainfall intensity
Temporal Resolution	High-resolution rainfall data	≤ 15-minute intervals	Accurate peak flow estimation
Areal Coverage	Rainfall representative of catchment	Gauge within or nearest to catchment	Ensures realistic runoff estimation
Spatial Variability	Considered for large urban catchments	Thiessen polygon / AWS network	Improves accuracy for large cities
Antecedent Moisture Condition	Assumed near saturation	AMCII to AMCIII	Reflects high runoff in impervious areas
Extreme Short-Burst Rainfall	Critical for urban flooding	Cloudburst-type events	Checks drain surcharge & flooding
Historical Rainfall Record	Minimum long-term data	≥ 20–30 years preferred	Frequency analysis and IDF development
Climate Change Factor	Increasingly recommended	1.1–1.3 multiplier on intensity	Future-proofing drainage systems
Data Source	Reliable, official sources	IMD, City AWS, Municipal records	Data credibility and audit compliance
Missing Data Handling	Mandatory data quality checks	Statistical infilling methods	Ensures continuity of rainfall series

Key Urban Drainage Design Notes

- Urban drainage relies more on rainfall intensity than rainfall depth.
- Short-duration storms control pipe sizing and inlet spacing.
- Peak runoff occurs very quickly due to high imperviousness.
- Under-designed rainfall inputs are a primary cause of urban flooding.

8.4 Intensity-duration-frequency relationships

8.4.1 Meaning of IDF relationships

An intensity-duration-frequency relationship expresses how rainfall intensity varies with storm duration and return period. It is one of the most important tools in drainage hydrology because it allows the engineer to select a rainfall intensity appropriate to the chosen design storm duration and level of risk.

8.4.2 General Behavior

Shorter-duration storms generally have higher rainfall intensity, while longer-duration storms have lower intensity but greater total depth. Rainfall associated with larger return periods is also generally more intense than rainfall associated with smaller return periods for the same duration. These relationships are commonly presented in the form of equations, curves, or tables derived from long-term rainfall records.

8.4.3 Relevance to Drainage Design

For small urban catchments, IDF data are especially important because design flows are often estimated using methods that depend directly on rainfall intensity. In village and peri-urban systems, IDF analysis also remains useful, although the critical storm duration may be longer because runoff concentration is slower and contributing areas may be larger.

8.4.4 Need for Consistency With Catchment Response

Proper use of IDF data requires consistency between rainfall duration and catchment response time. Selecting an intensity for an unrealistic duration can produce misleading discharge estimates. IDF analysis must therefore always be linked to time of concentration and catchment characteristics.

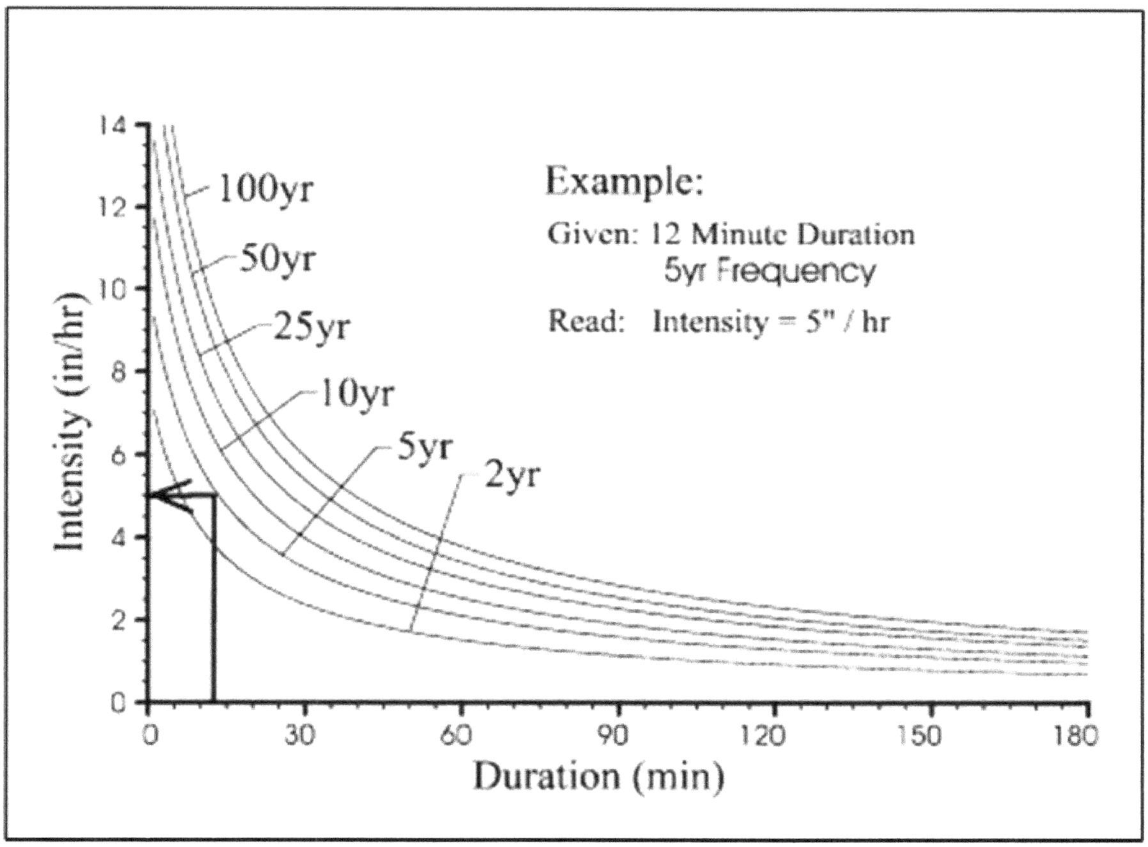

Figure 8.2 Typical IDF curve and its use in drainage design

8.5 Runoff Coefficients and Loss Estimation

8.5.1 Why Loss Estimation is Needed

Not all rainfall becomes runoff. A portion is lost through interception, infiltration, depression storage, and related abstractions. Hydrologic analysis therefore requires a method for estimating the fraction of rainfall that contributes to flow at the design point.

8.5.2 Runoff Coefficient Concept

One of the simplest ways to represent this effect is through a runoff coefficient, which expresses the proportion of rainfall expected to appear as direct runoff. Runoff coefficients depend on land use, imperviousness, soil condition, slope, vegetation cover, and surface storage. Paved roads, roofs, and compacted surfaces usually have high coefficients, while vegetated plots and permeable soils generally have lower values.

8.5.3 Average and Weighted Coefficients

In simple drainage design methods, a single average runoff coefficient may be used for the whole catchment. In more detailed studies, different coefficients may be assigned to different land-use zones and then combined by area weighting. This is often preferable in mixed urban or semi-urban catchments where surface conditions differ significantly.

8.5.4 Other Loss Concepts

Loss estimation may also be approached through infiltration indices, initial abstraction concepts, or method-specific parameters. Whatever the method, the core principle is that runoff estimation should reflect actual catchment behavior rather than assume that all rainfall becomes immediate surface flow.

Table 8.2 Indicative runoff coefficients for typical land-use classes

A. Built-Up / Urban Land Uses

Land-Use Class	**Typical Surface Characteristics**	**Indicative Runoff Coefficient (C)**
Central Business District (CBD)	Highly impervious (buildings, roads)	0.80 – 0.95
High-density residential	Buildings, paved roads, footpaths	0.60 – 0.80
Medium-density residential	Mixed paved & unpaved areas	0.45 – 0.60
Low-density residential	Plots with gardens/open areas	0.30 – 0.45
Industrial area	Large roofs, paved yards	0.60 – 0.85
Commercial area	Parking + built-up	0.70 – 0.90

B. Transportation & Infrastructure Surfaces

Land-Use Class	Surface Type	Indicative C
Asphalt / concrete roads	Fully impervious	0.80 – 0.95
Paved parking areas	Concrete / bituminous	0.75 – 0.90
Footpaths / plazas	Paved or tiled	0.60 – 0.85
Rail yards / terminals	Ballast + paved zones	0.40 – 0.65

C. Open & Semi-Permeable Areas

Land-Use Class	Surface Description	Indicative C
Parks / playgrounds	Grass, compacted soil	0.15 – 0.30
Institutional campuses	Mixed open & built-up	0.30 – 0.50
Urban vacant land	Bare soil / sparse cover	0.20 – 0.40
Compacted open ground	Hard soil surface	0.35 – 0.55

D. Agricultural & Rural Land Uses

Land-Use Class	Condition	Indicative C
Cultivated land	Flat to gentle slope	0.20 – 0.40
Agricultural fields	Steep slope / compacted	0.30 – 0.50
Orchards / plantations	With ground cover	0.15 – 0.30
Fallow land	Bare soil	0.25 – 0.45

E. Natural Land Cover

Land-Use Class	Vegetation / Condition	Indicative C
Dense forest	Good canopy, forest floor	0.05 – 0.15
Open forest	Moderate vegetation	0.10 – 0.25
Grassland	Natural grass cover	0.10 – 0.25
Scrub / bush	Sparse vegetation	0.20 – 0.40

F. Water Bodies & Wet Areas

Land-Use Class	Description	Indicative C
Lakes / reservoirs	Open water surface	1.00
Ponds / tanks	Standing water	0.90 – 1.00
Wetlands	Saturated soil & vegetation	0.40 – 0.70
Floodplains	Seasonally inundated	0.30 – 0.60

8.6 Rational Method for Small Catchments

8.6.1 Importance of the Rational Method

The rational method is one of the most widely used techniques for estimating peak runoff from small catchments. It is especially common in drainage design because it is simple, practical, and suitable for localized components such as roadside drains, inlets, branch drains, and small culverts.

8.6.2 Basic Idea

The method relates peak discharge to three principal factors: runoff coefficient, rainfall intensity, and catchment area. Its engineering basis is that peak discharge occurs when rainfall of selected intensity continues for a duration at least equal to the time of concentration, allowing the full catchment to contribute runoff at the outlet.

8.6.3 Typical Use and Strengths

Because of its simplicity, the rational method is frequently used for both preliminary and detailed design of small drainage areas. It is especially useful where the design objective is peak flow rather than the complete runoff hydrograph.

8.6.4 Limitations

The rational method does not describe the full runoff hydrograph and assumes relatively uniform rainfall over the catchment and selected duration. It is less suitable for large catchments, storage-dominated systems, and situations where runoff routing and timing must be represented in detail. Even so, it remains valuable when applied within its intended range using realistic coefficients, correct catchment delineation, and appropriate rainfall intensity.

Table 8.3 Application Steps of the Rational Method for a Small Drainage Catchment

Step No.	Application Step	Input / Data Required	Description / Purpose	Typical Output
1	Delineate Drainage Catchment	Topographic map / GIS data	Identify the contributing area draining to the design point	Catchment boundary
2	Determine Catchment Area (A)	GIS measurement / plan	Calculate the drainage area contributing runoff	Area (ha or km^2)
3	Estimate Time of Concentration (Tc)	Catchment length, slope, surface type	Determine travel time of runoff from farthest point to outlet	Tc (minutes)
4	Select Design Storm Return Period	Design guidelines / risk level	Choose appropriate return period based on importance of drain	Return period (years)
5	Select Storm Duration	Tc value	Storm duration equal to or slightly greater than Tc	Duration (minutes)
6	Obtain Rainfall Intensity (I)	IDF curves / rainfall data	Extract intensity corresponding to selected duration and return period	Intensity (mm/hr)
7	Determine Runoff Coefficient (C)	Land-use & surface data	Reflect runoff potential of catchment surfaces	Dimensionless C value
8	Apply Rational Formula	C, I, A	Compute peak runoff using Rational Method	Peak discharge (Q)
9	Compute Design Discharge	Peak runoff + safety margin	Final discharge for drain sizing	Design Q (m^3/s)
10	Drain Sizing & Design Check	Q, slope, roughness	Size drain section and verify capacity	Drain dimensions

Rational Method Equation (for reference)

$$Q = CIA$$

Where:

- **Q** = Peak runoff (m^3/s)
- **C** = Runoff coefficient
- **I** = Rainfall intensity (mm/hr)
- **A** = Catchment area (ha or km^2, as per unit convention)

Key Applicability Notes

- Best suited for small catchments (typically ≤ 50–200 ha)
- Assumes uniform rainfall intensity over the catchment
- Widely used for urban storm drains, street drains, and small culverts
- Not recommended for large watersheds or storage-dominant systems

8.7 Unit hydrograph and Hydrograph Methods

8.7.1 Why hydrograph Methods are Needed

For larger or more complex catchments, peak discharge alone may not be sufficient for drainage design. In such cases, it is necessary to estimate how runoff varies with time, which is represented by a hydrograph. A hydrograph shows the relationship between discharge and time during a rainfall-runoff event.

8.7.2 Applications

Hydrograph methods are useful for designing detention structures, retention systems, pump operations, trunk drainage, major culverts, and systems where runoff from multiple sub-catchments combines. These methods allow the engineer to consider not only peak flow but also runoff duration, storage, and routing effects.

8.7.3 Unit hydrograph Concept

One common approach is the unit hydrograph, which represents the runoff response of a catchment to a unit depth of effective rainfall occurring over a specified duration. Other approaches include synthetic hydrographs, empirical hydrograph development, and numerical hydrologic modeling. Method choice depends on project scale, data availability, and required level of accuracy.

8.7.4 Relevance to Larger Drainage Planning

Hydrograph analysis is especially useful in peri-urban and watershed-scale drainage planning where runoff from several sub-catchments may reach a common point at different times and where temporary storage plays an important role. This temporal dimension cannot be captured adequately by peak-flow-only methods.

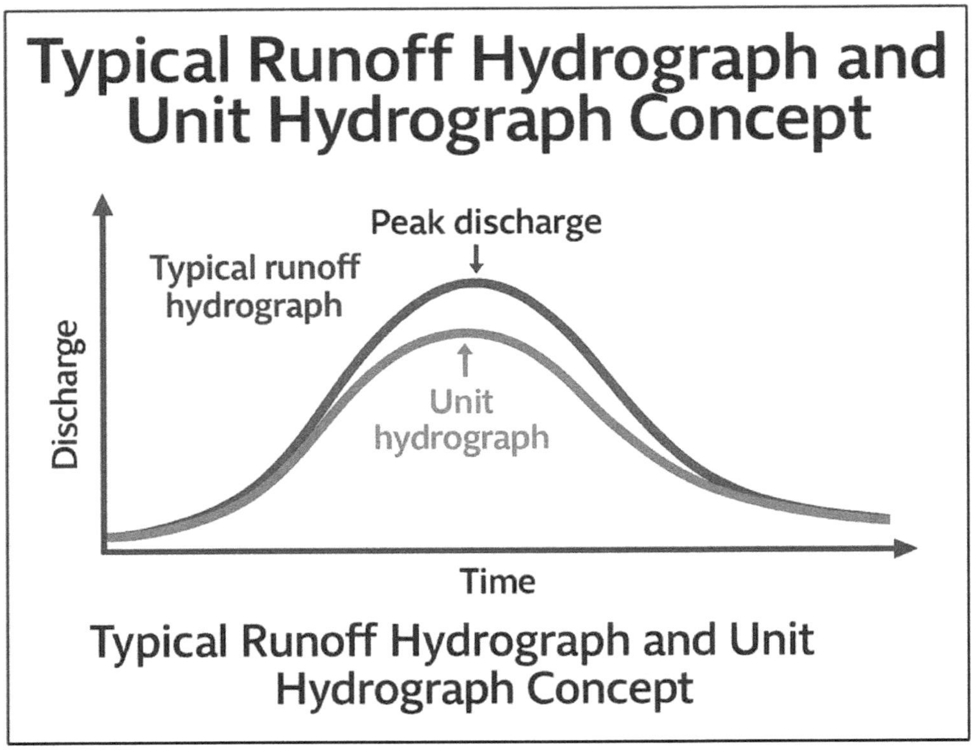

Figure 8.3 Typical runoff hydrograph and unit hydrograph concept

8.8 Design Storm Selection

8.8.1 Meaning of Design Storm

A design storm is a selected rainfall event used as the basis for hydrologic analysis and drainage design. It is defined by characteristics such as duration, return period, temporal distribution, and sometimes spatial pattern. Its purpose is to provide a rational and consistent basis for estimating design runoff.

8.8.2 Factors Affecting Design-Storm Choice

The appropriate design storm depends on the type of drainage component, the selected return period, catchment response time, and acceptable level of risk. A local roadside drain may be designed using a relatively frequent storm, while a trunk drain or flood-sensitive corridor may require a more severe event.

8.8.3 Relation to Hydrologic Method

Design storm selection should also match the analysis method. If the rational method is used, rainfall intensity is generally selected from IDF data for a duration related to time of concentration. If hydrograph methods are used, a full storm hyetograph may be needed to represent rainfall variation through time.

8.8.4 Importance of Balanced Selection

A poorly chosen design storm can distort the entire drainage design. If it is too mild, the system may flood frequently; if it is excessively conservative without practical justification, the resulting design may become uneconomical or impractical. Balanced selection requires hydrologic realism, acceptable risk, and alignment with project purpose.

Table 8.4 Design storm selection by drainage component and settlement type

A. Local Collection & Minor Components

Drainage Component	**Urban Areas**	**Semi-Urban / Peri-Urban**	**Village / Rural**	**Design Rationale**
Roof drains & downpipes	2–5 year storm	2–5 year storm	2 year storm	Localized surcharge acceptable
Curb inlets / gully pits	2–5 year storm	2–5 year storm	2 year storm	Short-term ponding tolerable
Local street drains	5 year storm	2–5 year storm	2–5 year storm	Limited surface disruption
Roadside open drains	5–10 year storm	5 year storm	2–5 year storm	Low consequence of failure

B. Secondary Conveyance Components (Sub-Catchment Level)

Drainage Component	**Urban Areas**	**Semi-Urban / Peri-Urban**	**Village / Rural**	**Design Rationale**
Secondary storm sewers	10 year storm	5–10 year storm	5 year storm	Neighborhood-scale flooding risk
Road culverts (minor crossings)	10–25 year storm	10 year storm	5–10 year storm	Traffic & connectivity protection
Junction chambers / manholes	10–25 year storm	10 year storm	5 year storm	Flow concentration points
Minor natural drains	10–25 year storm	10 year storm	5–10 year storm	Controlled overflow acceptable

C. Primary Trunk & Major System Components

Drainage Component	Urban Areas	Semi-Urban / Peri-Urban	Village / Rural	Design Rationale
Trunk storm drains	25–50 year storm	25 year storm	10–25 year storm	City-scale flooding risk
Major box drains	25–50 year storm	25 year storm	10–25 year storm	Critical conveyance
Major road / rail culverts	25–50 year storm	25 year storm	25 year storm	Strategic connectivity
Pumping stations (if any)	25–50 year storm	25 year storm	Rarely used	System-wide reliability

D. Storage, Flood Accommodation & Outfall Components

Drainage Component	Urban Areas	Semi-Urban / Peri-Urban	Village / Rural	Design Rationale
Detention ponds	25–50 year storm	25 year storm	10–25 year storm	Peak attenuation
Retention ponds	25–50 year storm	25 year storm	10–25 year storm	Volume control
Floodplains (planned)	50–100 year event	25–50 year event	Controlled flooding	Risk accommodation
Final outfall structures	50–100 year storm	25–50 year storm	25 year storm	Catastrophic failure avoidance

E. Contextual Adjustment by Settlement Risk Profile

Settlement Context	Adjustment to Design Storm Selection
Dense urban core	Use upper end of return-period ranges
Coastal settlements	Increase by one class + tide/surge check
Hill towns	Moderate return period + velocity control
Floodplain settlements	Moderate return period + adaptive flooding
Critical infrastructure zones	Apply next higher storm category

Key Planning Principles

- Higher return periods downstream where flows and consequences accumulate
- Urban systems demand higher serviceability and reliability
- Rural and village systems allow adaptive, lower-risk performance
- Storage and floodplains are deliberate design elements, not failures

8.9 Impact of Land Use Change on Runoff

8.9.1 Hydrologic Effect of Development

Land use change has a major effect on hydrologic response. As settlements expand, natural soil surfaces are replaced by roofs, roads, pavements, and compacted plots. This generally increases runoff volume, shortens time of concentration, reduces infiltration, and raises peak discharge.

8.9.2 Relevance Across Settlement Types

In urban areas, even moderate increases in imperviousness may significantly alter the drainage load. In semi-urban and peri-urban areas, the effect may be more dramatic because drainage networks often remain incomplete while land conversion proceeds rapidly. In village settings, road upgrading, plot raising, and reduction of open infiltration space can also modify runoff behavior.

8.9.3 Need to Account for Future Development

Hydrologic analysis should account not only for present land use but also for likely future development. A system designed only for current conditions may become inadequate within a short period if the catchment urbanizes rapidly. This is particularly important for trunk drains, culverts, and fixed outfall structures that are difficult to enlarge later.

8.9.4 Methods of Incorporating Change

Land-use change may be addressed through revised runoff coefficients, updated catchment boundaries, staged design assumptions, or scenario analysis. Whatever method is used, ignoring land transformation can result in serious underestimation of design flows.

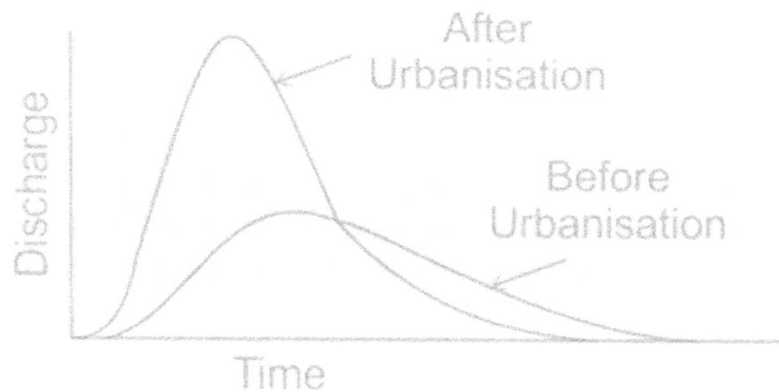

Figure 8.4 Effect of urbanisation on runoff volume and peak discharge

8.10 Climate Variability In Hydrologic Design

8.10.1 Climate Variability Matters

Climate variability and long-term climatic change introduce uncertainty into drainage design because rainfall behaviour may not remain the same as in the historical record. Changes in rainfall intensity, storm frequency, seasonal concentration, and extreme-event occurrence can affect the adequacy of drainage systems over their design life.

8.10.2 Effect on Drainage Performance

Even where total annual rainfall changes little, short-duration storm intensity may increase, which is especially important for urban drainage. More variable rainfall may also produce intense events separated by dry periods, affecting runoff generation and infrastructure performance. In low-lying settlements, intense rainfall combined with high downstream water levels may further reduce drainage efficiency.

8.10.3 Design Response to Uncertainty

Hydrologic design should consider whether historical records alone are sufficient or whether additional allowance should be made for uncertainty and changing conditions. This may involve using updated design guidance, applying safety margins, considering multiple design scenarios, or identifying future adaptation options.

8.10.4 Principle of Resilient Hydrologic Design

The aim is not to predict future climate with complete certainty, but to avoid a design that is rigidly tied to outdated assumptions. Resilient hydrologic design acknowledges uncertainty and prepares drainage systems to cope with a plausible range of future conditions.

8.11 Indian Application Notes

8.11.1 Relevance of Rainfall Design Methods

In Indian drainage practice, hydrologic design often depends strongly on rainfall frequency analysis, IDF relationships, and return-period selection suited to settlement context. Design rainfall estimates can vary depending on the method used, which reinforces the importance of method selection and local calibration.

8.11.2 Urbanisation Pressure

Rapid land conversion in Indian urban and peri-urban areas increases the risk that runoff estimates based only on current land use may quickly become outdated. Drainage systems designed without allowance for future imperviousness often prove inadequate within a short span of time.

8.11.3 Local Data Limitations

Many Indian cities and towns face challenges with incomplete or short rainfall records, limited automatic weather stations, and inconsistent data quality. Engineers often need to supplement official datasets with regional studies, empirical adjustments, or scenario-based design assumptions.

8.11.4 Adaptation to Settlement Context

Drainage design in India must adapt to diverse settlement types—from dense metropolitan cores to peri-urban growth corridors and rural villages. Each context requires different return periods, storm intensities, and allowances for land-use change.

8.11.5 Climate Change Considerations

Recent guidance increasingly recommends applying multipliers to rainfall intensity to account for climate change. In Indian practice, factors of 1.1–1.3 are often suggested to future-proof drainage systems against more intense short-duration storms.

8.12 Planning Implications

The transformed chapter shows that hydrologic analysis is the foundation of quantitative drainage design. Catchment delineation, rainfall interpretation, runoff estimation, design-storm selection, and future-condition assessment together determine the discharge values used throughout subsequent hydraulic design.

The chapter also demonstrates that no single hydrologic method suits every case. The rational method may be adequate for small catchments, while hydrograph methods are more suitable where storage, routing, and runoff timing must be represented explicitly. Good engineering practice therefore depends on matching method to context.

Finally, the chapter makes clear that hydrologic design should not be based only on the past. Land use change, evolving drainage conditions, and climate uncertainty require the designer to think beyond historical rainfall records and current catchment appearance.

8.13 Summary Points

1. Hydrologic analysis converts rainfall and catchment information into design runoff estimates.
2. Catchment delineation is the first and one of the most important steps in hydrologic design.
3. Built development may alter the effective drainage catchment substantially.
4. Rainfall analysis should focus on data relevant to the catchment response and design purpose.
5. IDF relationships link rainfall intensity to storm duration and return period.
6. Runoff coefficients represent the fraction of rainfall appearing as direct runoff
7. The rational method is useful for estimating peak runoff in small drainage catchments.
8. Hydrograph methods are more appropriate where runoff timing and storage effects matter.
9. Design storm selection must reflect risk, method, and settlement context.
10. Land use change generally increases runoff volume and peak discharge while shortening concentration time.
11. Climate variability introduces uncertainty that should be considered in resilient hydrologic design.

8.14 References

1. *Various authors. (n.d.). Textbooks on engineering hydrology and stormwater design.*
2. *Various authors. (n.d.). Drainage manuals describing rainfall analysis, runoff estimation, and design storm use.*
3. *Various authors. (n.d.). Studies and reviews on the rational method and its practical application in drainage.*
4. *Various authors. (n.d.). Literature on urbanisation impacts and unit hydrograph response in changing catchments.*

5. Various authors. (n.d.). *Publications on climate variability and resilient drainage design assumptions.*

6. Various authors. (n.d.). *Indian and related studies on rainfall analysis and design rainfall estimation for urban drainage.*

Chapter 9

Hydraulic Design of Open and Closed Drains

Learning Objectives

By the end of this chapter, the learner should be able to:

1. Design basic open and closed drainage elements.
2. Select suitable cross-sections and pipe sizes.
3. Evaluate hydraulic performance under normal and adverse conditions.
4. Incorporate key appurtenances into drainage system design.
5. Explain the hydraulic significance of inlets, manholes, surcharge, backwater, tailwater, and freeboard in drainage systems.

9.1 Introduction

Hydraulic design is the process of determining the size, shape, slope, and performance characteristics of drainage components so that runoff can be conveyed safely and efficiently. Once design discharge has been estimated through hydrologic analysis, hydraulic design translates that flow into physical infrastructure such as roadside drains, covered drains, stormwater pipes, manholes, junctions, and outfalls.

This chapter presents the basic principles for hydraulic design of open and closed drainage systems in urban, semi-urban, and village settings. It explains how drain type is selected, how section dimensions are determined, and how issues such as surcharge, backwater, inlet capacity, and hydraulic safety are addressed.

The chapter is therefore the practical link between estimated runoff and constructed drainage works. It emphasizes that hydraulic design is not only a matter of flow equations, but also of constructability, maintenance, operational behavior, and local site constraints.

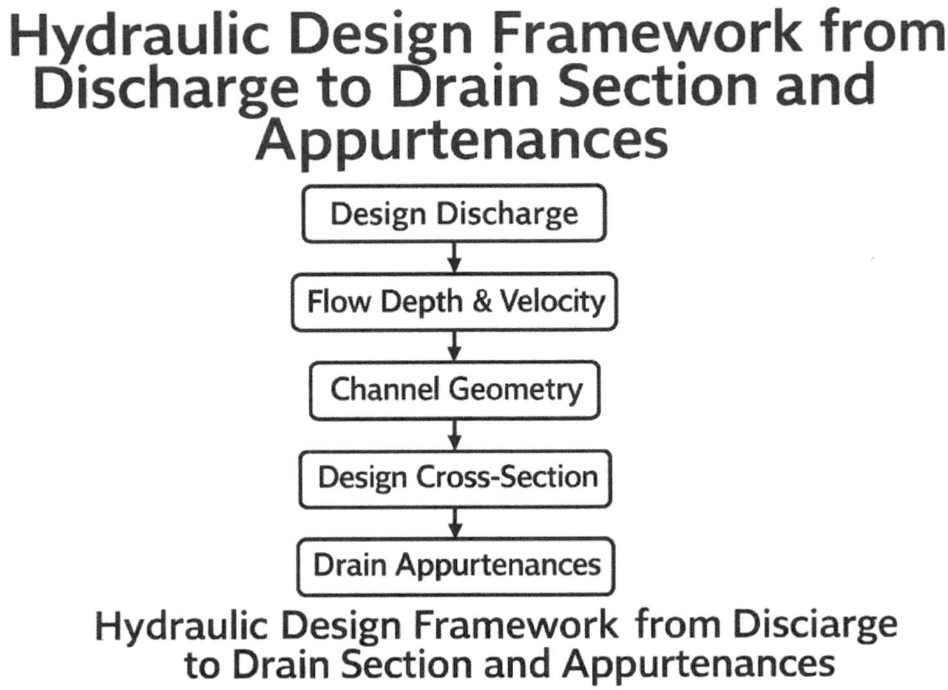

Figure 9.1 Hydraulic design framework from design discharge to drain section and appurtenances

9.2 Hydraulic Design Objectives

9.2.1 Primary Objective

The primary objective of hydraulic design is to ensure that the drainage system can carry the selected design flow without causing unacceptable flooding, erosion, surcharge, structural distress, or maintenance difficulty. A hydraulically sound system should not merely convey water, but should do so in a stable, safe, and maintainable manner.

9.2.2 Practical Design Aims

Hydraulic design seeks to provide adequate flow capacity, maintain suitable velocities for self-cleansing without causing erosion, ensure continuity between connected elements, avoid excessive head losses at inlets and junctions, protect adjacent land and structures from overtopping and scour, and accommodate downstream control and maintenance variation.

9.2.3 Settlement-specific Emphasis

Hydraulic objectives vary by settlement type. In urban systems, the emphasis may be on rapid and reliable conveyance under severe space constraints. In village systems, design may prioritize simple geometry, lower cost, easier maintenance, and safe discharge to local outfalls. In semi-urban contexts, the system may need to serve both current flow and future expansion.

9.3 Selection of Open Versus Closed Drains

9.3.1 Basic Decision

One of the first hydraulic decisions is whether runoff should be conveyed in an open drain or a closed drain. This choice affects capacity, cost, inspection, safety, maintenance, and land use.

9.3.2 Open Drains

Open drains are common in village roads, peri-urban corridors, and many roadside systems because they are relatively simple to construct, easy to inspect, and adaptable to varying flow conditions. They are particularly suitable where land is available and regular visual inspection is possible. However, they may create safety issues, interfere with access, attract solid-waste dumping, and reduce usable roadside space.

9.3.3 Closed Drains

Closed drains include covered channels and buried stormwater pipes. These are often preferred in dense urban settings where pedestrian and vehicle movement must be maintained and where open drains are undesirable for safety or land-use reasons. Closed systems can also provide a cleaner streetscape, but they are generally more expensive and more difficult to inspect and maintain.

9.3.4 Factors Controlling Selection

The choice depends on expected discharge, flow variation, right-of-way, traffic and pedestrian condition, maintenance capacity, public safety, aesthetics, utility congestion, and local construction feasibility. In many settlements, hybrid solutions are used, with open drains in less congested reaches and covered sections near crossings, market zones, or dense frontage.

Table 9.1 Comparison of open and closed drains for different settlement settings

A. SettlementSpecific Planning Context

Settlement Type	Dominant Planning Concerns
Urban Coastal Cities	High land value, tidal backflow, storm surge, dense infrastructure
Hill Resorts / Tourist Towns	Steep slopes, erosion, landslides, aesthetics
Riverine Towns	River flooding, backwater effects, sediment load, seasonal inundation

B. Suitability Overview

Settlement Type	Open Drains	Closed Drains	Recommended System Strategy
Urban coastal cities	⚠ Limited	☑ High	Closed + pumping + gated outfalls
Hill resorts	☑ High	⚠ Limited	Open lined drains with drops
Riverine towns	☑ High	⚠ Moderate	Open drains + flood routing

C. Hydraulic Performance Comparison

1. Urban Coastal Cities

Aspect	**Open Drains**	**Closed Drains**
Handling intense rainfall	Good	Good
Tidal / surge backflow control	Poor	☑ Effective (with flap gates)
Overflow behaviour	Visible, manageable	Risk of hidden surcharge
Adaptability to extremes	High	Moderate
Overall suitability	⚠ Moderate	☑ High

2. Hill Resorts / Hill Towns

Aspect	**Open Drains**	**Closed Drains**
Flow conveyance on steep slopes	☑ Excellent	⚠ Limited
Velocity control & erosion safety	☑ Can be engineered	✕ High risk
Debris handling	☑ Easy	✕ Frequent blockage
Flood warning visibility	☑ High	✕ Low
Overall suitability	☑ High	⚠ Low

3. Riverine Towns

Aspect	**Open Drains**	**Closed Drains**
Backwater tolerance	☑ Better	✕ Poor
Performance during long flooding	☑ Acceptable	✕ High failure risk
Sediment accommodation	☑ Good	✕ High siltation
System resilience	☑ High	⚠ Moderate
Overall suitability	☑ High	⚠ Moderate

D. Structural & Safety Considerations

Aspect	**Urban Coastal**	**Hill Resorts**	**Riverine Towns**
Open drain stability	Moderate (lined)	Critical	Good
Closed drain structural risk	Uplift & corrosion	Settlement risk	Sediment pressure
Public safety	High priority	Very high (tourists)	Moderate
Scour / erosion risk	Moderate	Very high	Low–moderate

E. Operation & Maintenance Implications

Aspect	**Urban Coastal Cities**	**Hill Resorts**	**Riverine Towns**
Open drain maintenance	Frequent	Frequent debris removal	Seasonal
Closed drain maintenance	Complex & costly	Difficult & risky	High silt removal
Failure detection	Easy (open)	Easy (open)	Easy (open)
Disaster response	Technologydriven	Immediate & manual	Seasonal preparedness

F. Environmental, Social & Tourism Aspects

Aspect	Urban Coastal Cities	Hill Resorts	Riverine Towns
Aesthetics	Critical	Very critical	Moderate
Odour & nuisance	Low tolerance	Zero tolerance	Moderate tolerance
Tourism sensitivity	High	Very high	Moderate
Environmental integration	Mangroves / creeks	Natural streams	Floodplains

G. Cost & Lifecycle Considerations

Aspect	Urban Coastal	Hill Resorts	Riverine Towns
Open drain cost	Moderate	Low	Low
Closed drain cost	Very high	High	Moderate
Lifecycle cost	High	Moderate	Low–moderate
Construction complexity	High	High	Low

H. Recommended Drainage System Typologies

Settlement Type	Recommended Drain System	Design Explanation
Urban coastal cities	**Closed drains + pumping + tide gates**	Land scarcity, hygiene, flood safety
Coastal CBD zones	**Fully closed storm sewer**	Zero surface disruption
Hill resorts	**Open lined drains with drops & chutes**	Velocity control & debris safety
Hill periurban areas	**Wide open drains**	Maintenance simplicity
Riverine towns	**Open drains with flood routing**	Adaptive flooding
Riverfront zones	**Open drains + raised outfalls**	Backwater accommodation

Executive Planning Insights

- **Urban coastal cities** require **engineered closed systems** but must retain **open/relief pathways downstream**.
- **Hill resorts** are best served by **open, visually inspectable drains** designed for erosion control.
- **Riverine towns** benefit from **accepting water visibly and routing it safely**, rather than hiding it.
- **Hybrid systems** are almost always superior to fully closed systems in hazardprone settings.

9.4 Design of Trapezoidal and Rectangular Drains

9.4.1 Common Open-drain Sections

Hydraulic design of open drains involves selecting a cross-section and dimensions capable of carrying design discharge at acceptable depth and velocity. Common shapes include rectangular, trapezoidal, triangular, and U-shaped sections. The selected form depends on required capacity, available space, side stability, construction method, and lining type.

9.4.2 Rectangular Sections

A rectangular drain is often used where space is limited and lining is provided. It is hydraulically convenient in built environments and can be integrated with roadside structures, but it may require stronger walls and more careful attention to maintenance access.

9.4.3 Trapezoidal Sections

A trapezoidal section is often suitable where more space is available, especially in semi-urban and village systems, because it can be more stable and economical. Side slopes can improve earth stability and reduce structural demand compared with vertical-walled sections.

9.4.4 Design Procedure

Design typically involves selecting trial dimensions and slope, then checking whether the section provides sufficient capacity under design discharge. Hydraulic evaluation includes flow area, wetted perimeter, hydraulic radius, roughness, and energy slope. Freeboard should also be provided where appropriate to allow for uncertainty, wave effects, and minor obstruction.

9.4.5 Field Practicality

Open-drain design must also account for side-wall stability, access for desilting, debris accumulation, and user safety. Hydraulic adequacy alone is therefore not sufficient; the section must also be workable in field conditions.

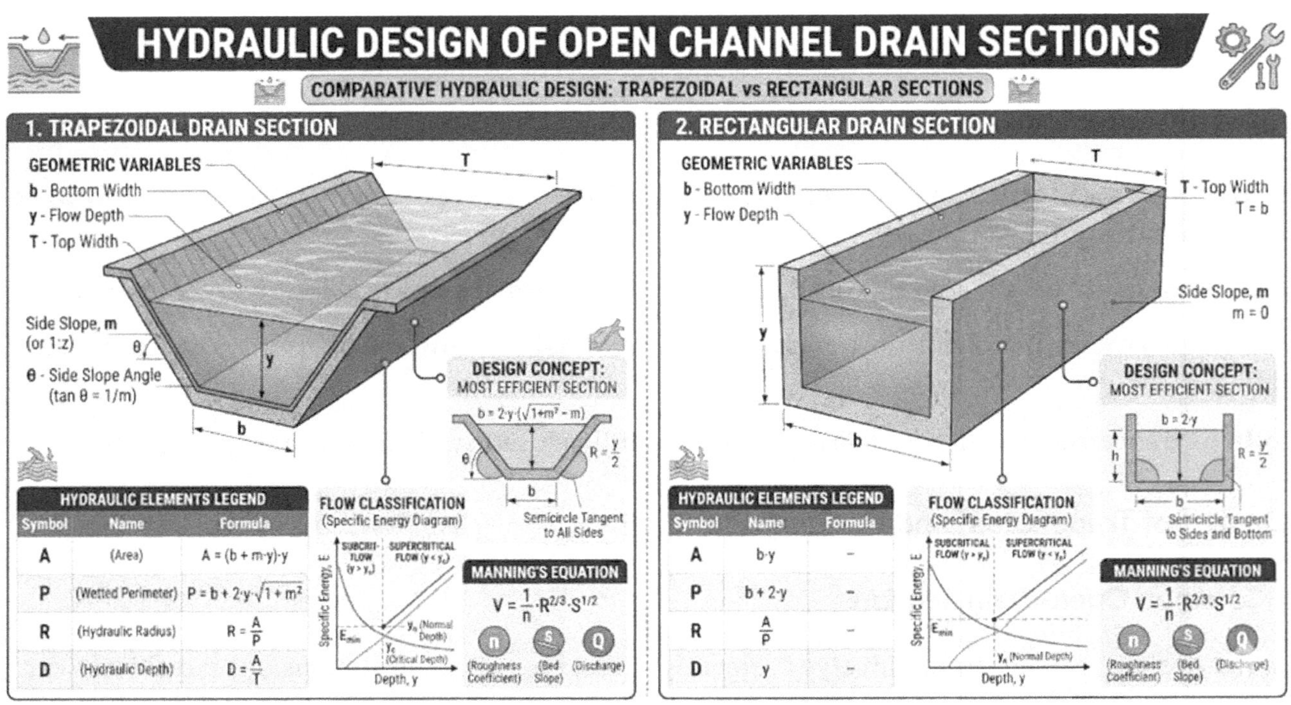

Figure 9.2 Trapezoidal and rectangular drain sections with hydraulic design variables

9.5 Lined and Unlined Drain Sections

9.5.1 Lined Drains

Open drains may be lined or unlined depending on flow condition, soil stability, maintenance requirements, and local economy. Lined drains use materials such as concrete, brick masonry, stone pitching, or precast elements. They offer better erosion resistance, more stable geometry, and generally lower hydraulic roughness than natural earth channels.

9.5.2 Unlined Drains

Unlined drains are usually earth channels or natural flow paths with little structural treatment. They are often suitable in village and low-density areas where discharge is modest, slope is gentle, and land is available. Their main limitations are greater vulnerability to erosion, vegetation growth, side collapse, and sediment deposition.

9.5.3 Performance Comparison

Because lined drains generally have lower roughness and more stable geometry, they usually convey more flow for a given section and slope. However, selection should reflect long-term performance rather than initial cost alone. A lined drain may be justified where velocities are high or side erosion is likely, while an unlined drain may be acceptable where natural drainage can be preserved and maintenance is manageable.

9.5.4 Partial Treatment

In some cases, partial lining or localized protection is sufficient instead of full lining. This can be appropriate where only short reaches are erosion-prone or where outlets, bends, and transitions need added protection.

9.6 Pipe Flow Design Principles

9.6.1 Closed Conduit Behavior

Closed stormwater systems often use circular pipes or similar conduit sections operating under gravity flow. Hydraulic design of such systems focuses on ensuring that the pipe network can carry design discharge with acceptable depth, velocity, and head loss.

9.6.2 Governing Variables

Pipe flow design depends on internal diameter, longitudinal slope, pipe roughness, flow depth relative to pipe size, inlet and outlet conditions, and local energy losses at bends and junctions.

9.6.3 Part-full and Full-Flow Conditions

In gravity drainage, storm pipes may flow partially full under ordinary conditions and approach full flow during larger storms. The design should ensure that selected size and slope can carry the target discharge without causing excessive surcharge, backwater, or upstream flooding at inlets. Where full-flow conditions are expected, hydraulic grade-line behavior becomes important.

9.6.4 Practical Constraints

Pipe design must also consider minimum cover, trench depth, utility crossings, manhole spacing, maintenance access, and the feasibility of achieving desired slope in flat terrain. In dense urban systems, hydraulic design often requires balancing ideal hydraulic capacity against severe physical constraints.

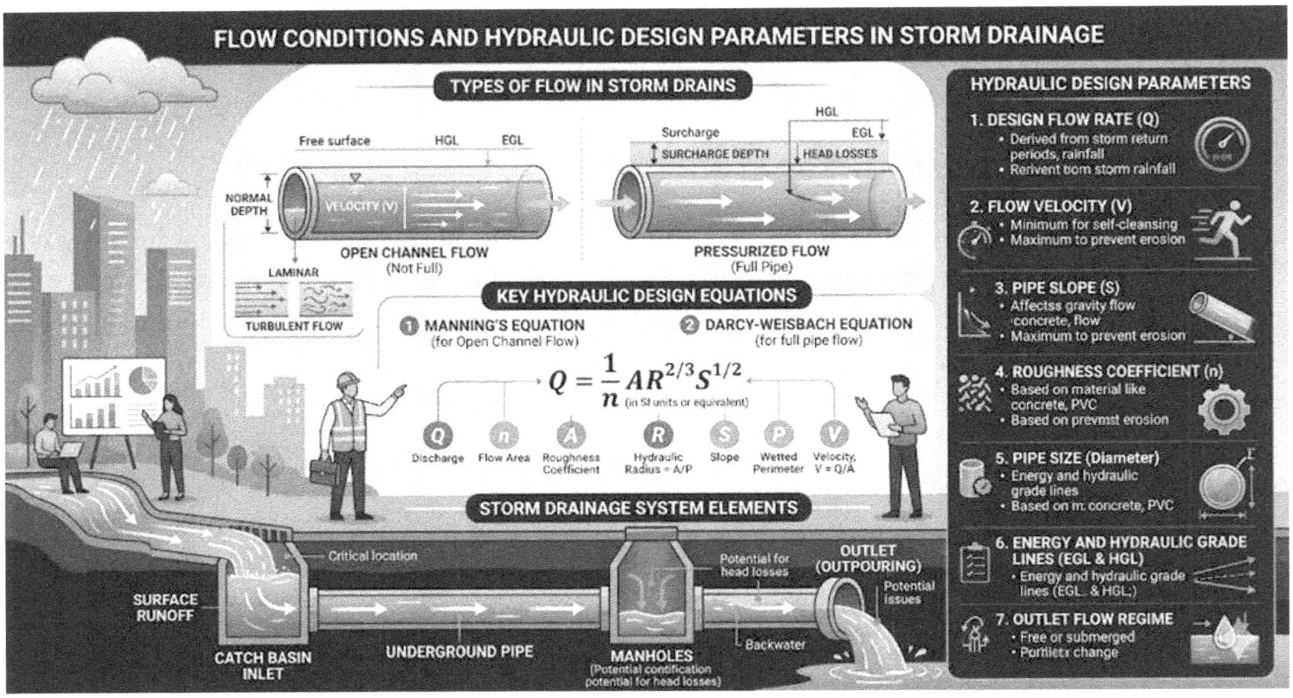

Figure 9.3 Pipe flow conditions and hydraulic design parameters in storm drainage

9.7 Diameter, Slope, and Flow Capacity

9.7.1 Core Design Variables

The most fundamental hydraulic variables in conduit and drain design are size, slope, and resulting capacity. In an open drain, capacity depends on section shape, flow depth, slope, and roughness. In a pipe, capacity depends on diameter, slope, and material roughness.

9.7.2 Interdependence of Variables

These variables are interrelated, so design usually involves iteration. A larger diameter or section increases capacity, but may also increase excavation, cost, land use, and structural demand. A steeper slope increases velocity and carrying capacity, but may not be available because of terrain limitation, downstream control, or utility conflict.

9.7.3 Velocity Limits

Resulting velocities must remain within acceptable limits. Very low velocities encourage siltation and blockage, while very high velocities may damage unlined sections, scour outlets, or create turbulence at transitions and structures.

9.7.4 Flat-terrain Challenge

In flat urban or peri-urban terrain, achieving suitable slope is often one of the greatest hydraulic challenges. In such cases, design may require larger conduits, local drop structures, pumping, storage, or revised alignment to maintain workable gradients.

Key Practical Relationships Derived from Drain Design Principles

- **Velocity is primarily driven by slope:** To achieve the minimum required self-cleansing velocity (usually 2.0–2.5 ft/s in sanitary sewers), slope must be adjusted.
- **Larger pipes need smaller slopes to reach minimum velocity:** As diameter increases, hydraulic radius increases, reducing friction. For example, an 8” pipe achieves 2 ft/s at 0.20% slope, whereas a 4” pipe needs 0.50%.
- **Capacity is driven primarily by diameter:** Capacity increases proportional to (D^{8/3}). Doubling pipe size from 4” to 8” boosts capacity by roughly six times at the same slope.
- **Steeper slope increases both velocity and capacity:** Velocity increases by the square root of slope ((S^{1/2})). Quadrupling slope doubles velocity, and capacity follows the same square-root relationship.

9.8 Design of Inlets and Interception Structures

9.8.1 Function ofInlets

Inlets are the points where surface runoff enters the drainage system. Their hydraulic role is critical because even a well-sized downstream drain will not perform effectively if runoff is not captured properly at the surface.

9.8.2 Common Inlet Forms

Common inlet types include kerb openings, grated inlets, gully pits, catch basins, side-entry inlets, and simple drop openings. The hydraulic design of inlets involves assessing how much runoff they can intercept under expected flow depth, spread, and road-slope conditions.

9.8.3 Inlet Spacing And Placement

In urban road drainage, inlet spacing is especially important. If inlets are too far apart, water may spread excessively along the carriageway or bypass the intended capture point. In peri-urban and village settings, simpler side openings may be sufficient, but they still need to be positioned so that runoff naturally reaches them.

9.8.4 Additional Features

Inlet design should also consider silt traps, sumps, debris screening, and safe connection to the downstream drain. Poor inlet design is a common cause of localized road flooding even where the main drainage line has adequate capacity.

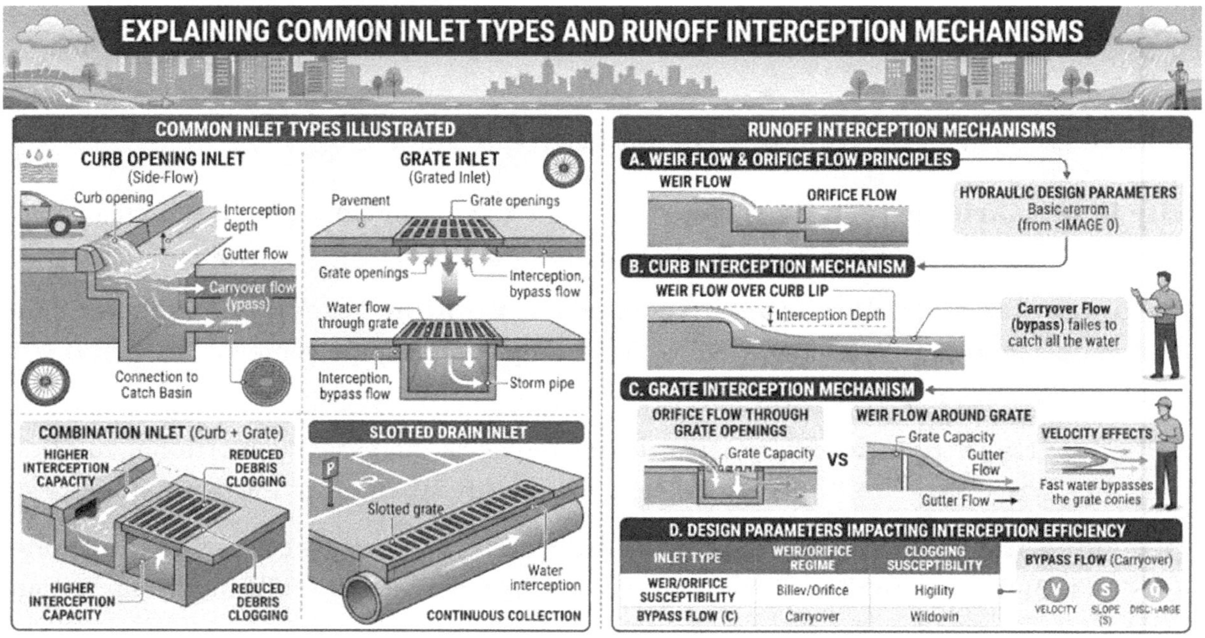

Figure 9.4 Common inlet types and runoff interception mechanisms

9.9 Manholes, Chambers, And Junctions

9.9.1 Role in Closed Systems

Manholes and junction chambers are essential components of closed drainage systems. They provide access for inspection, cleaning, direction change, slope change, and branch-pipe connection. From a hydraulic perspective, they also affect flow continuity and local head loss.

9.9.2 Basic Design Requirements

A well-designed manhole or chamber should allow safe access for maintenance, accommodate incoming and outgoing pipes at appropriate levels, provide a smooth flow path through the invert, minimize turbulence and sediment deposition, and withstand structural and traffic loads.

9.9.3 Junction Hydraulics

At junctions, hydraulic design must consider changes in direction, merging flow, and possible surcharge during storms. Sudden transitions or poorly aligned connections can create head loss, turbulence, and deposition. Where one pipe enters at a substantially higher level, a drop arrangement may be required to preserve both hydraulic and structural performance.

9.9.4 Spacing and Placement

Manhole spacing should reflect maintenance requirements, alignment changes, and pipe size. Although chambers are often treated as structural details, they are also hydraulic control points and must be designed with flow behavior in mind.

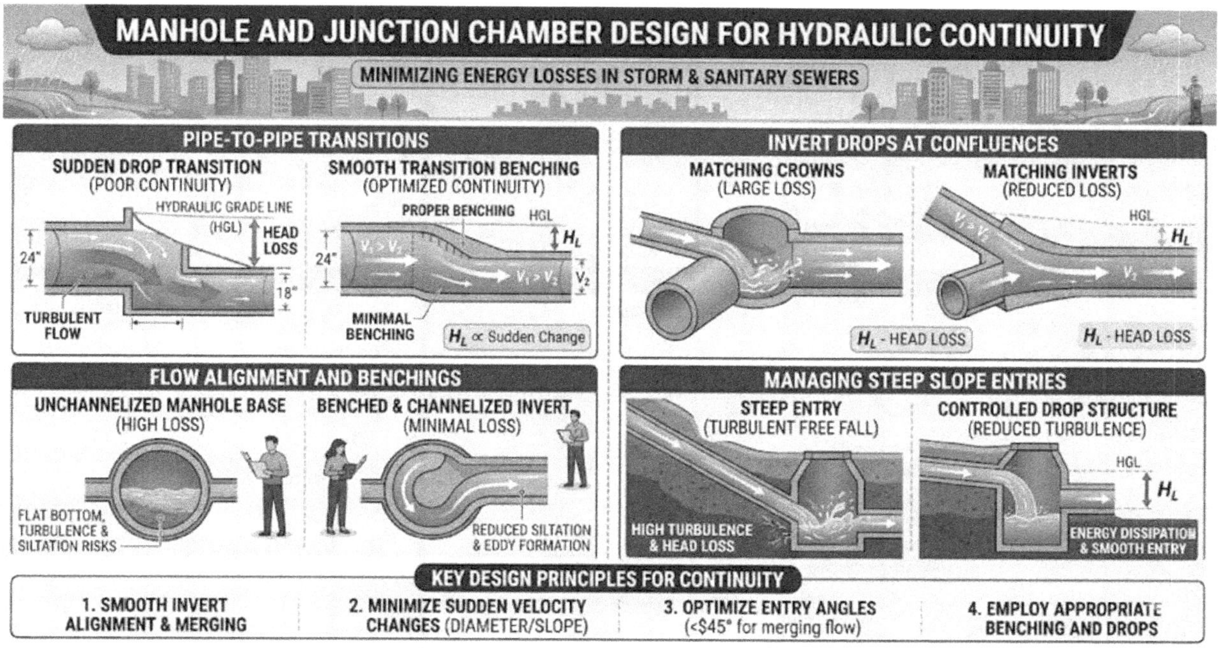

Figure 9.5 Manhole and junction chamber details affecting hydraulic continuity

9.10 Surcharge, Backwater, and Tailwater Conditions

9.10.1 Non-ideal Hydraulic Conditions

Drainage systems do not always operate under ideal free-flow conditions. During intense rainfall or when downstream levels rise, the system may experience surcharge, backwater, or tailwater effects that reduce capacity and cause upstream flooding.

9.10.2 Surcharge

Surcharge occurs when a closed conduit or chamber carries flow under pressure because the water level rises above the crown of the pipe. This can force water upward through inlets or manholes and reduce the efficiency of the upstream system.

9.10.3 Backwater and Tailwater

Backwater is the upstream rise in water level caused by a downstream control such as a narrow culvert, flat outlet, obstruction, or high receiving-water level. Tailwater is the water level at the outlet that influences discharge from the upstream structure. These effects are especially important in flat terrain, downstream-congested systems, tidal or river-influenced outfalls, and common receiving points.

9.10.4 Design Implication

Hydraulic design must evaluate not only local conveyance capacity but also downstream controls and system interaction. In some cases, this requires checking hydraulic grade lines, outlet submergence, and flow transitions under storm conditions.

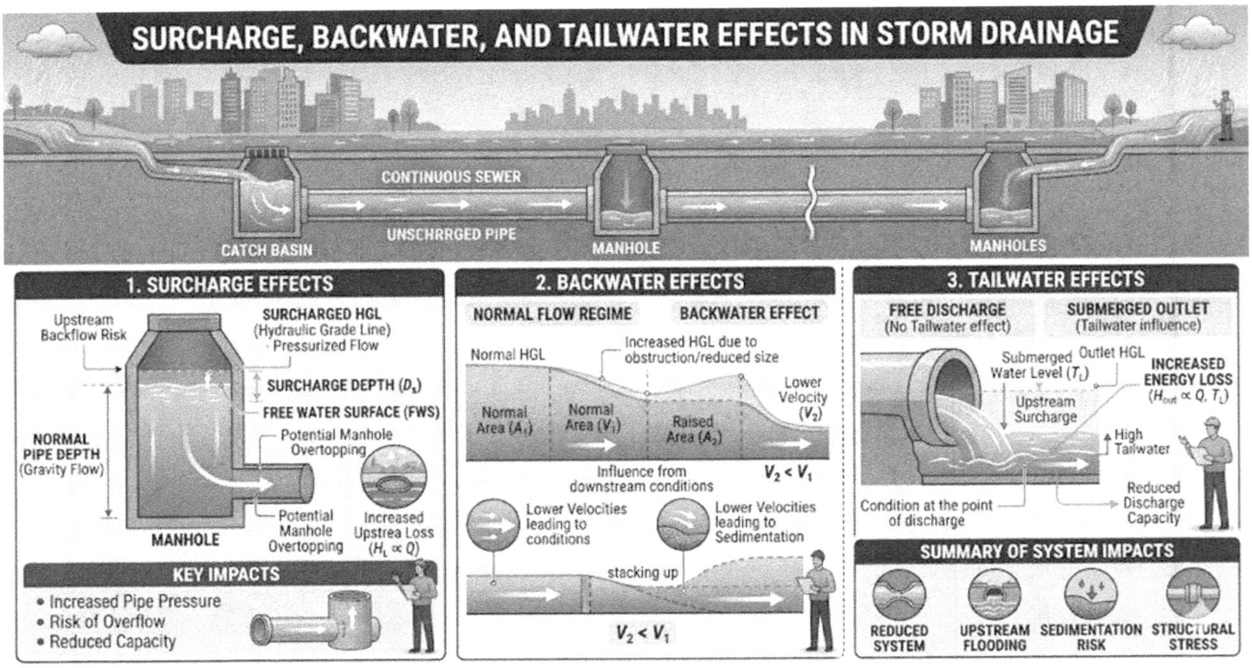

Figure 9.6 Surcharge, backwater, and tailwater effects in storm drainage systems

9.11 Hydraulic Checks for Safety and Efficiency

9.11.1 Need for Post-selection Checks

Once a drain or conduit section is selected, it must be checked for overall hydraulic performance, safety, and practical efficiency. These checks help ensure that the system will function under real field conditions rather than only under simplified theoretical assumptions.

9.11.2 Typical Checks

Important checks include adequacy of flow capacity under design discharge, acceptable flow depth and freeboard, minimum self-cleansing velocity, maximum permissible velocity for erosion control, continuity across bends and junctions, inlet and outlet stability, absence of adverse surcharge or backwater, and compatibility with maintenance access and debris management.

9.11.3 Open and Closed System Specifics

For open drains, additional checks may include side stability, overtopping risk, and pedestrian safety. For pipe systems, checks often include hydraulic grade-line behavior, chamber performance, and access spacing.

9.11.4 Hydraulic Efficiency And Operational Realism

Hydraulic efficiency should always be judged together with operational reality. A section that is hydraulically adequate on paper may still fail if it is too narrow to clean, too shallow to maintain slope continuity, or too vulnerable to blockage. Good hydraulic design therefore combines calculation, judgment, and constructability.

Table 9.2 Hydraulic design checks for open drains, pipes, inlets, and junctions

Component	**Hydraulic Design Check**	**Rationale/Purpose**	**Typical Design Criteria / Checks**
Open Drains (Channels)	Freeboard Check	Prevent overtopping of the channel during the design event.	Verify a minimum vertical clearance between the Design Water Level (DWL) and the channel top (e.g., 0.5–1.0 ft or % of depth).
	Velocity Checks (Min & Max)	Ensure sufficient velocity for self-cleansing and limit velocity to prevent erosion.	Min Velocity: ~2–3 fps (0.6–0.9 m/s). Max Velocity: 3–5 fps (grass), 10–15 fps (concrete).
	Depth/Capacity Check	Verify sufficient cross-sectional area to pass design flow at allowable depth.	Flow depth < permissible depth; capacity verified using Manning's equation.
	Hydraulic Jump Check	For steep channels transitioning to flatter slopes, ensure energy dissipation occurs.	Location and stability of hydraulic jump must be calculated and managed.
Pipes (Sewers/ Culverts)	Velocity Checks (Full & Partial Flow)	Manage sedimentation and erosion.	Min Velocity: ~2–3 fps. Max Velocity: 10–15 fps.
	Capacity Check (Full Flow)	Ensure pipe diameter can pass design peak flow.	Calculated using Manning's or Darcy-Weisbach equations.
	Surcharge/HGL Check	Prevent excessive HGL rise, upstream flooding, and manhole overtopping.	HGL < pipe crown; freeboard below street level.

(Contd.)

Component	**Hydraulic Design Check**	**Rationale/Purpose**	**Typical Design Criteria / Checks**
	Tailwater/Backwater Effects	Manage outlet conditions that reduce capacity.	Analyze performance under high tailwater conditions.
Inlets	Interception Capacity/ Efficiency	Ensure inlets capture desired flow rate.	(Q_{captured} \geq \text{Target Interception}). Efficiency (E = Q_{captured}/Q_{total}).
	Gutter Flow/Spread Check	Limit roadway water spread for traffic safety.	Check maximum water spread vs. road classification limits.
	Clogging Factor	Account for partial blockage.	Apply safety factor (e.g., check against 50% capacity).
	Depth & Hydraulic Regime Check	Ensure inlet operates within safe hydraulic regime.	Verify depth and flow regime under design storm.
Junctions (Manholes/ Chambers)	Minor Losses (Form Losses) Check	Determine and minimize energy losses caused by changes in flow direction, merging flows, and diameter transitions.	Calculate losses using $h_L = K(V^2/2g)$. Ensure losses do not cause excessive upstream HGL rise. Use smooth shaping and benching.
	HGL Transition Check	Ensure the Hydraulic Grade Line moves smoothly through the junction without sudden spikes.	Review invert drops and matching crown/invert alignments. Minimize sudden velocity changes.
	Energy Dissipation (if drop structure)	Manage high velocity entries from steep pipes.	Design a plunge pool or suitable drop structure to reduce turbulence and erosion within the junction chamber.

Understanding the Interconnection of Checks

These checks are not performed in isolation. The design process is iterative:

- A change in **Pipe Slope** directly affects **Velocity**, **Capacity**, and the **HGL**.
- Selecting a pipe to maximize **Capacity** (larger diameter) might result in low **Velocity**, leading to a **Sedimentation Risk**.
- The **Inlet Type** and **Location** determine the **Inlet Capacity**, which must match the upstream gutter flow (controlled by pavement slope and roughness) to prevent excessive **Gutter Spread**.
- **Minor Losses** calculated for a complex junction (high K) increase the upstream **Head Loss** (H_L), directly contributing to **Upstream Surcharge**.
- A **High Tailwater** condition will raise the **Outlet HGL**, potentially causing **Backwater Effects** and reducing the entire system's discharge capacity, triggering the need to redesign pipes for increased size.

9.12 Indian Application Notes

9.12.1 Open and Covered Storm Drains

Indian urban drainage practice commonly uses both open and covered storm drains depending on settlement density, available space, safety concerns, and access needs. This makes the choice between open and closed systems an important planning and hydraulic decision rather than a purely structural one

9.12.2 Maintenance-sensitive Design

In many Indian field conditions, hydraulic adequacy alone is not enough. Drain sections, inlet openings, and chamber layouts must also reflect actual desilting practice, debris load, and ease of inspection if they are to perform reliably over time.

9.12.3 Relevance in Dense Corridors

In congested urban corridors with utility interference and limited slope availability, pipe sizing, chamber design, backwater assessment, and inlet placement become especially important. These are practical hydraulic challenges that strongly influence whether a theoretically adequate design will function in the field

9.13 Planning Implications

The transformed chapter shows that hydraulic design is the stage where estimated runoff becomes physical infrastructure. Decisions about drain type, section shape, lining, pipe size, slope, inlet arrangement, and chamber details determine whether the system can perform safely and efficiently under field conditions.

The chapter also demonstrates that hydraulic design must consider system interaction rather than isolated components. A well-sized drain can still fail if inlet capture is poor, downstream tailwater is high, junction losses are excessive, or maintenance access is inadequate.

Finally, the chapter makes clear that good hydraulic design balances theory and practicality. Manning-based capacity checks, velocity limits, and hydraulic grade-line considerations are essential, but they must be applied with attention to local terrain, maintenance reality, safety, and constructability.

9.14 Summary Points

1. Hydraulic design converts design runoff into physical drainage infrastructure.
2. The main objectives are adequate capacity, acceptable velocity, continuity, stability, and maintainability.
3. Open and closed drains each have advantages and limitations depending on context.
4. Rectangular and trapezoidal drains are common open-section choices.
5. Lined drains offer more erosion resistance and geometric stability than unlined drains.

6. Pipe flow design depends on diameter, slope, roughness, and hydraulic grade-line behaviour.
7. Size and slope must always be considered together in hydraulic design.
8. Inlet design is critical because poor capture can cause flooding even when downstream capacity is adequate.
9. Manholes and junctions are hydraulic as well as maintenance elements.
10. Surcharge, backwater, and tailwater effects can strongly reduce upstream drainage performance.
11. Hydraulic checks must include both efficiency and operational practicality.

9.15 References

1. *Various authors. (n.d.). Textbooks on open-channel hydraulics and stormwater pipe design.*
2. *Various authors. (n.d.). Guidance on Manning-based hydraulic design for open channels and conduits.*
3. *Various authors. (n.d.). Urban drainage and roadway storm-drain design manuals covering inlets, street drainage, and storm sewers.*
4. *Various authors. (n.d.). Standards and criteria related to manholes, junctions, hydraulic grade lines, and surcharge.*
5. *Various authors. (n.d.). Guidance on open-drain stability and lined versus unlined channel treatment.*
6. *Various authors. (n.d.). Indian urban drainage guidance recognising open and covered storm drains in municipal practice.*

Chapter 10

Road Drainage and Cross-Drainage Structures

Learning Objectives

By the end of this chapter, the learner should be able to:

1. Explain the function of road drainage components.
2. Design or size minor cross-drainage structures at a conceptual level.
3. Assess drainage needs for road corridors in settlements.
4. Recognize common failures at roadway drainage interfaces.
5. Distinguish among side drains, gutters, kerb inlets, culverts, causeways, and small bridges in road drainage systems.

10.1 Introduction

Road drainage is one of the most critical elements of settlement infrastructure because roads act both as transport corridors and as artificial flow pathways. During rainfall, road surfaces collect and concentrate runoff, while embankments, kerbs, shoulders, and crossings influence how water is intercepted, conveyed, and discharged. If road drainage is inadequate, the result may include pavement damage, traffic disruption, access failure, waterlogging, and accelerated deterioration of surrounding infrastructure.

This chapter explains the engineering principles of road drainage and the role of cross-drainage structures in urban, semi-urban, and village contexts. It covers surface drainage along roads, kerb and side drains, catch basins, culverts, causeways, small bridges, drainage at intersections and depressions, and protection of embankments and side slopes.

The chapter highlights that road drainage is not merely a pavement concern. It is also a settlement drainage, accessibility, and flood-management issue, especially where roads double as runoff paths or barriers to natural flow.

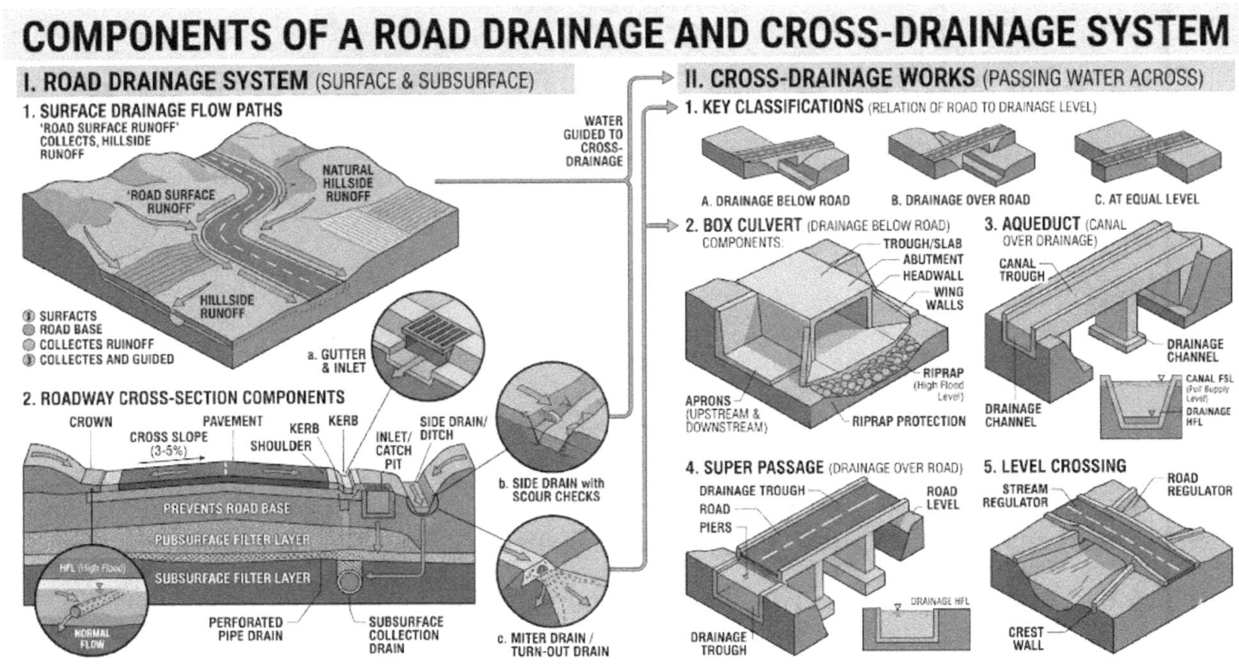

Figure 10.1 Components of a road drainage and cross-drainage system

10.2 Importance of Road Drainage

10.2.1 Protection of Pavement And Subgrade

Road drainage is essential because roads are exposed, sloping surfaces that rapidly generate runoff during rainfall. If water is allowed to remain on the carriageway, shoulders, or adjoining land, it can weaken pavement layers, damage subgrades, cause potholes, and reduce traffic safety. Proper drainage therefore preserves the structural integrity and service life of the road.

10.2.2 Importance in Settlements Of All Sizes

The need for road drainage is not limited to highways or major streets. In villages and semi-urban settlements, even low-volume roads require adequate drainage because they often serve as the only access route to homes, schools, markets, farms, and health facilities. A single blocked culvert or roadside drain can isolate part of a settlement during heavy rainfall.

10.2.3 Effects on Adjacent Land

Road drainage also influences surrounding land and buildings. Water flowing uncontrolled from roads can flood adjacent plots, erode roadside edges, and create stagnant pools near habitations. The design of road drainage must therefore protect both the road and the settlement through which it passes.

10.2.4 Surface and Cross-drainage

In engineering terms, surface drainage removes water from the road and its immediate margins, while cross-drainage conveys water across the road alignment so that natural or planned flow is not obstructed. Both functions are necessary for a complete road drainage system.

10.3 Surface Drainage Along Roads

10.3.1 Meaning of Surface Drainage

Surface drainage along roads refers to the collection and removal of runoff from the carriageway, shoulders, medians, footpaths, and adjoining edges. Its purpose is to prevent excessive water accumulation on the pavement and to guide runoff into suitable drains or inlets before it creates damage or traffic hazards.

10.3.2 Influence of Road Geometry

The effectiveness of surface drainage depends heavily on road geometry. Cross fall, camber, longitudinal slope, shoulder grading, kerb alignment, and pavement-edge condition all influence how quickly water leaves the road surface. Even a well-designed side drain will perform poorly if road levels do not direct runoff toward it.

10.3.3 Urban and Non-urban Conditions

In urban streets, surface drainage often depends on kerb-guided flow toward inlets, gullies, or covered edge drains. In semi-urban and village roads, runoff may drain to earthen shoulders, roadside channels, or open side drains. The required treatment depends on traffic level, road type, built frontage, and maintenance capacity.

10.3.4 Basic Design Principle

Road drainage must begin at the pavement surface itself. If water is not removed efficiently from the road, downstream structures cannot fully compensate for poor surface-flow control.

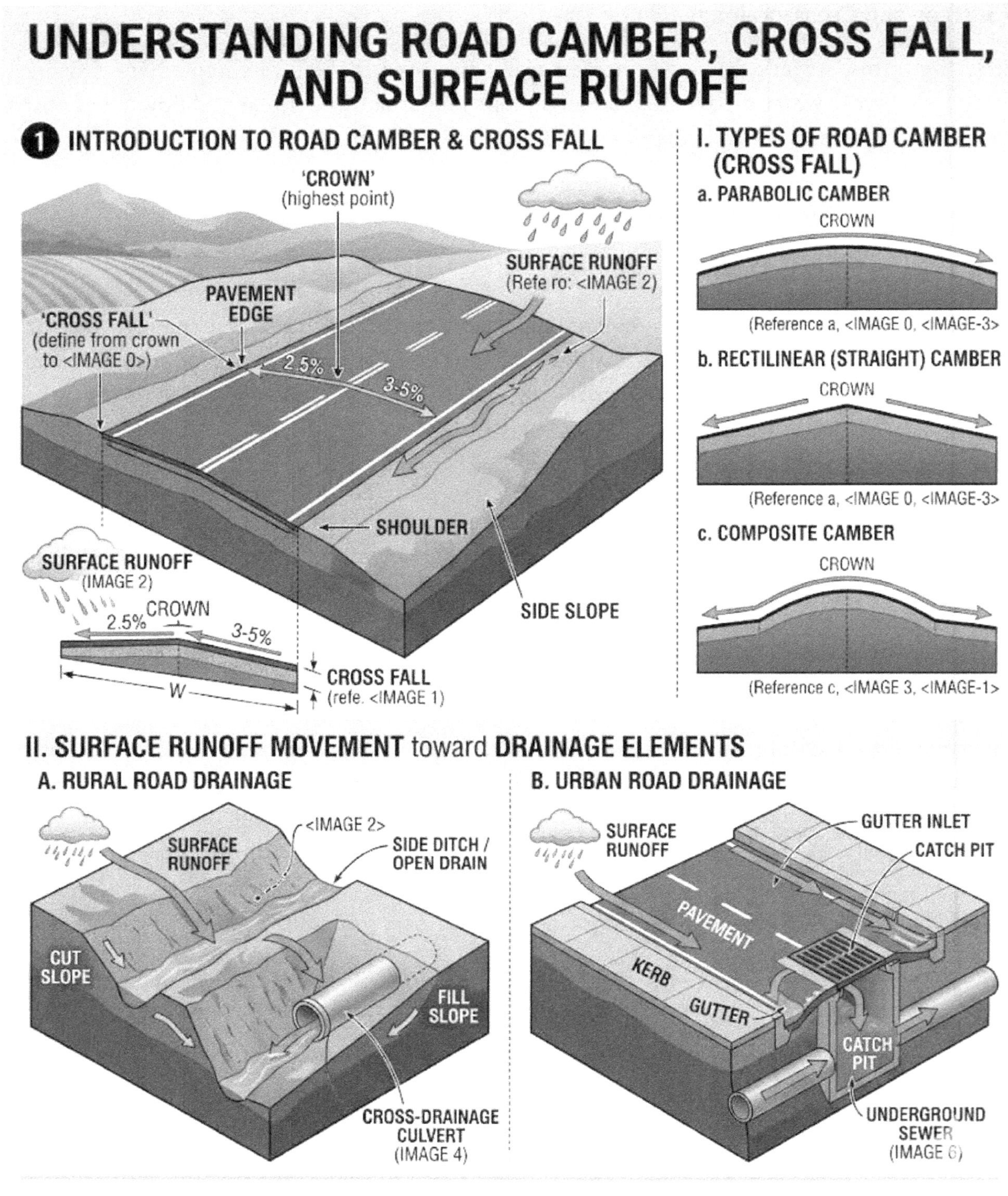

Figure 10.2 Road camber, cross fall, and surface runoff movement toward drainage elements

10.4 Kerb Drains, Gutters, and Side Drains

10.4.1 Kerb Drains

Kerb drains are common in urban streets where defined pavement edges guide runoff toward side-entry inlets or gullies. They are useful where land is limited and where pedestrian movement, frontage access, and underground services require a compact drainage form. However, they require proper slope, regular cleaning, and effective inlet placement to avoid ponding.

10.4.2 Gutters

Gutters are shallow channels formed along the edge of the pavement, often integrated with kerb systems. They carry runoff longitudinally toward collection points. Their hydraulic performance depends on gutter slope, roughness, road cross fall, and allowable spread of water on the pavement.

10.4.3 Side Drains

Side drains are open or lined channels located beside the road. They are common in rural roads, village streets, peri-urban corridors, and some urban-edge roads. Depending on context, they may be trapezoidal, rectangular, or shallow V-shaped. Side drains are generally easier to inspect and clean than buried systems, but they occupy more space and may interfere with access if not designed properly.

10.4.4 Combined Use

The choice among these elements depends on road type, surrounding development, available width, drainage discharge, and maintenance practice. In many cases, combined solutions are used, such as kerb-and-gutter flow discharging into side drains or covered roadside channels.

Table 10.1 Comparison of kerb drains, gutters, and side drains in settlement road drainage

Comparison Feature	Combined/Integrated Kerb Drains	Surface Gutter / Fluted Gutter (Channel)	Side Drain / Open Ditch (Trapezoidal/ V-shape)
Common Location	The edge of paved roads in urban and high-density residential/commercial areas.	Pavement edge, often in shared-space streets, low-density areas, or large parking areas.	Alongside lower-order roads, rural-urban boundaries, or unpaved shoulders of low-density settlements.
Structure & Material	Typically precast concrete or cast-in-place concrete. Modern integrated units are a single product combining the kerb profile and a continuous drainage inlet and channel.	Shallow channel formed within the pavement or using precast components. Can be lined (concrete, stone) or unlined (fluted grass, less common in settlements).	An open longitudinal trench or ditch, typically with sloped sides (trapezoidal or V-shaped). Can be unlined (earth/grass) or lined (concrete, stone).

(Contd.)

Comparison Feature	Combined/Integrated Kerb Drains	Surface Gutter / Fluted Gutter (Channel)	Side Drain / Open Ditch (Trapezoidal/ V-shape)
Primary Function	1. Defines road edge and protects pavement. 2. Acts as a physical barrier for pedestrians. 3. Collects and directs stormwater to gully inlets or integrated channels.	Collects surface runoff directly from the pavement and directs it to a disposal point (sewer inlet or side drain). Allows vehicles to cross easily.	Collects runoff from the road surface and adjacent landscape. Carries water longitudinal along the road to a major cross-drainage structure or water body.
Water Capacity	High to Very High (especially combined kerb & gutter systems connected to underground sewers).	Low to Medium (suited for minor runoff volumes due to limited depth and width).	High to Very High (capacity can be adjusted by changing the width, depth, and slope).
Installation Cost (Relative)	High (requires specialized forms, labor-intensive construction, and connection to underground sewer networks).	Medium (simpler construction than kerb systems, especially if formed with pavement).	Low (if unlined; Medium if lined with concrete or stone).
Maintenance Requirement	Frequent & Specialized (requires regular street sweeping and vacuuming of gully pots/integrated channels to prevent blockage by debris).	Frequent (prone to accumulation of sediment and leaves; easily cleaned by standard sweeping).	Periodic & Labor-Intensive (requires regular clearing of silt, vegetation, and dumped refuse to maintain flow capacity).
Pedestrian/Cyclist Interaction	Barrier (the kerb face acts as a barrier, directing pedestrians to designated crossings).	Accessible (designed to be easily crossed; fluted gutters are low-profile).	Hazard (uncovered ditches are a significant tripping and falling hazard; often require culverts at property entrances).
Best Use Case in Settlements	High-traffic urban/commercial streets; dense residential areas needing clear pedestrian separation and efficient storm sewer connection.	Shared-space "Woonerf" streets; pedestrian plazas; low-density residential pavement edges; car parks with mild slopes.	Peripheral settlement areas where space is available and underground sewer networks are not feasible/justified; managing high runoff from slopes.

10.5 Catch Basins and Gully Pits

10.5.1 Function

Catch basins and gully pits are structures designed to intercept runoff from road surfaces and transfer it into underground or covered drainage networks. They are especially important in urban road drainage where water must be removed at frequent intervals without obstructing traffic or pedestrian movement.

10.5.2 Typical Arrangement

A gully pit usually receives runoff through a grate or kerb opening and may include a sump below outlet level to trap sediment and debris. A catch basin performs a similar function and often includes a chamber with an outlet pipe leading to the drainage network. These structures help reduce direct entry of heavy silt into downstream conduits.

10.5.3 Key Design Factors

Design must consider location, spacing, inlet capacity, sump depth, access for cleaning, and connection to the downstream drain. If spacing is too large, runoff may bypass the inlet and accumulate on the road. If sump capacity is too small or cleaning is infrequent, blockage may occur and interception efficiency may fall sharply.

10.5.4 Maintenance Relevance

In dense commercial or traffic-heavy areas, inlet blockage is a common problem because litter and silt accumulate quickly. Hydraulic design must therefore be supported by practical maintenance planning. A structurally sound gully pit is ineffective if it cannot be cleaned or if it is positioned below a poorly graded road surface.

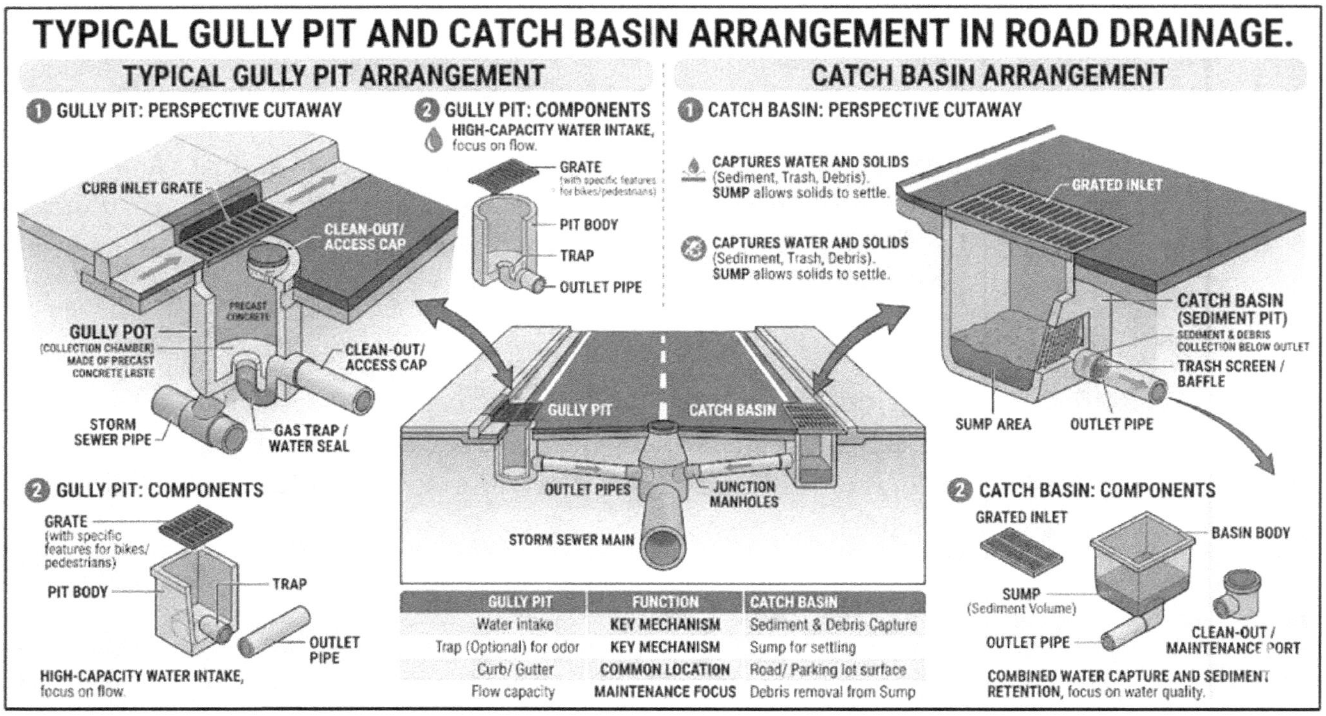

Figure 10.3 Typical gully pit and catch basin arrangement in road drainage

10.6 Culverts and Pipe Crossings

10.6.1 Role in Cross-drainage

Culverts and pipe crossings are cross-drainage structures that convey water beneath a road embankment or carriageway. Their purpose is to preserve the natural or designed flow path across the road while maintaining road continuity above.

10.6.2 Common Forms

Culverts may be circular pipes, pipe arches, box sections, slab culverts, or other forms depending on discharge, span, headroom, roadway width, and local construction practice. Small roads and village roads often use pipe culverts, while larger flows or wider crossings may require box culverts or minor bridge structures.

10.6.3 Hydraulic Design Considerations

Hydraulic design of culverts must consider inlet capacity, barrel capacity, outlet condition, headwater level, tailwater influence, and potential blockage. A culvert may fail not because the barrel is structurally weak, but because the inlet is undersized, the approach is poorly aligned, or sediment and debris reduce the effective opening. Likewise, an adequate culvert may still create flooding if the downstream channel is obstructed.

10.6.4 Structural and Maintenance Considerations

Culvert design must also account for structural loading, bedding, scour protection, embankment stability, and maintenance access. Since culverts often become critical bottlenecks in settlement drainage, their design and upkeep require special attention.

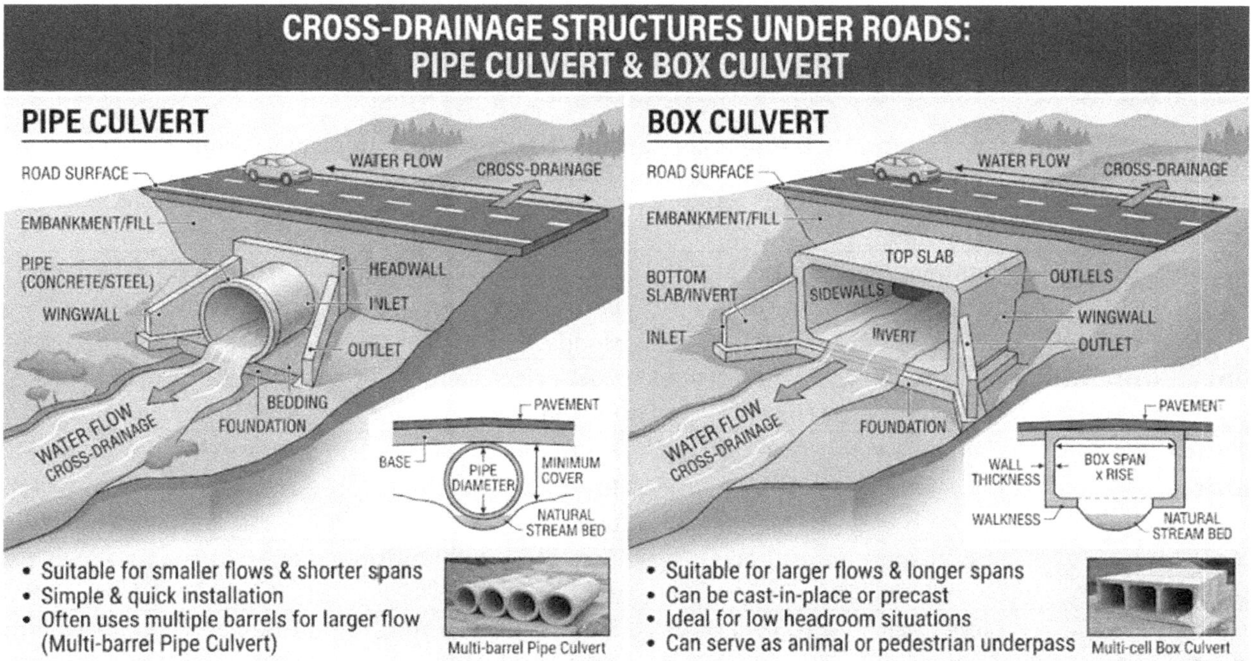

Figure 10.4 Pipe culvert and box culvert as cross-drainage structures under roads

10.7 Causeways and Small Bridges

10.7.1 When Larger Structures Are Needed

Where drainage flow is too wide, too variable, or too hydraulically demanding for a simple culvert, causeways or small bridges may be more appropriate. These structures provide road continuity while allowing water to pass beneath or, in some cases, over the crossing during specified conditions.

10.7.2 Causeways

A causeway is a low-level crossing that may permit overtopping during higher flows. It is often used in low-volume rural or village roads where traffic interruption for limited periods is acceptable and where a full bridge would be uneconomical. Causeways must be designed so that overtopping occurs safely and does not rapidly damage the structure or the road approaches.

10.7.3 Small Bridges

A small bridge provides a more permanent and higher-capacity crossing. It is appropriate where flow width, flood depth, debris load, or traffic importance makes a culvert or causeway inadequate. Small bridges are common where drainage channels are wider, sediment transport is significant, or road functionality must be maintained during larger storm events.

10.7.4 Selection Criteria

The choice among culvert, causeway, and small bridge depends on discharge magnitude, hydraulic behavior, acceptable risk, cost, maintenance needs, and the strategic importance of the road. The decision should therefore be based on both engineering analysis and service requirements.

Table 10.2 Conceptual comparison of culverts, causeways, and small bridges

Comparison Feature	Culvert	Causeway (Low-Water Crossing)	Small Bridge
Core Concept	Self-contained drainage structure beneath embankment	Low-profile crossing, overtopped during floods	Structured passage carrying road over waterway
Primary Function	Continuous drainage without interrupting traffic	Cost-effective ford for seasonal rivers	Reliable crossing for significant gaps
Hydraulic Handling	Channels water completely beneath roadway	Allows minor flow through pipes, flood overtops	Spans channel, flood passes beneath deck
Typical Span/ Length	< 6 m span; multi-barrel possible	Variable, small multiple openings	6–30 m span typical
Foundation & Support	Embedded in soil, concrete floor	Stabilized riverbed, small footings	Independent substructure with piers/abutments
Structure Complexity	Low	Low–Medium	High
Road Interaction	Seamless continuation over culvert	Traffic slows, closed during floods	Raised deck, uninterrupted travel
Best Application	Minor streams, manageable flows	Seasonal rivers, low traffic	Large perennial streams, permanent crossings
Relative Cost (Construction)	Lowest	Medium	Highest
Relative Cost (Maintenance)	Low	Medium	High

10.8 Drainage at Intersections and Depressions

10.8.1 Why These Locations are Critical

Road intersections and depressions are among the most common locations for local flooding because they act as collection points for runoff from multiple directions. Drainage design at such locations requires special attention even when adjacent road sections appear satisfactory.

10.8.2 Intersections

At intersections, runoff may converge from several streets, and pavement geometry may create shallow basins if grades are not coordinated properly. In dense urban areas, blocked or undersized inlets at intersections often cause rapid ponding that disrupts both traffic and pedestrian movement. In peri-urban and village settings, the problem may be aggravated by uneven surfacing, roadside obstruction, or lack of clearly defined flow paths.

10.8.3 Depressions and Sag Points

Depressed road sections, underpasses, sag points, and low crossings are especially vulnerable because they depend on drainage structures operating efficiently under concentrated inflow. Such locations may require greater inlet capacity, duplicate inlets, local sump arrangements, or pumped drainage where gravity discharge is not feasible.

10.8.4 Design Lesson

Road drainage design cannot rely only on uniform longitudinal sections. Critical nodes such as intersections and depressions require localized hydraulic checks and often need stronger design provisions than ordinary road reaches.

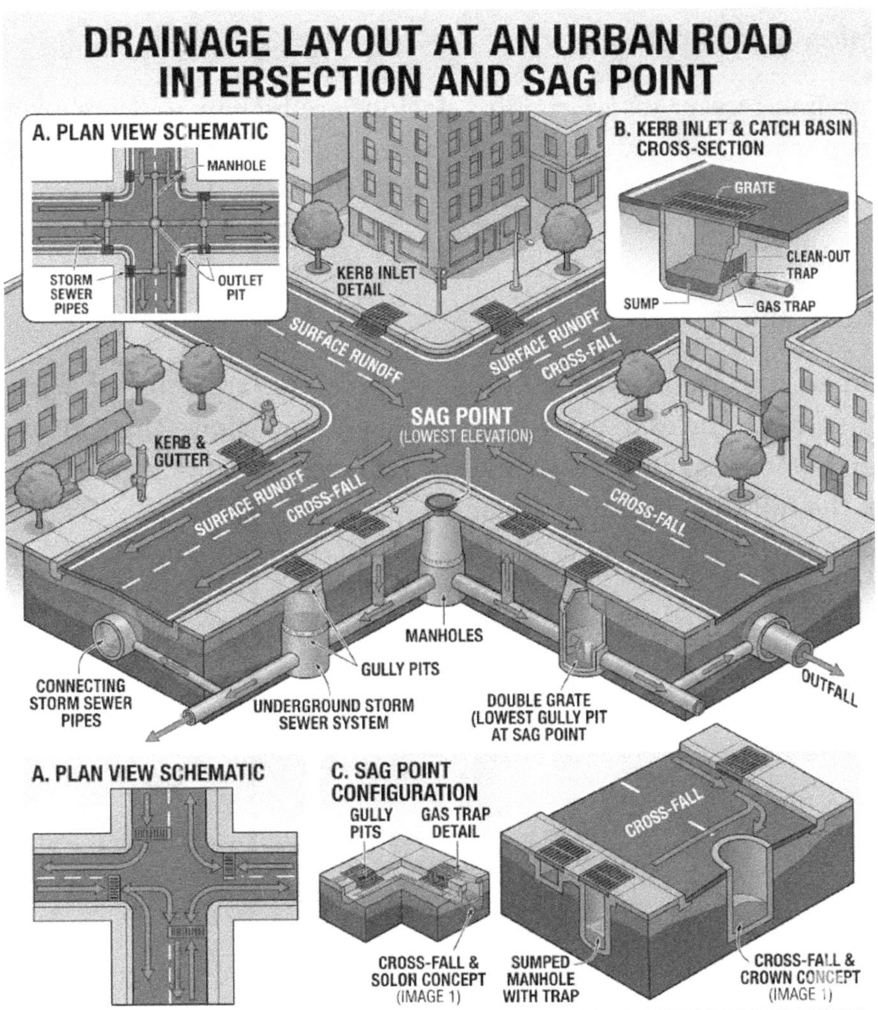

Figure 10.5 Drainage layout at an urban road intersection and sag point

10.9 Embankment Drainage and Slope Protection

10.9.1 Embankments as Drainage Barriers

Road embankments alter natural drainage by raising the road above surrounding terrain. While this may improve trafficability, it also creates the risk of intercepting or obstructing natural flow, concentrating runoff, and causing erosion along side slopes and at outlets.

10.9.2 Two Main Drainage Tasks

Two key issues in embankment drainage are:

1. Water from the road surface and side drains must be conveyed safely without damaging the embankment.
2. External runoff approaching the road must be allowed to cross or bypass the embankment through culverts or side drainage so that upstream flooding does not occur.

10.9.3 Slope Protection Measures

Slope protection is often necessary where runoff velocities are high or soils are erodible. Measures may include turfing, stone pitching, riprap, chute drains, lined side drains, toe walls, and energy dissipation at discharge points. Without such protection, even a correctly sized drain may cause gullying and slope instability.

10.9.4 Common Field Problem

In village and peri-urban roads, embankment drainage problems are common because roads are often raised incrementally without corresponding cross-drainage provision. This can trap water upstream, damage adjoining land, and weaken road shoulders. Embankment drainage must therefore always be considered as part of the full road drainage system.

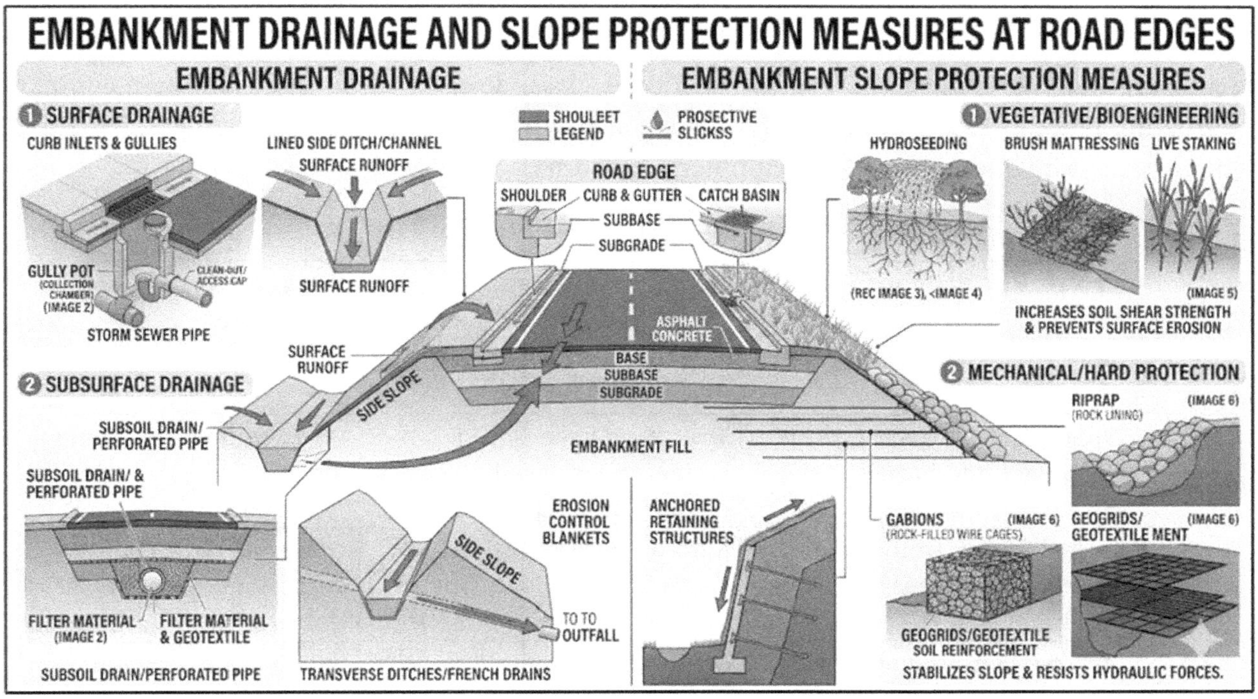

Figure 10.6 Embankment drainage and slope protection measures at road edges

10.10 Urban and Rural Road Drainage Considerations

10.10.1 Urban Roads

Road drainage design differs significantly between urban and rural or village conditions. In urban areas, road space is constrained, imperviousness is high, underground utilities are present, and runoff concentrates quickly. This often leads to reliance on kerb inlets, covered drains, storm sewers, and closely spaced collection structures. Aesthetic, safety, and traffic requirements are also stronger.

10.10.2 Rural and Village Roads

In rural and village roads, open side drains, earthen channels, culverts, and simple cross-drainage structures are more common. The hydraulic principles remain the same, but cost, simplicity, ease of maintenance, and compatibility with local land use become more important. Agricultural drainage, access crossings, and roadside encroachment must also be considered.

10.10.3 Semi-urban Roads

Semi-urban roads often combine the challenges of both systems. They may have increasing paved area and traffic demand, but still rely on partially developed roadside drains and limited maintenance. Such roads require flexible solutions that anticipate future urbanization while remaining practical in the present.

10.10.4 Context-based Design

Road drainage design should reflect the road's functional class, settlement setting, land-use context, and available maintenance system rather than applying a single standard form everywhere. This is a particularly important principle in mixed-development corridors.

Table 10.3 Urban, semi-urban, and village road drainage characteristics

Comparison Feature	Urban Road Drainage	Semi-urban Road Drainage	Village (Rural) Road Drainage
System Classification	Closed / "Urban Section"	Combined / Transitional Section	Open / "Rural Section"
Dominant Flow Path	Collected on surface, quickly channeled into buried storm sewer	Transition from closed systems near development to open systems further away	Directed into longitudinal ditches
Primary Components	Kerb & Gutter, Gully Pits, Underground Pipes, Manholes, Detention Basins	Kerb & Gutter (clusters), Catch Basins, Pipes, Open ditches, Culverts	Open Side Ditches, Vegetated Swales, Culverts, Turn-out Drains
Surface Pavement Type	Impervious (bitumen, concrete)	Mixed (impervious + pervious shoulders)	Varied (paved primary, unpaved minor roads)
Runoff Characteristics	Rapid concentration, high peaks	Mixed regimes	Slower concentration, infiltration encouraged
Land Availability	Extremely limited	Transitional	Ample
Materials Used	Concrete, cast iron, bitumen	Concrete, stone, native soil	Native soil, vegetation, stone, concrete culverts
Construction Cost	Highest	Medium	Lowest
Maintenance & Operation	Specialized & costly	Mixed	Periodic & labor-intensive
Connection to Networks	Centralized municipal storm sewers	Local pipe networks to open channels	Direct to natural water bodies
Environmental Concerns	Heavy metals, hydrocarbons, litter	Transition pollutants	Sediment, agricultural runoff
Aesthetics	Clean, structured	Transitional	Natural, unobtrusive

Summary of Key Transitions

- **Closed to Open:** Urban systems emphasize speed; rural systems emphasize volume handling and infiltration.
- **Infrastructure to Landscape:** Urban drainage is engineered; rural drainage integrates with natural catchments.
- **Speed to Storage:** Cities remove water quickly; rural areas store and infiltrate water.

10.11 Indian Application Notes

10.11.1 Relevance in Indian Settlements

In Indian settlement conditions, road drainage often serves a wider role than simple pavement protection because roads frequently act as both access routes and major runoff pathways during monsoon rainfall. This makes roadside drains, culverts, and low-point drainage structures especially important in both urban and village contexts.

10.11.2 Raised Roads and Blocked Cross-Drainage

Embankment drainage issues are particularly relevant where roads are raised incrementally but cross-drainage provision does not keep pace. Such conditions can trap water upstream, affect adjoining plots and agricultural land, and create recurring local flooding.

10.11.3 Need for Maintainable Solutions

Indian field conditions underline the need for maintainable design solutions. Roadside drains, culverts, and cross-drainage structures must be designed not only for hydraulic adequacy but also for ease of cleaning, desilting, and repair under local resource constraints.

10.12 Planning Implications

The transformed chapter shows that road drainage is a critical intersection between transportation engineering and settlement drainage management. The road corridor must not only remain trafficable but must also receive, guide, and discharge runoff safely without damaging adjoining land or obstructing natural flow.

10.13 Indian Application Notes

10.13.1 Relevance in Indian Settlements

In Indian settlement conditions, road drainage often serves a wider role than simple pavement protection because roads frequently act as both access routes and major runoff pathways during monsoon rainfall. This makes roadside drains, culverts, and low-point drainage structures especially important in both urban and village contexts.

10.13.2 Raised Roads And Blocked Cross-Drainage

The source chapter's observations on embankment drainage are particularly relevant where roads are raised incrementally but cross-drainage provision does not keep pace. Such conditions can trap water upstream, affect adjoining plots and agricultural land, and create recurring local flooding.

10.13.3 Need for Maintainable Solutions

Indian field conditions also underline the need for maintainable designs. Gully pits, side drains, culverts, and slope-protection measures must be designed so that they can be cleaned, inspected, and repaired under real local maintenance systems.

10.14 Planning Implications

The road drainage is a critical intersection between transportation engineering and settlement drainage management. The road corridor must not only remain trafficable but must also receive, guide, and discharge runoff safely without damaging adjoining land or obstructing natural flow.

The local failures at inlets, culverts, intersections, and embankments can produce effects that extend beyond the road itself. This is why road drainage should be planned as part of the larger drainage network rather than as an isolated roadside detail.

The road drainage solutions must be adapted to context. Urban streets, semi-urban corridors, and village roads differ in geometry, land use, maintenance capacity, and hydraulic behaviour, so appropriate drainage forms will also differ.

10.15 Summary Points

1. Road drainage protects pavements, traffic movement, and surrounding settlement areas.
2. Roads function as both transport corridors and artificial runoff pathways.
3. Surface drainage begins with correct road geometry, including cross fall, camber, and longitudinal slope.
4. Kerb drains, gutters, and side drains are the first controlled drainage elements along roads.
5. Catch basins and gully pits intercept runoff and transfer it to covered or underground drains.
6. Culverts and pipe crossings preserve flow paths beneath roads.
7. Causeways and small bridges are needed where culverts are hydraulically or functionally insufficient.
8. Intersections, sag points, and depressed road sections require special drainage attention.
9. Embankment drainage must address both road runoff and intercepted external flow.
10. Slope protection is essential where runoff concentration causes erosion risk.
11. Urban, semi-urban, and village roads require different drainage treatments even though the hydraulic principles are shared.

10.16 References

1. *Various authors. (n.d.). Textbooks on highway drainage and transportation infrastructure.*
2. *Various authors. (n.d.). Road drainage design manuals and culvert design guidance.*
3. *Various authors. (n.d.). Municipal references on stormwater inlets, gutters, and roadway drainage.*
4. *Various authors. (n.d.). Rural and low-volume road drainage references, including roadside and cross-drainage guidance.*
5. *Various authors. (n.d.). Case-based technical guidance on culvert and cross-drainage design in settlement and corridor applications.*

Chapter 11

Urban Drainage Systems

Learning Objectives

1. Describe the engineering features of urban drainage systems.
2. Identify constraints unique to dense urban environments.
3. Propose design strategies for flood-prone and congested urban areas.
4. Evaluate the need for pumping and advanced control measures.

11.1 Introduction

Urban drainage systems are designed to manage runoff in densely built environments where impervious surfaces, limited open space, underground utilities, and high infrastructure concentration create complex hydraulic conditions. Unlike rural or low-density settings, cities generate rapid runoff, offer little natural storage, and often contain aging or overloaded drainage networks. As a result, urban drainage engineering requires careful integration of hydrology, hydraulics, land-use planning, road drainage, utility coordination, and flood management.

This chapter examines the structure and functioning of urban drainage systems and the engineering issues that shape their design, operation, and improvement. It focuses on the characteristics of urban catchments, storm sewer networks, covered drains, trunk drainage, utility conflicts, low-lying area drainage, flood hotspots, and retrofitting of existing systems.

11.2 Characteristics of Urban Drainage Catchments

Urban drainage catchments differ significantly from natural or rural catchments because built development alters the way rainfall is converted into runoff. Roofs, pavements, roads, parking areas, and compacted surfaces reduce infiltration and increase the proportion of rainfall that becomes direct surface flow. As a result, urban catchments generally produce higher runoff volumes and sharper peak discharges than undeveloped land.

Another key characteristic of urban catchments is hydraulic connectivity. In cities, runoff is quickly transferred from roofs and paved areas to streets, gutters, inlets, and storm drains. This shortens the time of concentration and causes water to reach drainage structures more rapidly. Even small storms can therefore create intense local drainage demand, especially where the network is already stressed.

Urban catchments are also affected by artificial grading, plot filling, basements, road embankments, boundary walls, and utility corridors. These features may modify or disconnect natural drainage paths. In many older cities, urban drainage patterns no longer reflect the original terrain, making careful investigation essential before design or rehabilitation.

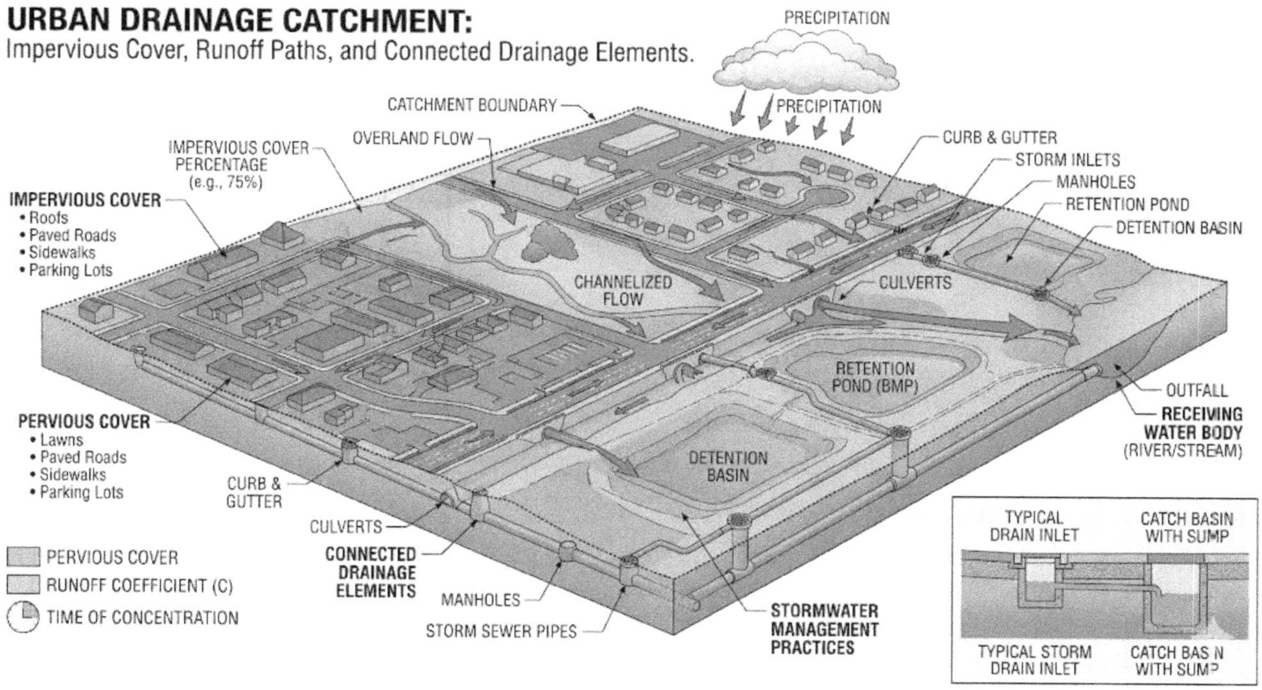

Figure 11.1 Urban drainage catchment showing impervious cover, runoff paths, and connected drainage elements.

11.3 Impervious Surfaces and Runoff Response

Imperviousness is one of the defining features of urban drainage systems. Surfaces such as concrete roads, asphalt pavements, tiled courtyards, rooftops, and compacted parking areas allow very little infiltration. Consequently, a large fraction of rainfall becomes surface runoff almost immediately.

As imperviousness increases, several hydrologic changes occur: runoff volume increases, peak discharge rises, time of concentration decreases, surface detention decreases, and flow becomes more concentrated and rapid.

These changes place heavy demand on drainage infrastructure, especially in older urban areas where the system may have been designed under lower development intensity. Local flooding often occurs not only because rainfall is extreme, but because the built environment has amplified runoff beyond the capacity of existing drains.

The effect of imperviousness is also uneven across the city. Commercial areas, transport corridors, industrial plots, and dense residential zones may generate very different runoff patterns. Therefore, urban hydrologic analysis should account for spatial variation in impervious cover rather than treating the city as a uniform surface.

Table 11.1 Impervious surface types and their effect on runoff response.

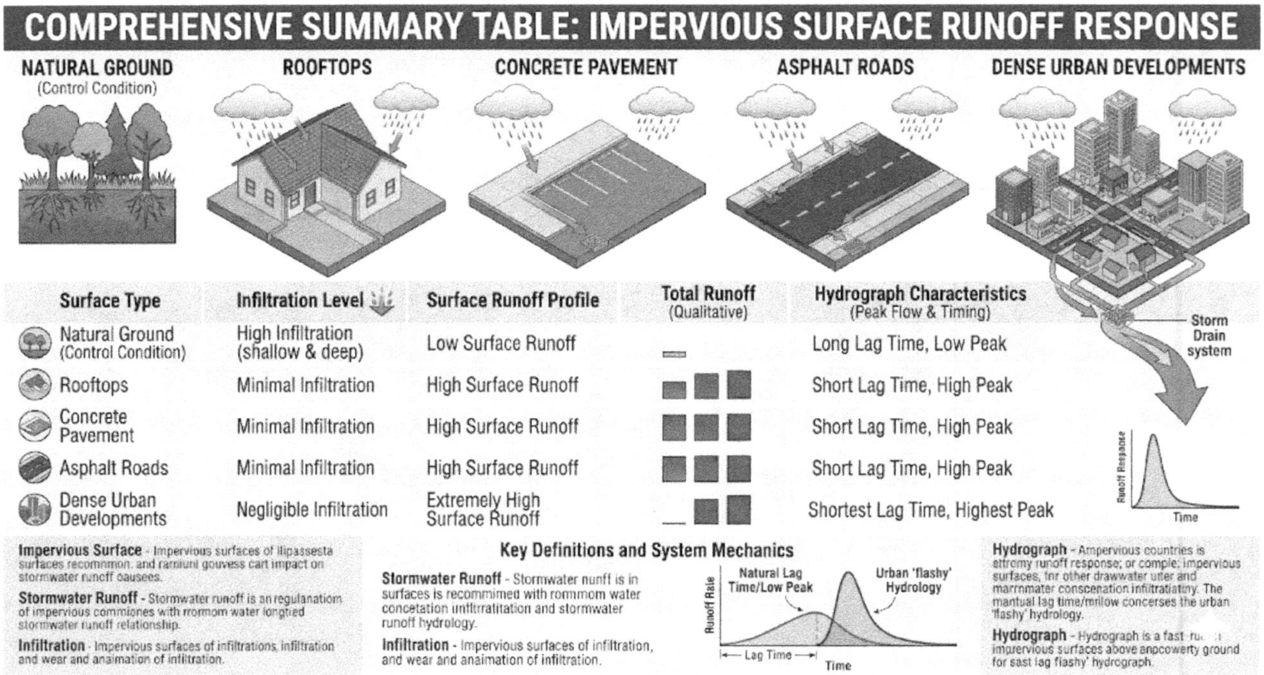

11.4 Urban Storm Sewer Systems

Storm sewer systems are closed drainage networks designed to collect and convey runoff through buried pipes, inlets, manholes, and outfall structures. They are widely used in urban areas where land is scarce, road space is intensively used, and open drains are impractical or undesirable.

A typical urban storm sewer network includes surface inlets or catch basins, lateral and branch pipes, manholes and junction chambers, trunk sewers, and outfalls to receiving drains, channels, rivers, or storage systems.

The design of storm sewer systems must account for pipe capacity, inlet spacing, hydraulic grade line, surcharge behavior, maintenance access, and interaction with road drainage. In dense built-up areas, these systems often operate near capacity during major rainfall events, making them sensitive to blockage, sedimentation, and tailwater effects.

Storm sewers offer several advantages, including space efficiency, reduced exposure of wastewater or stagnant water, and compatibility with dense streetscapes. However, they also require skilled construction, accurate grade control, and regular inspection. If maintenance access is poor, hidden defects can remain undetected until severe flooding occurs.

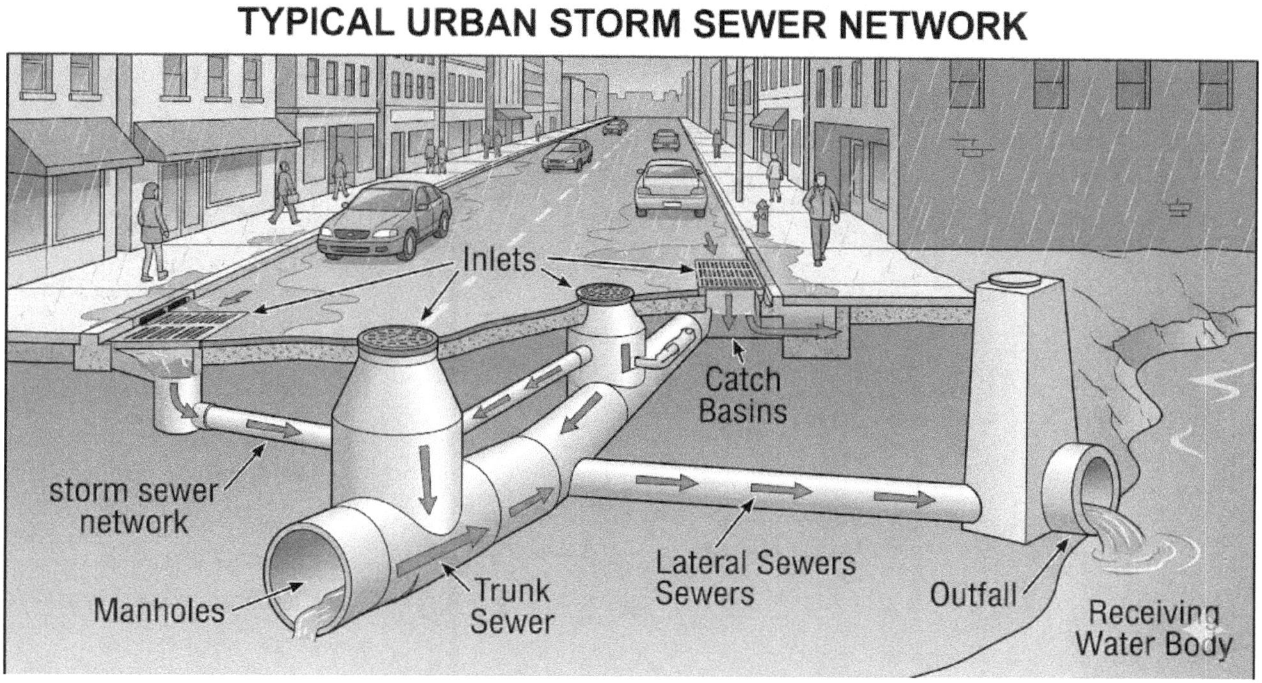

Figure 11.2 Typical urban storm sewer network with inlets, manholes, trunk sewer, and outfall.

11.5 Covered Drains and Underground Networks

Covered drains occupy an intermediate position between open drains and fully buried storm sewers. They are commonly used in urban areas where drainage must be accommodated along roadsides or building lines without leaving an open channel exposed at the surface. Such systems may consist of masonry or concrete drains roofed with removable or fixed slabs.

Covered drains can be useful where pedestrian movement, frontage access, or safety concerns make open drains undesirable, but where a full pipe sewer is unnecessary or too expensive. They allow linear drainage along urban roads while preserving some access for cleaning, especially if inspection openings or removable covers are provided.

Underground networks may include both covered roadside drains and larger buried pipe systems. In many cities, these networks evolve over time rather than being built as a single planned system. Older covered drains may connect to newer storm sewers, culverts, or natural channels, often creating discontinuities in capacity or invert level.

The main engineering concern is that covered systems can conceal problems. Sediment accumulation, structural deterioration, illegal wastewater inflow, and collapse are less visible than in open drains. Therefore, covered urban drainage must be designed with inspection and maintenance in mind, not just with hydraulic capacity alone.

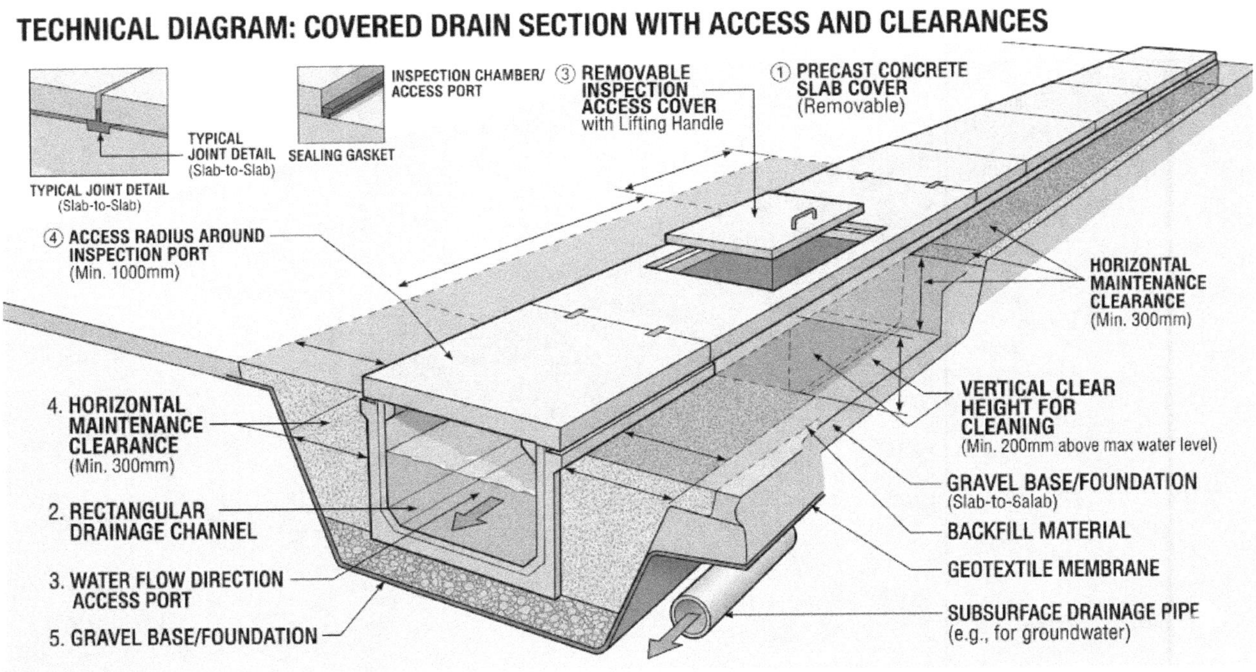

Figure 11.3 Covered drain section showing slab cover, channel, inspection access, and maintenance clearances.

11.6 Trunk Drains and Outfall Systems

Trunk drains are the main conveyance routes that receive runoff from branch drains, local sewers, and intermediate drainage lines. In urban drainage systems, they carry large volumes of water toward the final outfall, which may be a river, canal, lake, detention basin, wetland, pumping station, or other receiving facility.

The performance of the entire urban drainage network often depends on the adequacy of trunk drains. Local drains may be properly sized, but if the trunk system is undersized, encroached upon, or poorly maintained, upstream flooding will still occur. This is why many urban flood problems are system-level failures rather than local drain defects.

Outfall systems are equally critical. A city may possess an extensive drainage network, yet still experience flooding if the outfall is submerged, obstructed, tide-affected, or hydraulically disconnected from the main drainage path. In low-lying cities, downstream river levels or tidal conditions may prevent gravity discharge at certain times.

Urban drainage planning must therefore treat trunk drains and outfalls as strategic infrastructure. Their protection, rehabilitation, and capacity assessment should receive special priority, especially where urban growth is increasing runoff faster than the main system is being upgraded.

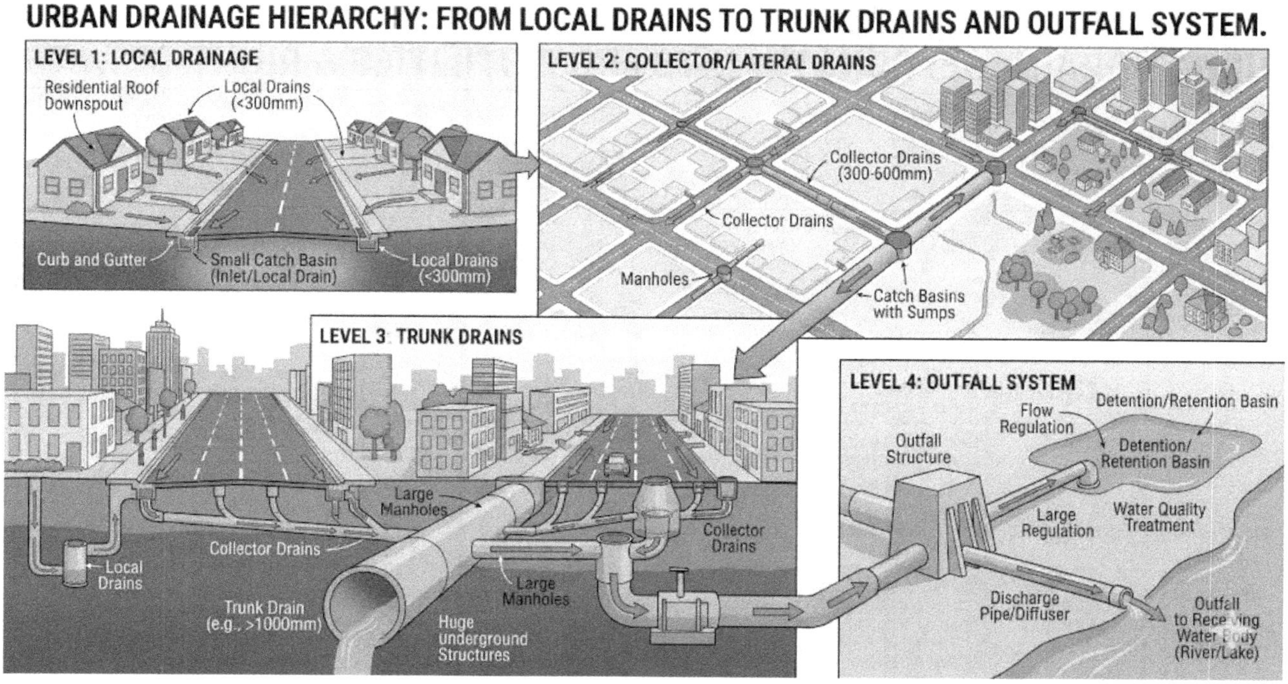

Figure 11.4 Urban drainage hierarchy from local drains to trunk drain and outfall system.

11.7 Utility Conflicts and Right-of-Way Issues

One of the greatest practical challenges in urban drainage design is the presence of existing utilities and limited right-of-way. Water supply lines, sewer mains, electric ducts, telecommunications cables, gas lines, and transport infrastructure often compete for the same underground corridor needed for drainage.

Utility conflicts can force drainage alignments to be deeper, narrower, offset, or segmented in ways that reduce hydraulic efficiency and increase construction complexity. In rehabilitation projects, it may be difficult to replace or enlarge an existing drain because other services have been installed above, beside, or across it. In some cases, the available corridor is so constrained that only partial improvements are feasible.

Right-of-way problems are equally severe. Buildings may encroach upon former drainage channels, informal developments may occupy natural flow corridors, and roads may lack sufficient shoulder width for proper side drains. In older urban areas, the original drainage corridor may have been progressively lost to unplanned expansion.

As a result, urban drainage design often involves compromise and coordination. Engineering solutions must be technically sound, but also feasible within the built fabric of the city. Utility mapping, staged construction, protective works, and inter-agency coordination are therefore central to urban drainage practice.

Table 11.2 Common urban drainage constraints related to utilities and right-of-way.

URBAN DRAINAGE: CONSTRAINTS FROM UTILITIES & RIGHT-OF-WAY

CONSTRAINT CATEGORY	DESCRIPTION / KEY ISSUES		ILLUSTRATION / EXAMPLES	MITIGATION & ADAPTATION STRATEGIES
UTILITIES (UNDERGROUND & OVERHEAD)	CONFLICTING ELEVATIONS	Drainage needs to go where other pipes are (classh ar drainage need of clases (gas) ander water conflicts)		UTILITY RELOCATION or DRAINAGE VERTICAL CURVES
	DENSE NETWORKS	High number of pipes and wires are comp more pipes and maze states around the utilility chemical networks.		• MULTI-LEVEL UTILITY DUCTS • SHARED TRENCHES
	CRITICAL INFRASTRUCTURE	Constraints around major trunk lines discusses, a lurod arge buried around a high-voltage line and high-voltage tower.	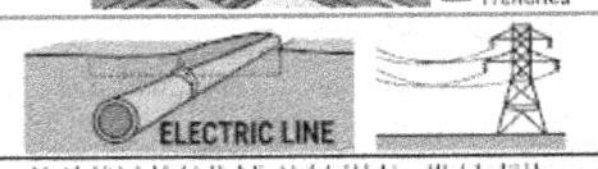	• SETBACK DISTANCES • SPECIALIZED CROSSINGS
	LIMITED HORIZONTAL SPACE	Narrow streets are snow.eret and drainage elements from numbered road inorizontal streets.		• SMALLER, DEEPER DRAINAGE ELEMENTS • BIORETENTION / PERMEABLE PAVEMENT
RIGHT-OF-WAY (ROW) & SURFACE CONSTRAINTS	NARROW / LIMITED ROW	Limited surface area, street plan, for of boundary lines, boundary and in limited street.	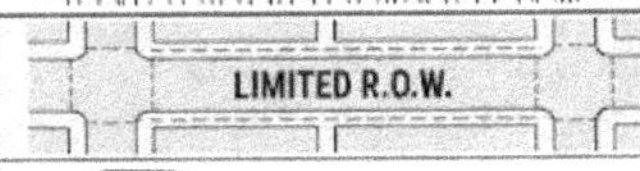	• LINEAR DRAINAGE SYSTEMS (e.g., TRENCHES) • BIOFILTRATION ALONG CURBS
	STEEP SLOPES	Excessive speed of water. Excessive water to speed of wate escesitive drainage of water.	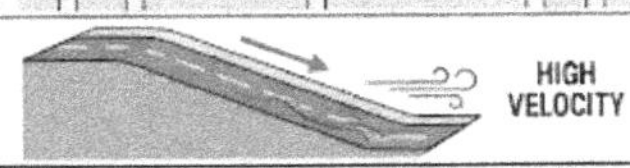	• CHECK DAMS • STEP DRAINAGE • DROP STRUCTURES
	EXISTING STRUCTURES & LANDMARKS	Inability to move things, on historic retaining wall and existing potential drainage locations		• RE-ROUTING DRAINAGE PATHS • SPECIAL DESIGN ADJUSTMENTS
	MULTIPLE USERS	Competting needs (sidewalks, bike lanes) and separated zone snes zones about competing fallewalks.		• INTEGRATED GREEN INFRASTRUCTURE • MULTIPLE-USE DESIGN

11.8 Pumped Drainage for Low-Lying Urban Areas

Gravity flow is generally preferred in drainage systems because it is simple and reliable, but in many urban areas gravity alone is insufficient. Low-lying districts, reclaimed land, underpasses, basements, and areas affected by high downstream water levels may require pumped drainage to remove stormwater effectively.

Pumped systems are used where the drainage outfall is higher than the flood-prone area, where tidal or river backwater prevents gravity discharge, or where localized depressions trap water. These systems may include sump structures, wet wells, pumping stations, rising mains, control gates, standby power, and automated control systems.

The use of pumps introduces a major operational dimension into urban drainage management. Unlike gravity systems, pumped drainage depends on mechanical reliability, energy supply, operator response, and maintenance discipline. Failure of a pump during a storm can quickly lead to serious flooding, especially in enclosed or heavily urbanized zones.

Therefore, pumped drainage should be applied carefully. It is often necessary in certain contexts, but it should not be used as a substitute for poor planning where gravity alternatives still exist. Where pumps are unavoidable, redundancy, emergency operation procedures, and maintenance planning become essential parts of the design.

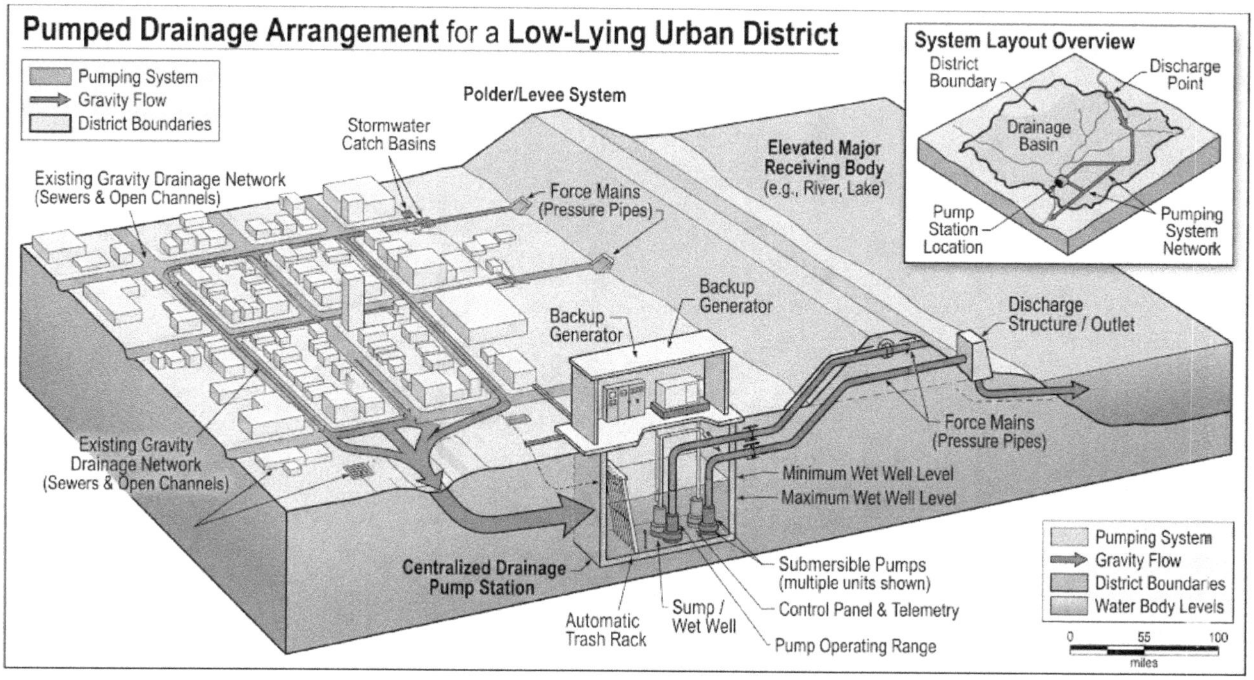

Figure 11.5 Pumped drainage arrangement for a low-lying urban district.

11.9 Flood Hotspots and Drainage Bottlenecks

Urban flood hotspots are locations where flooding recurs repeatedly because local hydraulic conditions, catchment concentration, infrastructure limitations, or downstream constraints combine to produce failure. Common hotspot locations include low intersections, underpasses, market areas, transport terminals, basement access points, narrow culvert crossings, and densely built low-lying neighborhoods.

A drainage bottleneck is a point in the system where flow is restricted and upstream performance is controlled by a local deficiency. This may be caused by an undersized drain, blocked inlet, narrowed crossing, encroached channel, flat gradient, utility obstruction, or submerged outfall. Although bottlenecks are often physically small, their influence can extend over a large upstream area.

The identification of hotspots and bottlenecks is a major task in urban drainage management. These locations often deserve priority because improvement at a single control point may significantly reduce flooding across a wider area. However, not all visible hotspots are local problems; some are symptoms of larger trunk or outfall deficiencies.

Urban drainage analysis should therefore evaluate both the local geometry of hotspots and their position within the wider network. Effective intervention depends on knowing whether the true cause is nearby or elsewhere in the system.

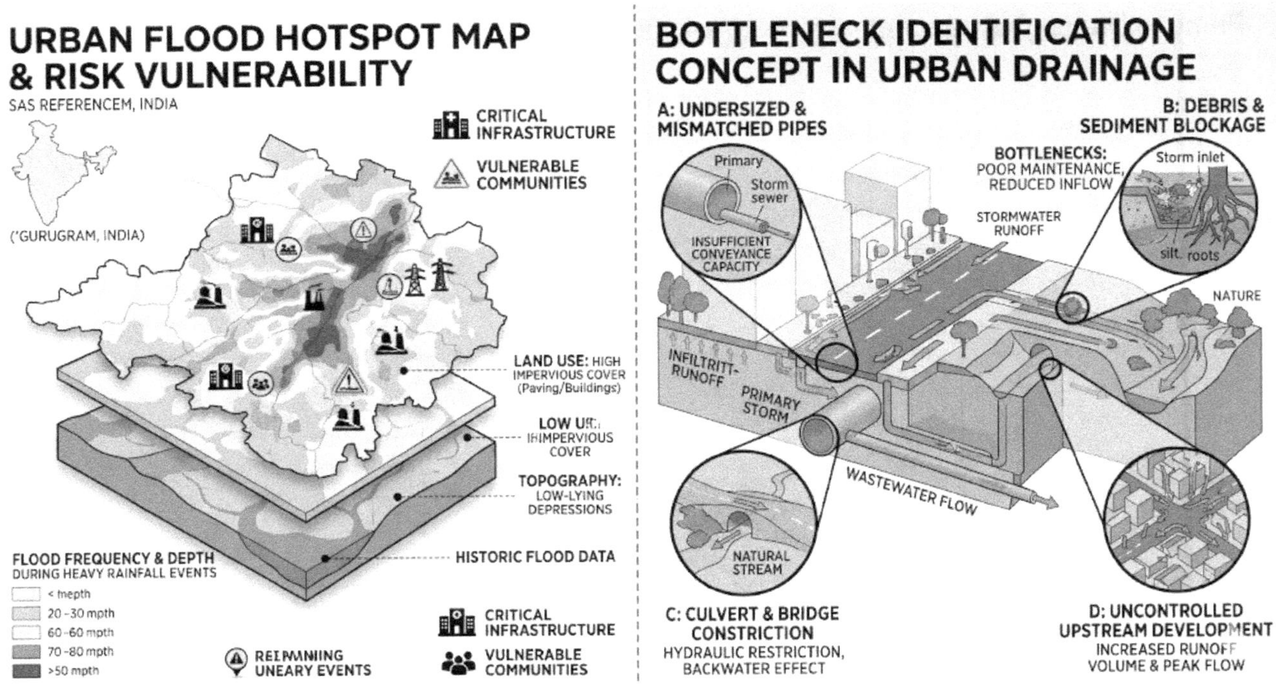

Figure 11.6 Urban flood hotspot map and bottleneck identification concept.

11.10 Retrofitting Existing Urban Systems

Most cities must improve drainage under existing built conditions rather than through complete redevelopment. Retrofitting refers to the modification, upgrading, or rehabilitation of existing drainage infrastructure so that it performs better under current and future demand.

Retrofitting may involve widening or deepening open drains, replacing undersized pipes, adding inlets or catch basins, reconstructing collapsed covered drains, removing obstructions and encroachments, regrading road surfaces, improving culvert openings, adding storage or pumping, and introducing sustainable drainage features where feasible.

The challenge of retrofitting lies in working within an occupied, utility-congested, and often space-constrained environment. Construction must often proceed in stages, with traffic management, temporary drainage, utility protection, and minimal disruption to residents and businesses. In older urban cores, even small upgrades may require significant coordination.

Despite these difficulties, retrofitting is often the only realistic path toward urban drainage improvement. It is therefore important that retrofit design be practical, phased, and compatible with the long-term drainage strategy of the city.

Table 11.3 Typical retrofit measures for upgrading urban drainage systems.

TYPICAL RETROFIT MEASURES FOR UPGRADING URBAN DRAINAGE SYSTEMS

MEASURE TYPE / CATEGORY	SPECIFIC RETROFIT MEASURE	DESCRIPTION / MECHANISM	KEY UPGRADE / FUNCTION	IMPLEMENTATION EXAMPLE
Nature-Based Solutions (NBS) & Green Infrastructure (GI)	• Permeable Pavement • Rain Gardens (Bioretention Cell) • Green Roofs	Replaces impervious surfaces; engineered soils and plants for infiltration and evapotranspiration; vegetation on rooftops	Peak flow reduction, volume reduction, quality treatment (filtration, biological uptake)	Full parking lot retrofit; landscaped area in a park; entire industrial roof
Grey Infrastructure & Conveyance Improvements	• Pipe Upsizing • Parallel Pipes (Twinning) • Installation of Culvert Upgrades • Online Storage	Replaces small pipes; adds additional capacity alongside existing; increases flow; uses enlarged pipe sections for temporary	Localized capacity increase, flow bottleneck removal, temporary storage	Trunk sewer on Main St.; redundant pipe added; narrow crossing replaced
Flow Control & Peak Shaving	• Install/Retrofit Flow Restrictors (Orifices, Weirs) • Add Smart Controllers (RTC) • Diversion Structures	Controls downstream flow rate; uses real-time data to control valves; redirects excessive flows to storage	System-wide peak reduction, optimization of storage, relief of downstream capacity	Storage tank outfall; logic implemented in detention large-scale river flow division
Water Quality Treatment & Pollutant Removal	• Hydrodynamic Separators • Screening Devices • Biofiltration Swales	Uses vortex or settling to remove solids; physical barriers (bars, nets); long swales with specific plants and soils	Removal of trash, sediment, heavy metals, nutrients	Pre-treatment for an outfall; end-of-pipe net; wide roadside landscape feature
Operation & Maintenance (Essential Retrofit)	• Specialized Cleaning (Jetting, Vacuum) • Sediment and Debris Removal (Silt/Root Cutters) • System-wide CCTV Inspection	High-pressure water, powerful vacuums, mechanical tools; uses cameras to map and inspect	Restoration of original capacity, extends system life, proactive maintenance, identifies defects	Full system flushing; large culvert desilting; 5-year cyclical inspection program
RETROFIT SYSTEM INTEGRATION				

11.11 Summary Points

1. Urban drainage systems manage runoff in dense cities where impervious surfaces and limited open space intensify flooding.
2. Urban catchments produce higher runoff volumes and sharper peaks due to reduced infiltration and rapid hydraulic connectivity.
3. Impervious surfaces like roads and rooftops increase runoff, shorten concentration time, and strain drainage networks.
4. Storm sewer systems, with inlets, pipes, manholes, and outfalls, are common in cities but sensitive to blockages and capacity limits.
5. Covered drains provide roadside drainage in dense areas but conceal problems like sedimentation and collapse.
6. Trunk drains and outfalls are strategic elements; failures here often cause system-wide flooding.
7. Utility conflicts and right-of-way issues complicate drainage design, requiring coordination with other infrastructure.
8. Pumped drainage is essential in low-lying or tide-affected areas but introduces risks of mechanical failure and energy dependence.
9. Flood hotspots and bottlenecks recur at intersections, underpasses, and culverts, often controlling upstream performance.
10. Retrofitting existing systems—through upgrades, regrading, or added storage—is often the only feasible improvement path in older cities.

11.12 References

1. Butler, D., Digman, C., Makropoulos, C., & Davies, J. W. (2024/2025). *Urban drainage* (5th ed.). Routledge.
2. Butler, D., Digman, C., Makropoulos, C., & Davies, J. W. (2018). *Urban drainage* (4th ed.). CRC Press / Routledge.
3. Butler, D., & Davies, J. W. (2010). *Urban drainage* (3rd ed.). Taylor & Francis.
4. Butler, D., & Davies, J. W. (n.d.). *Urban drainage* (2nd ed.). CRC Press / Taylor & Francis.
5. American Society of Civil Engineers/Environmental and Water Resources Institute. (2006). *Standard guidelines for the design of urban subsurface drainage systems (ASCE/EWRI 12-05, with related installation and O&M guidance 13-05 and 14-05)*. American Society of Civil Engineers.
6. Drainage Services Department. (n.d.). *Stormwater drainage manual*. Government of Hong Kong.
7. Federal Highway Administration. (2001). *Urban drainage design manual (Hydraulic Engineering Circular No. 22, 2nd ed., FHWA-NHI-01-021)*. U.S. Department of Transportation.
8. Federal Highway Administration. (n.d.). *Urban drainage design manual (HEC-22, 3rd ed.)*. U.S. Department of Transportation.
9. Chartered Institution of Water and Environmental Management Urban Drainage Group. (2021). *Integrated urban drainage modelling guide* (2nd ed.).
10. International Journal of Innovative Research in Science, Engineering and Technology. (n.d.). *Stormwater drainage solutions for urban area*.

Chapter 12

Semi-Urban and Peri-Urban Drainage Systems

Learning Objectives

By the end of this chapter, the learner should be able to:

1. Explain the physical and hydrologic characteristics of semi-urban and peri-urban drainage environments.
2. Distinguish transitional drainage conditions from those of fully urban and fully rural systems.
3. Assess the effects of mixed land use, land conversion, and partial infrastructure on runoff generation and drainage performance.
4. Describe methods for integrating natural drains with engineered drainage works.
5. Develop staged, flexible, and cost-effective drainage strategies for expanding settlement areas.
6. Evaluate drainage issues arising at the urban-rural interface.

12.1 Introduction

Semi-urban and peri-urban areas represent dynamic zones of transition where rural land is progressively converted into urban uses, but the supporting infrastructure often develops slowly, unevenly, or in fragments. In such locations, drainage engineering becomes more complex than in either established cities or stable rural landscapes because the catchment itself is changing while runoff pathways, surface conditions, and land ownership patterns remain unsettled.

These areas commonly occur at the outer edge of cities, along highway corridors, around industrial expansion belts, near newly plotted residential colonies, and within villages gradually absorbed by urban growth. Their drainage systems are rarely planned as complete networks from the beginning; instead, they emerge piecemeal through culverts, roadside drains, open channels, temporary ditches, modified natural depressions, and isolated drainage improvements introduced in response to development pressure.

For this reason, semi-urban drainage should not be treated merely as a smaller version of urban drainage. It is a distinct planning and engineering challenge defined by uncertainty, mixed land use, interrupted natural hydrology, limited budgets, and the need to anticipate future intensification before the final urban form is known.

12.2 Nature of Semi-Urban and Peri-Urban Settlements

A semi-urban or peri-urban settlement may be understood as a zone in transition from predominantly rural land use to increasingly urban occupation and function. These areas may include suburban extensions, ribbon development along major roads, peripheral municipal wards, institutional campuses, warehouse clusters, industrial fringes, resettlement colonies, and old villages now surrounded by new construction.

These settlements often display a visible mixture of development stages. One stretch may contain paved roads, drains, and formal building plots, while a neighboring stretch may remain agricultural or open, drained only by field channels or natural depressions. This juxtaposition of formal and informal land transformation creates hydraulic inconsistency across the catchment.

Figure 12.1 Typical semi-urban and peri-urban drainage setting showing mixed development, open land, roads, and natural drainage paths.

In older rural settings, the drainage system may have evolved naturally over time, and in established urban settings, it may have been formalized into storm sewers, roadside drains, or trunk channels. In semi-urban settings, however, both systems often overlap imperfectly, so the resulting drainage network is incomplete, discontinuous, and vulnerable to failure.

From an engineering viewpoint, the principal difficulty lies in predicting how current conditions will change. Areas that are still partly permeable may soon be paved, local depressions may be filled, natural channels may be encroached upon, and traffic corridors may require crossing structures that alter flow direction and concentration. Thus, the drainage designer must plan with both present and future catchment states in mind.

12.3 Transitional Drainage Characteristics

The drainage characteristics of semi-urban and peri-urban areas are transitional in both physical and hydrologic terms. Surface slopes may still reflect the original rural terrain, but runoff generation increasingly resembles urban behavior as roofs, pavements, compacted earth, boundary walls, and roads accumulate.

A common feature is the presence of discontinuous drainage lines. A roadside drain may begin in a developed colony, terminate abruptly at an undeveloped parcel, reappear near a culvert crossing, and finally discharge into a pond, a natural nallah, or a shallow low-lying depression. Such discontinuity weakens hydraulic efficiency and makes system diagnosis difficult because flooding may originate not at the location where water accumulates, but at an upstream break or downstream restriction.

Figure 12.2 Discontinuous drainage pattern in a transitional catchment. illustrating fragmented flow paths, intermittent streams, and areas where surface runoff vanishes into the ground.

Another feature is variability in cross-section, material, and condition. Some drain reaches may be earthen, others brick-lined, some concrete, and others merely informal channels cut by repeated runoff. Maintenance levels also vary sharply, with some sections cleaned regularly while adjacent reaches are obstructed by solid waste, vegetation, silt, or construction debris.

Hydrologically, these areas exhibit mixed response. Impervious surfaces generate rapid runoff, but nearby open or agricultural areas may still absorb, detain, or delay part of the flow. As a result, runoff timing and concentration are irregular, and local flooding often depends on micro-topography, temporary obstructions, and the degree to which natural storage areas remain available.

Conventional urban drainage assumptions may therefore be only partly valid. At the same time, a purely rural approach may underestimate future runoff, especially where development is proceeding rapidly. The engineer must interpret the drainage system as a hybrid condition requiring hybrid planning logic.

Table 12.1 Comparison of rural, semi-urban, peri-urban, and urban drainage characteristics.

Aspect	Rural	Semi-Urban	Peri-Urban	Urban
Land Use	Predominantly agricultural and open land	Mixed residential and agricultural	Transitional mix of rural and urban land uses	Dense built-up areas with minimal open land
Surface Cover	Natural soil and vegetation	Partially paved, with open drains	Increasing impervious surfaces, fragmented vegetation	Highly impervious surfaces (roads, rooftops, pavements)
Drainage Type	Natural streams and field channels	Combination of natural and constructed drains	Discontinuous and modified natural channels	Fully engineered stormwater networks
Flow Pattern	Continuous, slow runoff	Seasonal or intermittent flow	Discontinuous, with infiltration and ponding	Rapid surface runoff, minimal infiltration
Catchment Response	High infiltration, low peak discharge	Moderate infiltration, localized flooding	Variable infiltration, frequent flash runoff	Low infiltration, high peak discharge and flash flooding
Maintenance	Minimal, community-based	Limited, often informal	Partially managed, transitional responsibility	Regular, institutionalized maintenance
Water Quality	Generally good, low pollution	Moderate pollution from domestic sources	Increasing contamination from mixed land use	High pollution from industrial and domestic sources
Drainage Challenges	Sedimentation, erosion	Poor connectivity, clogging	Fragmented flow paths, encroachment	Overloading, blockage, and urban flooding

12.4 Influence of Mixed Land Use Patterns

Mixed land use is one of the most defining and challenging features of peri-urban drainage design. Within a single catchment, there may be residential colonies, workshops, schools, godowns, informal markets, roads, vacant land, fields, institutional campuses, and remnant water bodies. Each of these land uses contributes runoff in a different way.

Commercial and industrial areas usually increase runoff rapidly because they introduce large roof areas, paved circulation surfaces, boundary walls, loading platforms, and reduced infiltration. Residential colonies alter grading, create roadside drainage demands, concentrate roof discharge, and often fill natural low points to improve plot usability. Agricultural land and open ground may still provide storage and infiltration, but their contribution is unstable because they may be developed later or cut off by road embankments and plot filling.

Because of this diversity, it is rarely accurate to represent the catchment using a single runoff coefficient or a single land surface assumption. Catchment subdivision becomes necessary so that runoff from different land parcels can be estimated and routed more realistically. This is especially important where high-runoff zones discharge toward lower-capacity open channels or into areas still functioning with rural drainage logic.

Mixed land use also introduces competing design expectations. A new commercial strip may demand curbside drainage and quick runoff removal, while adjacent agricultural land may tolerate shallow ponding or slower field drainage. The challenge is not simply to design individual drains, but to connect differing local standards into a coherent settlement-scale system.

In planning terms, the engineer must recognize that mixed land use is not a temporary inconvenience but a structural feature of semi-urban development. The drainage network should therefore be adaptable, segmented logically, and coordinated with probable future land conversion.

12.5 Integration of Natural and Engineered Drainage

One of the most important principles of peri-urban drainage engineering is the integration of natural drainage paths with engineered drainage structures. Natural channels, field drains, swales, depressions, village ponds, and seasonal flow routes often continue to carry runoff effectively even after built development begins to spread around them.

If these natural features are retained and connected properly, they can perform valuable hydraulic functions. They may convey stormwater, provide temporary storage, preserve watershed continuity, reduce the need for expensive underground systems, and limit the concentration of runoff into a few artificial outlets. They can also serve as the backbone around which future branch and trunk drainage systems are organized.

However, these pathways are often the first elements to be disturbed during land conversion. Developers may fill depressions, narrow natural channels, block local inlets, or cut across seasonal drainage routes without providing equivalent conveyance. Roads may sever drainage continuity if cross-drainage structures are absent or undersized. Once this occurs, runoff seeks new paths, frequently causing localized flooding, erosion, backwater effects, and waterlogging in adjacent low-lying settlements.

Figure 12.3 Integration of natural drains and engineered drainage elements.

The objective of integration is not to leave every natural feature untouched. Some channels may require widening, slope improvement, selective lining, bank stabilization, inlet control, or protection against erosion and encroachment. The crucial principle is that their hydraulic role should be understood and preserved even when physical improvement is necessary.

Engineered drains in semi-urban areas should therefore be conceived as connectors within a larger drainage landscape rather than isolated works serving only nearby roads or plots. A branch drain should discharge logically into a protected channel, a culvert should preserve continuity across embankments, and outlets should not be located where downstream erosion or obstruction is inevitable.

12.6 Drainage Impacts of Rapid Land Conversion

Rapid land conversion is among the most significant causes of drainage dysfunction in expanding fringe areas. Agricultural land, wetlands, village commons, orchard zones, and open depressions are often transformed into roads, residential layouts, industrial plots, commercial strips, and institutional developments without proportional drainage planning.

This process changes the hydrologic behavior of the catchment in several important ways. Infiltration declines as natural soil surfaces are covered or compacted. Runoff volume increases because more rainfall becomes direct surface flow. Peak discharge rises and time of concentration shortens because the catchment begins to convey water more rapidly toward low points and outlets. Local storage in ponds, depressions, and vegetated areas is lost, and natural flow paths are interrupted, diverted, or obliterated.

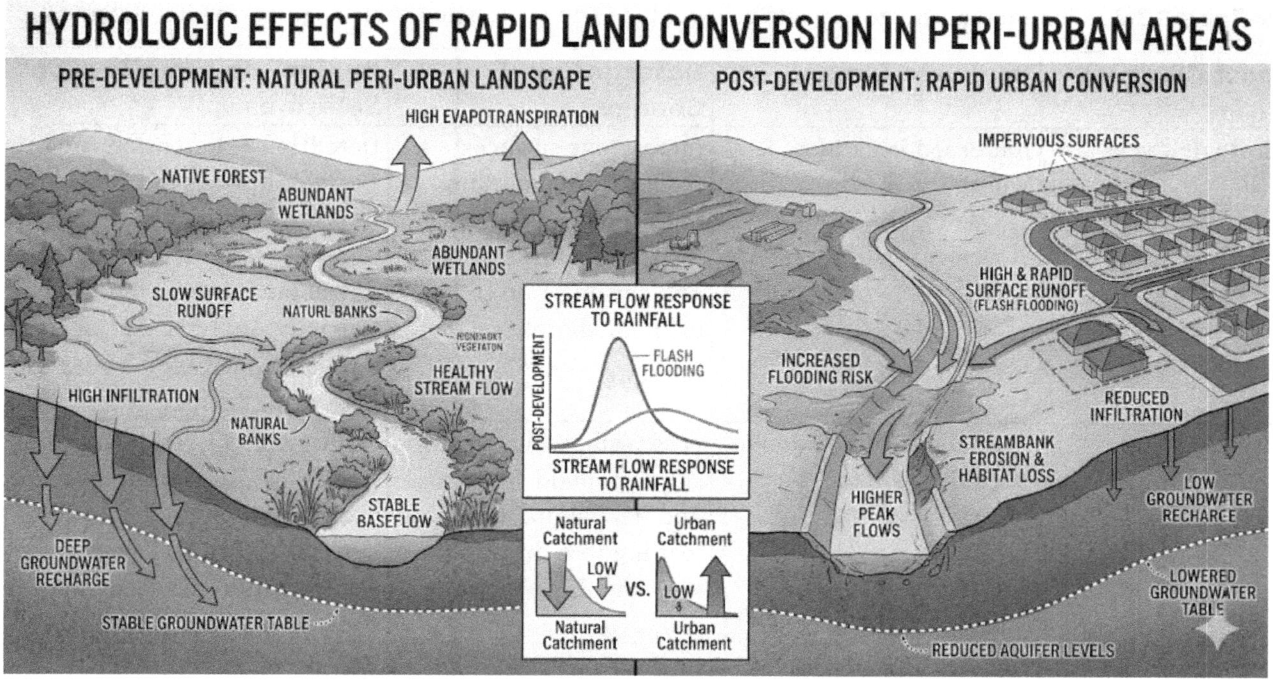

Figure 12.4 Hydrologic effects of rapid land conversion in peri-urban areas.

Among the most disruptive interventions are road embankments, indiscriminate plot filling, and boundary walls. A raised road without adequate culverts may act as a barrier that traps water on the upstream side. A filled colony may redirect runoff into older neighborhoods, agricultural fields, or village streets that sit at lower elevations. Compound walls may interrupt shallow sheet flow and force water into narrow concentrated routes that were never intended to function as principal drains.

The cumulative result is often a fragmented catchment in which traditional drainage pathways no longer operate, yet a formal urban system has not been provided. Flooding then becomes chronic, especially during intense storms, because each incremental development transfers its runoff burden to the surrounding landscape.

For this reason, peri-urban drainage planning must be proactive. Once development is complete, land acquisition becomes difficult, natural channels may be permanently encroached upon, and retrofitting costs rise sharply. Early recognition of drainage corridors, low points, and outfall requirements is therefore essential.

Table 12.2 Common drainage impacts of rapid land conversion and their engineering implications.

Impact Category	**Drainage Impact**	**Description of Change**	**Engineering Implications & Solutions**
Hydrologic Cycle	Increased Peak Flow & Volume	Impervious surfaces (roads, roofs, parking lots) replace vegetation, preventing infiltration and causing rain to run off immediately.	Detention & Retention Facilities: Design of ponds or underground tanks to capture excess runoff and release it slowly at pre-development rates. Storage Capacity Analysis: Recalculating system capacity for 2-year to 100-year storm events.
Hydrologic Cycle	Reduced Groundwater Recharge	Less water infiltrates the soil to replenish shallow and deep aquifers, leading to a drop in the water table.	Infiltration Systems: Installing dry wells, infiltration trenches, or managed aquifer recharge systems. Pervious Pavement: Using porous surfaces for parking lots and sidewalks.
Hydrologic Cycle	Decreased Evapotranspiration	Removal of trees and plants reduces the amount of water returned to the atmosphere through natural biological processes.	Green Infrastructure/ Urban Greening: Large-scale implementation of green roofs, tree canopies, and landscaped areas to mimic natural water loss.
Water Quality	Increased Pollutant Loading (Non-Point)	Runoff collects oil, heavy metals, fertilizers, pesticides, sediment, and trash from developed surfaces.	Stormwater Quality Controls: Installation of oil-grit separators, hydrodynamic separators, and sediment traps before discharge. Bioretention: Using rain gardens and bioswales to filter pollutants naturally.
Stream Morphology	Channel Erosion & Geomorphic Change	Frequent, high-velocity "flashy" flows act "like sandpaper" on streambanks, widening and deepening natural channels.	Stream Restoration: Re-engineering channels with riffle-pool sequences, natural rip-rap, or vegetation to stabilize banks. Flow Control: Utilizing bioretention to reduce the duration of erosive flows.
Stream Morphology	Downstream Flooding	Rapid conveyance of water through engineered pipes and concrete channels overwhelms natural waterways downstream.	Regional Stormwater Planning: Implementing watershed-scale storage solutions rather than site-specific ones alone. Culvert & Bridge Sizing: Upgrading downstream infrastructure to handle new flow regimes.

Impact Category	Drainage Impact	Description of Change	Engineering Implications & Solutions
Agricultural/Soil	Subsidence and Oxidation	Drainage of wetlands or organic (peat) soils for agriculture leads to soil compaction, decomposition, and land sinking.	Water Table Management: Controlled drainage systems (e.g., weir structures) to maintain necessary moisture and reduce oxidation rates. Land Grading: Re-leveling fields to maintain effective drainage patterns.
Ecological	Habitat Degradation/ Loss	Reduced baseflow (from low groundwater), increased water temperature (from heated pavement runoff), and segmentation.	Constructed Wetlands: Designing artificial wetland systems to treat water, manage flow, and provide compensatory habitat. Riparian Buffers: Engineering setback zones of native vegetation along waterways.

12.7 Staged Development of Drainage Infrastructure

Because peri-urban areas change over time, their drainage systems often need to be built in stages rather than as one final comprehensive network. A staged approach allows infrastructure investment to match current demand while preserving the framework needed for future expansion.

In the earliest stage, the main priority is usually protection of the natural drainage structure of the area. This may include preserving flow corridors, preventing obstruction of ponds and channels, constructing key roadside drains, and providing culverts where roads interrupt runoff paths. Basic outfalls should be identified and protected from erosion, blockage, and encroachment.

As development intensifies, intermediate stages may add branch drains, improve inlet arrangements, line erosion-prone reaches, widen critical sections, and strengthen the continuity of trunk conveyance routes. These interventions should be designed so that they support rather than constrain later upgrading.

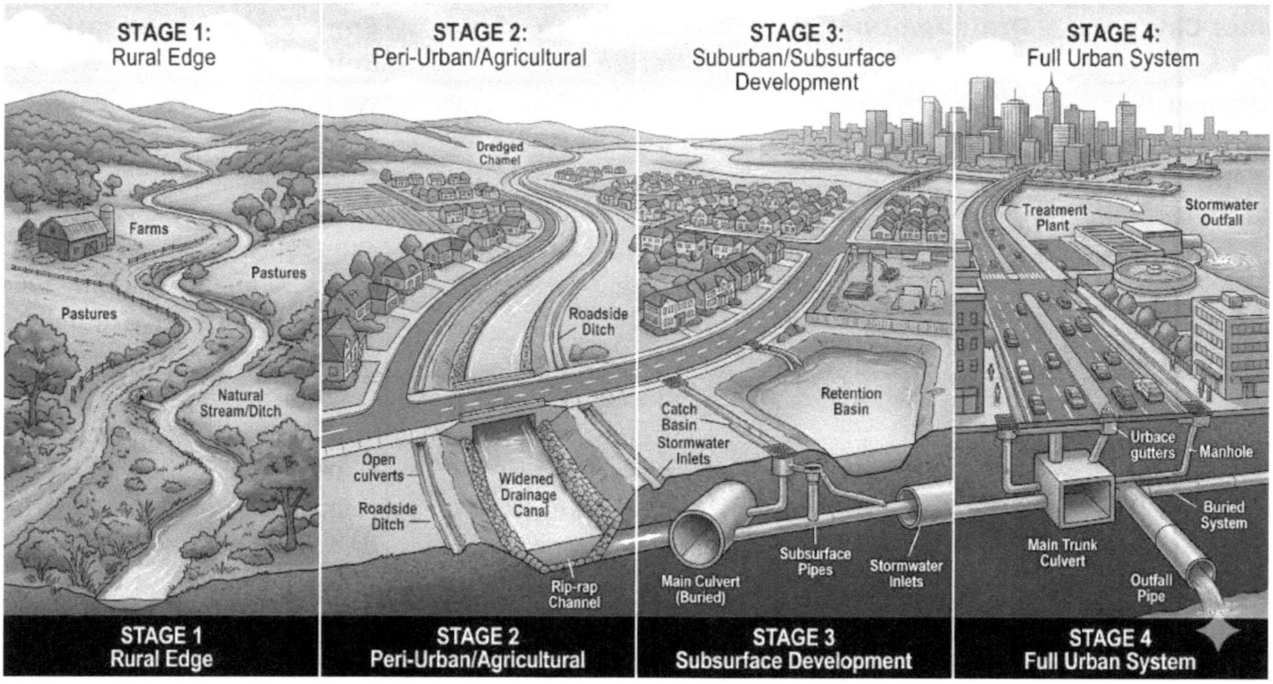

Figure 12.5 Staged development of drainage infrastructure from rural edge to urbanized system.

At more advanced stages of urbanization, the drainage system may need to evolve toward covered drains, storm sewers, detention facilities, pumping in low-lying areas, or more formal trunk drainage corridors. Even then, the efficiency of such improvements depends heavily on whether the earlier stages preserved corridor space and maintained logical system alignment.

The great advantage of staged development is practical feasibility. Many peri-urban areas do not have the budget, governance capacity, or final land-use clarity needed to justify a complete urban drainage system immediately. A staged strategy allows the system to remain functional in the present while avoiding the mistakes that make future improvements difficult.

12.8 Low-Cost and Flexible Design Approaches

Cost and uncertainty are central realities of semi-urban drainage planning. As a result, many effective systems in these areas must rely on low-cost and flexible design measures rather than expensive fully enclosed infrastructure from the outset.

Examples include well-graded open roadside drains, selectively lined reaches where erosion risk is high, simple culverts at crossings, locally constructed drop structures for steep slopes, and the use of ponds or depressions for temporary storage. Modular widening of drains, low-cost silt traps, and basic inlet improvements can also enhance performance substantially without requiring major investment.

Figure 12.6 Low-cost drainage measures for expanding settlements.

The value of flexibility lies in the uncertainty of future layout. A fully rigid enclosed drain system may become misaligned with later road widening, plot reorganization, or utility placement. In contrast, protected open corridors and staged channel improvements are often easier to adapt as the settlement matures.

Yet low cost should never be confused with weak design. Even simple drainage works require proper longitudinal slope, sufficient hydraulic capacity, continuity between reaches, stable outfalls, protection against scour, and provision for inspection and maintenance. Poorly designed low-cost systems often fail quickly and can become more expensive over time than a modest but well-engineered initial investment.

Therefore, the design objective should be affordability with hydraulic discipline. A simple system that is logically aligned, maintainable, and expandable is usually preferable to an expensive but incomplete system introduced without catchment-level planning.

Table 12.3 Low-cost and flexible drainage options for transitional settlements.

Drainage Option	Description	Materials / Construction	Advantages	Limitations / Maintenance Needs
Earthen Side Drains	Shallow, unlined channels along roads or plot boundaries	Compacted soil, shaped manually	Very low cost, easy to modify	Prone to erosion and clogging; requires periodic reshaping
Stone-Pitched Drains	Semi-permanent open drains with stone lining	Local stone, mortar (optional)	Durable, resists erosion, blends with natural terrain	Labor-intensive; may need re-pointing after heavy rains

(*Contd.*)

Drainage Option	Description	Materials / Construction	Advantages	Limitations / Maintenance Needs
Vegetated Swales	Shallow, grass-lined channels for infiltration and slow runoff	Native grasses, compacted subgrade	Enhances infiltration, improves water quality	Needs vegetation upkeep; less effective in high-flow areas
Perforated Pipe Systems	Subsurface flexible drainage for small plots or lanes	PVC or HDPE perforated pipes	Compact, adaptable, reduces surface clutter	Requires careful installation and periodic flushing
Modular Concrete Channels	Precast units for rapid deployment in developing areas	Precast concrete segments	Quick installation, scalable	Higher initial cost; joint sealing needed
Rainwater Harvesting Trenches	Combined drainage and recharge system	Gravel, sand, perforated pipe	Reduces runoff, supports groundwater recharge	Needs sediment removal and inspection
Reinforced Earth Bunds	Low embankments guiding runoff to safe outlets	Compacted earth, geotextile reinforcement	Controls sheet flow, prevents erosion	Vulnerable to overtopping; periodic compaction required

12.9 Managing Urban-Rural Drainage Interfaces

The boundary between urbanizing land and rural receiving areas is one of the most sensitive zones in drainage engineering. New development often concentrates runoff and discharges it into rural channels, ponds, agricultural drains, or village pathways that were never intended to carry urban-style peak flows.

This mismatch produces several common problems. Rural drains may overtop because their cross-sections are small and irregular. Increased discharge may cause downstream erosion, bank failure, and sediment movement. Agricultural land may flood more frequently, village roads may become impassable, and local ponds may silt up or receive higher inflows than their embankments can safely manage. In irrigation-dependent areas, conflict may also arise between storm drainage and existing water distribution pathways.

Figure 12.7 Urban-rural drainage interface and downstream impacts.

Managing this interface requires more than simply enlarging an outlet at the edge of development. The engineer must assess whether the receiving system has adequate capacity, whether storage or detention should be introduced upstream, whether channel protection is required, and whether downstream improvements are needed to maintain system balance.

Institutional coordination is equally important because the drainage boundary usually extends beyond the administrative boundary of the developed area. A system that functions well inside a new colony may still be unacceptable if it transfers flooding, erosion, or maintenance burdens to the rural fringe. Thus, the design domain must include the receiving environment, not merely the built-up parcel.

Table 12.4 Typical problems and management responses at the urban-rural drainage interface.

Problem	Description	Primary Causes	Management Responses	Expected Outcomes
Rapid Urban Runoff	Excessive stormwater discharge from paved surfaces	High imperviousness, inadequate detention	Construct retention ponds, green roofs, and permeable pavements	Reduced peak flow and downstream flooding
Pollution Transfer	Contaminants from urban areas entering rural streams	Industrial effluents, domestic waste, road runoff	Install sediment traps, vegetated buffers, and decentralized treatment units	Improved water quality and ecological health

(Contd.)

Problem	Description	Primary Causes	Management Responses	Expected Outcomes
Channel Erosion	Scouring and bank instability in receiving rural streams	Sudden high-velocity inflows from urban drains	Use energy dissipation structures, riprap, and vegetative stabilization	Controlled erosion and stable channel geometry
Sediment Deposition	Accumulation of silt and debris downstream	Construction runoff, eroded urban surfaces	Implement silt fences, sediment basins, and upstream check dams	Reduced sediment load and improved conveyance
Flooding in Transitional Zones	Localized inundation at the urban–rural boundary	Poor connectivity between systems, blocked culverts	Integrate drainage master plans, maintain culverts, and enhance cross-drainage	Improved flow continuity and reduced flood risk
Habitat Degradation	Loss of aquatic and riparian ecosystems	Altered flow regime, pollution, encroachment	Restore riparian vegetation, regulate discharge timing, and protect wetlands	Enhanced biodiversity and ecological resilience
Institutional Gaps	Lack of coordination between urban and rural authorities	Fragmented jurisdiction and planning	Establish joint watershed committees and integrated management frameworks	Harmonized governance and sustainable drainage planning

12.10 Planning Considerations for Engineers

In semi-urban drainage planning, the engineer must work under conditions of incomplete information and rapid change. Design cannot depend solely on present land use maps because the catchment may transform significantly within a few years.

A practical planning approach should therefore:

1. Identify existing natural drainage corridors, depressions, ponds, and outfalls before new development obscures them.
2. Distinguish present and likely future impervious zones within the catchment.
3. Protect corridors needed for future branch and trunk drains.
4. Provide cross-drainage wherever roads or embankments interrupt natural runoff movement.
5. Avoid transferring the drainage burden to adjacent rural or low-income settlements without downstream assessment.

The engineer should also recognize that drainage in transitional areas is closely linked with land management decisions. Plot filling, road formation, construction phasing, utility trenches, and encroachments can all have hydraulic consequences greater than their apparent local scale. Good drainage planning therefore requires coordination with layout approval, road design, and broader settlement development control.

12.11 Summary Point

1. Semi-urban and peri-urban areas are transitional zones between rural and urban landscapes, with fragmented drainage systems.
2. Drainage networks here are incomplete, discontinuous, and vulnerable due to mixed land use and uneven development.
3. Transitional catchments show irregular runoff behavior, combining rural infiltration with urban-style rapid discharge.
4. Mixed land use—residential, commercial, agricultural—creates diverse runoff responses, requiring catchment subdivision for realistic design.
5. Integration of natural drainage paths with engineered structures is vital to preserve hydraulic continuity and reduce costs.
6. Rapid land conversion increases runoff, reduces infiltration, and disrupts natural storage, often leading to chronic flooding.
7. Staged development of drainage infrastructure allows gradual adaptation to intensifying urbanization while preserving future corridors.
8. Low-cost, flexible designs (open drains, swales, modular channels) are practical but must maintain hydraulic discipline.
9. Urban-rural interfaces pose challenges like downstream flooding, erosion, pollution transfer, and institutional gaps.
10. Effective planning requires foresight, coordination, and protection of natural corridors to ensure adaptable and sustainable drainage systems.

12.12 References

1. Ministry of Housing and Urban Affairs. (2019). *Manual on storm water drainage systems*. Government of India.
2. Ball, J., et al. (2019). *Australian rainfall and runoff: A guide to flood estimation*.
3. Government of Western Australia. (n.d.). *Stormwater management manual for Western Australia*.
4. Engineers Australia, Institute of Public Works Engineering Australia, & Stormwater Western Australia. (2023). *Code of practice: Urban and peri-urban drainage modelling*.
5. *Hydrology of peri-urban catchments: Processes and modelling*. (2013). *Journal of Hydrology*.

6. Federal Highway Administration. (2024). *Storm drainage systems design guidance for transportation facilities* (Publication No. FHWA-HIF-24-006).
7. Federal Highway Administration. (2001). *Urban drainage design manual.*
8. Drainage Services Department. (n.d.). *Stormwater drainage manual.* Government of Hong Kong.
9. Department of Human Settlements. (n.d.). *Red book: Section L—Stormwater.*
10. CIRIA. (2015). *The SuDS manual (C753).*
11. CIRIA. (2000). *Sustainable urban drainage systems: Design manual for England and Wales (C522).*

Chapter 13

Village and Rural Settlement Drainage Systems

Learning Objectives

By the end of this chapter, the learner should be able to:

1. Describe the main features of drainage systems in villages.
2. Explain how rural settlement form and local topography influence drainage behavior
3. Assess the role of roadside drains, open channels, and local outfalls in village drainage systems.
4. Evaluate drainage issues around homes, public spaces, ponds, and agricultural land
5. Recommend low-cost and locally manageable engineering measures for rural drainage improvement.
6. Explain the importance of community participation in the long-term functioning of village drainage systems.

13.1 Introduction

Village and rural settlement drainage systems are generally simpler in physical form than urban drainage systems, yet they are no less important from an engineering and public health perspective. In many villages, drainage determines whether streets remain passable during monsoon periods, whether water stagnates near dwellings, whether sanitation conditions deteriorate, and whether local roads, ponds, and common spaces remain functional throughout the year.

Because rural settlements often depend on low-cost infrastructure, local materials, and informal or community-based maintenance, drainage design in such settings must balance technical adequacy with affordability, simplicity, and ease of upkeep. The challenge is not merely to remove stormwater, but to do so in a way that fits local topography, settlement pattern, agricultural surroundings, and institutional capacity

Unlike large urban networks with buried sewers and formally designed trunk systems, village drainage commonly depends on open surface paths, roadside channels, shallow drains, natural depressions, ponds, field drains, and small culvert crossings. These elements often function as part of a combined local drainage environment rather than as a strictly separated engineered network

For this reason, village drainage should not be viewed as a reduced version of urban drainage. It is a distinct field condition shaped by visible surface flow, strong dependence on minor level differences, close interaction with habitation and agriculture, and the crucial role of residents in maintenance and system continuity.

13.2 Characteristics of Village Drainage Systems[2]

Village drainage systems differ from urban systems in scale, layout, infrastructure type, and operational logic. In many rural settlements, runoff is conveyed through open drains, road edges, shallow channels, natural low lines, ponds, and local watercourses rather than through underground pipe systems. This gives village drainage a more direct dependence on visible surface grading and continuity of flow path

A key characteristic of rural drainage is the dominant influence of local topography. Even small differences in road level, courtyard level, house plinth height, pond embankment level, and field elevation may determine whether runoff drains away smoothly or remains trapped near homes and lanes. In flat or low-gradient areas, these small level variations become especially critical

Another major characteristic is the close spatial relationship between drainage and everyday village activities. Roads, courtyards, cattle sheds, handpump platforms, ponds, public buildings, and agricultural plots often lie close together, so drainage cannot be treated as a narrowly defined stormwater problem alone. Instead, it forms part of the broader rural environmental and sanitary system.

Figure 13.1 Typical village drainage settings, showing lanes, roadside drains, houses, pond outfall, and nearby agricultural fields

Village systems also tend to evolve incrementally. A drain may be improved in one street but remain earthen in another; a section may be covered locally for access; an embankment may be raised without corresponding outfall adjustment; and residents may modify side entries or crossings over time. These piecemeal changes can create discontinuity, reverse slope, bottlenecks, and maintenance difficulties.

From an engineering perspective, the village drainage system must therefore be understood as a practical surface-flow network that is strongly shaped by local construction practices, settlement growth, and community behavior. Good design depends less on complexity and more on continuity, correct levels, safe outfalls, and ease of maintenance.

13.3 Surface Drainage Patterns in Rural Settlements[2]

Surface drainage in rural settlements generally occurs through gravity flow over the visible ground surface, along lane edges, road margins, courtyards, shallow channels, and natural depressions leading toward low points or receiving bodies. Because underground drainage systems are uncommon, the surface itself often acts as the primary drainage framework.

The drainage pattern is frequently influenced by settlement morphology. In compact villages, closely spaced houses and narrow lanes may rapidly concentrate runoff into roadside drains or local depressions. In dispersed settlements, water may spread over open land, pass around homesteads, and reach a pond, field edge, or natural channel only after moving over a broader area.

Continuity of slope is therefore essential. A village that originally drained through shallow natural gradients may become waterlogged when roads are raised, plots are filled, access ramps are built without hydraulic consideration, or drains are blocked by silt and waste. Such changes may appear minor locally but can interrupt the overall drainage pattern of the settlement.

Assessment of rural drainage must pay close attention to micro-topography. In many villages, flooding problems cannot be understood from general contour patterns alone; they depend on subtle level differences between road crowns, door thresholds, courtyard depressions, and local outfall points. For that reason, field observation is often as important as mapped data

Engineering planning should guide runoff along safe and predictable routes that align with existing roads, lanes, and natural low lines. This approach helps drainage improvements work with the settlement form rather than against it, reducing both construction difficulty and future maintenance burden.

Table 13.1 Typical surface drainage patterns in rural settlements and their engineering implications.

Type of Rural Settlement	Typical Surface Drainage Pattern	Engineering Implications
Compact Village Form	Centralized drainage paths converging toward a common outfall or pond; limited infiltration due to dense built-up area	Requires well-designed surface channels and culverts; risk of localized flooding if outfall capacity is inadequate; easier maintenance due to concentrated layout
Linear Settlement (along roads or streams)	Drainage aligned parallel to the main road or stream; runoff directed through side drains and cross-drainage structures	Needs coordinated design of roadside drains and culverts; potential erosion along linear paths; easier hydraulic connectivity but higher maintenance along length
Dispersed Settlement Form	Multiple independent drainage paths from scattered houses; runoff infiltrates locally or drains into small ditches and creeks	Promotes natural infiltration and groundwater recharge; requires decentralized drainage planning; difficult to maintain uniform discharge control
Clustered Hamlet Form	Small clusters with semi-connected drainage paths leading to local depressions or natural channels	Moderate risk of ponding; suitable for small-scale surface drains and infiltration trenches; requires local coordination for shared outfalls
Agricultural Homestead Form	Surface runoff guided through field contours and shallow ditches; often integrated with irrigation channels	Enhances soil moisture management; requires contour alignment and erosion control; dual-purpose drainage-irrigation design beneficial for sustainability

13.4 Roadside Drains and Open Channels[2]

Roadside drains and open channels are the most common engineered drainage structures in villages and rural settlements. They are widely used because they are comparatively easy to construct, inspect, clean, and repair using local labor and materials. Where alignment, slope, and outfall are properly designed, they can perform effectively at low cost.

These drains may be earthen, brick-lined, stone-lined, or concrete-lined depending on local soil conditions, material availability, traffic exposure, and budget constraints. In many villages, simple trapezoidal or rectangular channels are provided along roads and lanes to intercept runoff from the carriageway and adjoining house fronts. Small slab covers or crossing elements may be added at entrances and lane intersections to preserve access.

The hydraulic performance of roadside drains depends on section continuity, sufficient longitudinal slope, absence of obstruction, and sound outlet connection. Common failures include dead ends without outfall, depressions with stagnant water, local narrowing due to private crossings, sediment accumulation, weed growth, and waste dumping.

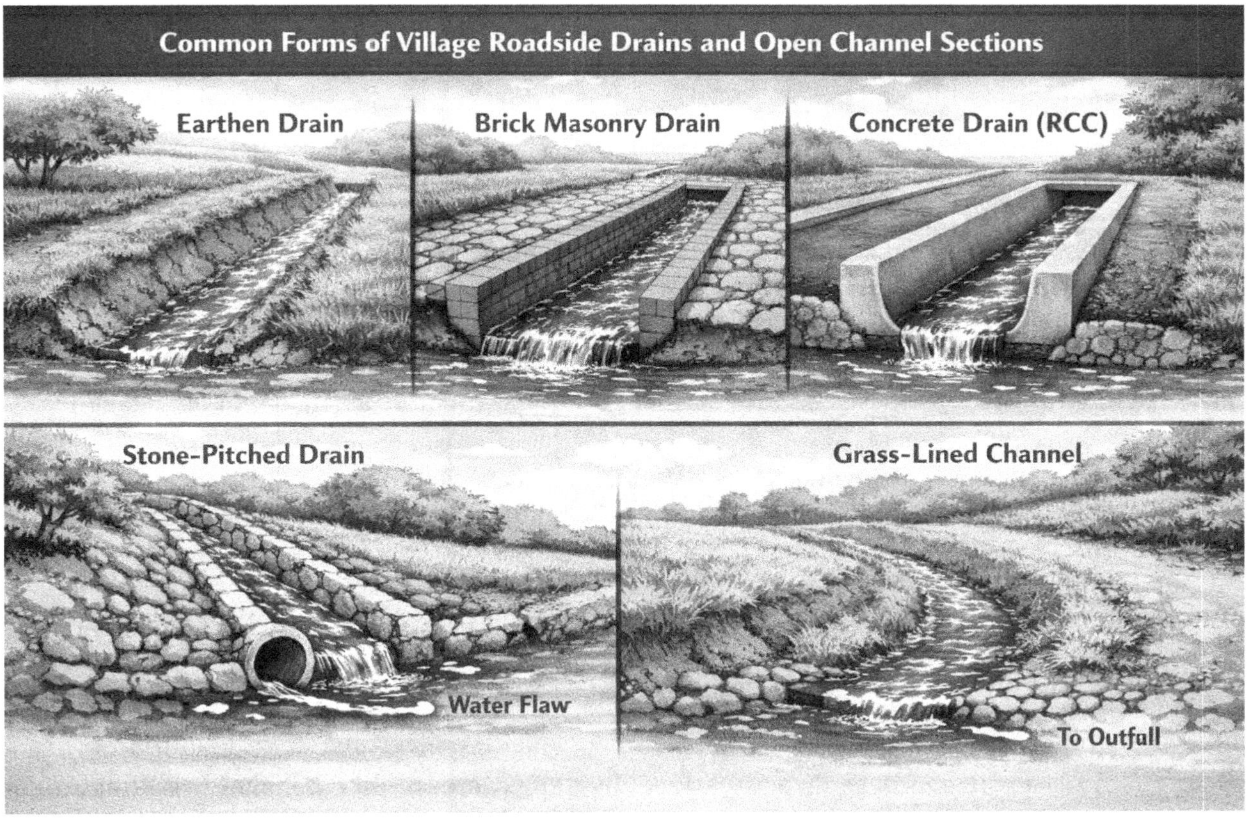

Figure 13.2 Common forms of village roadside drains and open channel sections.

Open channels remain especially appropriate in rural settings because their condition is visible and maintenance needs can be identified quickly. However, visibility alone does not guarantee performance. The channel must still have stable side slopes, manageable sediment deposition, practical cleaning access, and reasonable safety for children, pedestrians, and animals

The choice between lined and unlined drains should be made carefully. Lining may be justified in erosion-prone reaches, high-velocity sections, or village cores with frequent traffic and concentrated runoff, while earthen channels may remain acceptable in lower-flow reaches where soil is stable and routine reshaping is feasible.

Table 13.2 Advantages, limitations, and suitable applications of common village drain types.

Type of Village Drain	Advantages	Limitations	Suitable Applications
Earthen Drains	Simple and low-cost; easy to construct with local materials; promote infiltration	Prone to erosion and silting; require frequent maintenance; limited hydraulic capacity	Small rural lanes, agricultural plots, temporary drainage channels
Brick Masonry Drains	Durable and stable; moderate flow capacity; suitable for lined sections	Costlier than earthen drains; limited flexibility for alignment changes	Village streets, market areas, and semi-permanent drainage networks

(*Contd.*)

Type of Village Drain	Advantages	Limitations	Suitable Applications
Concrete Drains (RCC / Precast)	High strength and durability; smooth flow; low maintenance	High initial cost; requires skilled labor; limited adaptability to uneven terrain	Main village drains, approach roads, and high-flow zones
Stone Pitching Drains	Resistant to erosion; blend with rural aesthetics; good for steep gradients	Labor-intensive; less watertight; may require periodic repair	Hilly villages, roadside channels, and erosion-prone slopes
Covered Drains (Slab or Grated)	Prevents clogging by debris; safe for pedestrians; aesthetic improvement	Difficult to inspect and clean; higher construction cost	Densely populated village centers, marketplaces, and pedestrian zones
Vegetated / Grass-Lined Drains	Eco-friendly; enhance infiltration and water quality; reduce flow velocity	Limited capacity during heavy rainfall; require vegetation upkeep	Peripheral rural areas, low-flow zones, and sustainable drainage corridors

13.5 Drainage Around Homes and Public Spaces[2]

Household plots and public spaces are among the most sensitive locations in village drainage planning because stagnant water near homes directly affects hygiene, accessibility, comfort, and day-to-day life. Even where a main roadside drain exists, local drainage around courtyards, plinths, entrances, schools, health centers, panchayat buildings, and market spaces may still be inadequate

Drainage around homes depends on several interconnected factors, including plinth level, courtyard slope, roof runoff disposal, wastewater discharge pattern, and connection to nearby drains. If individual house platforms are raised without regard to neighboring plots, runoff may be diverted toward adjoining dwellings or trapped in shared spaces. Similarly, direct roof discharge into narrow lanes can cause erosion, splash, and muddy conditions.

Public spaces require especially careful design because they experience repeated use and often serve the broader community. Poor drainage near school entrances, public buildings, water collection points, or marketplaces affects not only water removal but also pedestrian safety, sanitation, and the social utility of the space

Engineering measures in such areas may include careful surface grading, short connecting drains, roof-water discharge control, inlet points at low spots, and safe connection to a downstream drain or outfall. The objective is to prevent ponding near high-use areas and to ensure that local improvements do not transfer water problems to adjacent plots

Where wastewater and stormwater are both present, the risk of mixed stagnation becomes more serious. In such cases, the drainage system must be planned not only for conveyance, but also for hygiene and safe movement within shared village spaces.

13.6 Outfalls to Ponds, Tanks, and Natural Drains[2]

Most village drainage systems eventually discharge into local outfalls such as ponds, tanks, streams, field channels, natural drains, or shallow depressions. The quality and adequacy of these receiving points are crucial to the overall success of the drainage system. A well-built drain with an inadequate outfall will still fail hydraulically.

Ponds and tanks are often traditional receiving bodies for village runoff, but their usefulness depends on available storage, seasonal water level, siltation condition, embankment stability, and the geometry of connection at the point of inflow. If they are already encroached upon, permanently waterlogged, or heavily silted, they may not be able to receive additional flow effectively.

Natural drains may also become compromised by vegetation growth, debris accumulation, poorly designed road crossings, or unauthorized filling. In such cases, upstream village drains may back up during rainfall events, producing flooding far from the actual point of obstruction.

Figure 13.3 Village outfall connections to pond, tank, and natural drain with basic protection works.

Outfall design should consider invert level, approach alignment, erosion control, and the possibility of backflow during high water conditions. Outlet protection may require brickwork, stone pitching, apron treatment, or small energy dissipation measures to prevent scour and local bank failure.

Where receiving water bodies serve multiple purposes, such as storage, livestock use, recharge, or ecological function, the quality of drained water may also need consideration. Selection and protection of outfalls should therefore be treated as a central element of rural drainage planning rather than as a final detail added after internal drains are laid out.

Table 13.3 Suitability criteria for village drainage outfalls and common causes of outfall failure.

Suitability Criteria for Village Drainage Outfalls	Common Causes of Outfall Failure
Adequate capacity to handle peak runoff volumes	Undersized outlet leading to overtopping and flooding
Stable location with minimal erosion risk	Scouring and erosion at discharge point
Free-flowing connection to natural watercourse or pond	Blockage due to siltation, debris, or vegetation growth
Accessibility for inspection and maintenance	Neglect of routine cleaning and structural upkeep
Proper alignment with village drainage network	Misalignment causing backflow or ponding
Safe discharge velocity to prevent downstream damage	Excessive discharge velocity leading to channel instability
Environmental compatibility (avoiding sensitive habitats)	Pollution or ecological disturbance from untreated runoff
Provision for energy dissipation (stilling basin, apron, etc.)	Structural failure of outlet works due to poor design or construction

13.7 Interaction with Agricultural Drainage[2]

Village settlement drainage often interacts directly with agricultural drainage because houses, roads, ponds, and fields are closely linked within the rural landscape. Runoff from habitation areas may enter field channels, while agricultural runoff may pass near or even through the village during heavy rainfall periods.

This interaction can be beneficial where field drains or natural channels provide downstream conveyance for settlement runoff. However, conflict arises when habitation drainage overloads agricultural channels, when road embankments block farm drainage, or when runoff from cropped land enters residential areas unexpectedly.

Seasonal variation is especially important in this relationship. During the monsoon, agricultural land may already be saturated, reducing its ability to receive additional runoff from the settlement. In low-lying terrain, ponding in fields can raise downstream water levels and slow or block drainage from the village itself.

Poorly planned village drains may also carry sediment or waste into channels used for irrigation or field drainage, creating operational and social conflict. This shows that village drainage must be planned not as an isolated settlement problem, but as part of a broader local hydrologic system that includes production land and seasonal water movement.

Coordination between habitation drainage and field drainage is therefore essential, especially in low-gradient regions where flow paths are sensitive to small level changes and temporary obstructions.

Table 13.4 Common interactions between village drainage and agricultural drainage, with design responses.

Interaction Type	Nature of Interaction	Engineering Implications	Design Responses
Shared Outfalls	Village drains and agricultural field drains discharge into the same outlet	Risk of overloading during peak rainfall; backflow into fields or village streets	Provide enlarged or reinforced outfall structures; include energy dissipation and erosion control measures
Cross-Connections	Village surface drains intersect agricultural contour drains	Potential mixing of polluted village runoff with irrigation water; hydraulic imbalance	Introduce check structures or diversion channels; separate clean irrigation flows from polluted runoff
Parallel Alignment	Village roadside drains run parallel to agricultural field drains	Increased cumulative discharge; potential erosion along parallel channels	Design adequate spacing and lining; stabilize banks with vegetation or stone pitching
Field-to-Village Runoff	Agricultural runoff enters village drains during heavy rainfall	Localized flooding in built-up areas; sediment deposition in village drains	Install silt traps and sedimentation basins; provide cutoff drains at field boundaries
Village-to-Field Runoff	Village stormwater drains discharge into agricultural fields	Waterlogging and crop damage; contamination of soil	Redirect village drains to natural streams; use controlled outlets with flap gates
Seasonal Irrigation-Drainage Overlap	Irrigation channels double as drainage paths in monsoon	Conflict between irrigation supply and drainage needs	Provide dual-purpose channels with adjustable gates; schedule irrigation to avoid peak drainage periods

13.8 Low-Cost Engineering Solutions

Low-cost engineering solutions are central to village drainage because financial resources, construction equipment, and institutional support are often limited. The objective is not to lower engineering standards, but to achieve effective drainage through practical and locally manageable means.

Examples of suitable low-cost measures include properly graded earthen or brick-lined roadside drains, local lining in erosion-prone reaches, small culverts beneath village roads, slab-covered crossings at house entrances, simple silt traps near inlets, improved outlet protection with brick or stone, and restoration of traditional drainage ponds or depressions.

These measures are successful not because they are inexpensive, but because they can be fitted to local conditions with hydraulic logic. A simple open drain with correct gradient, uninterrupted section, and reliable outfall will often perform better than a more expensive but poorly aligned drain with blocked discharge or inconsistent levels.

Ease of repair is another major criterion. In rural settings, systems that can be maintained, reshaped, desilted, or reconstructed using local labor and materials are often more sustainable over time than advanced but fragile installations requiring specialized support.

Low-cost design must therefore still respect sound engineering principles, including slope control, continuity, stable outfall treatment, manageable maintenance, and compatibility with future improvement. Affordability should support performance, not replace it.

13.9 Community Role in Rural Drainage Management

Community participation plays a larger role in village drainage management than in many urban systems because formal maintenance services are often limited or absent. Local residents directly influence drainage performance through cleaning practices, waste disposal behavior, maintenance of house-front drains, protection of outfalls, and informal alteration of drainage paths.

Community knowledge is also an important planning resource. Residents often know where water stagnates, which lanes become impassable, how long ponding lasts, which drains overflow first, and how drainage patterns have changed over time. This local information can significantly improve diagnosis and prioritization of works

At the same time, community-managed systems face clear challenges. Private crossings, local closures of drain sections, dumping of refuse, and blocking of channels for short-term convenience can reduce continuity and create disputes between households or neighborhoods.[2]

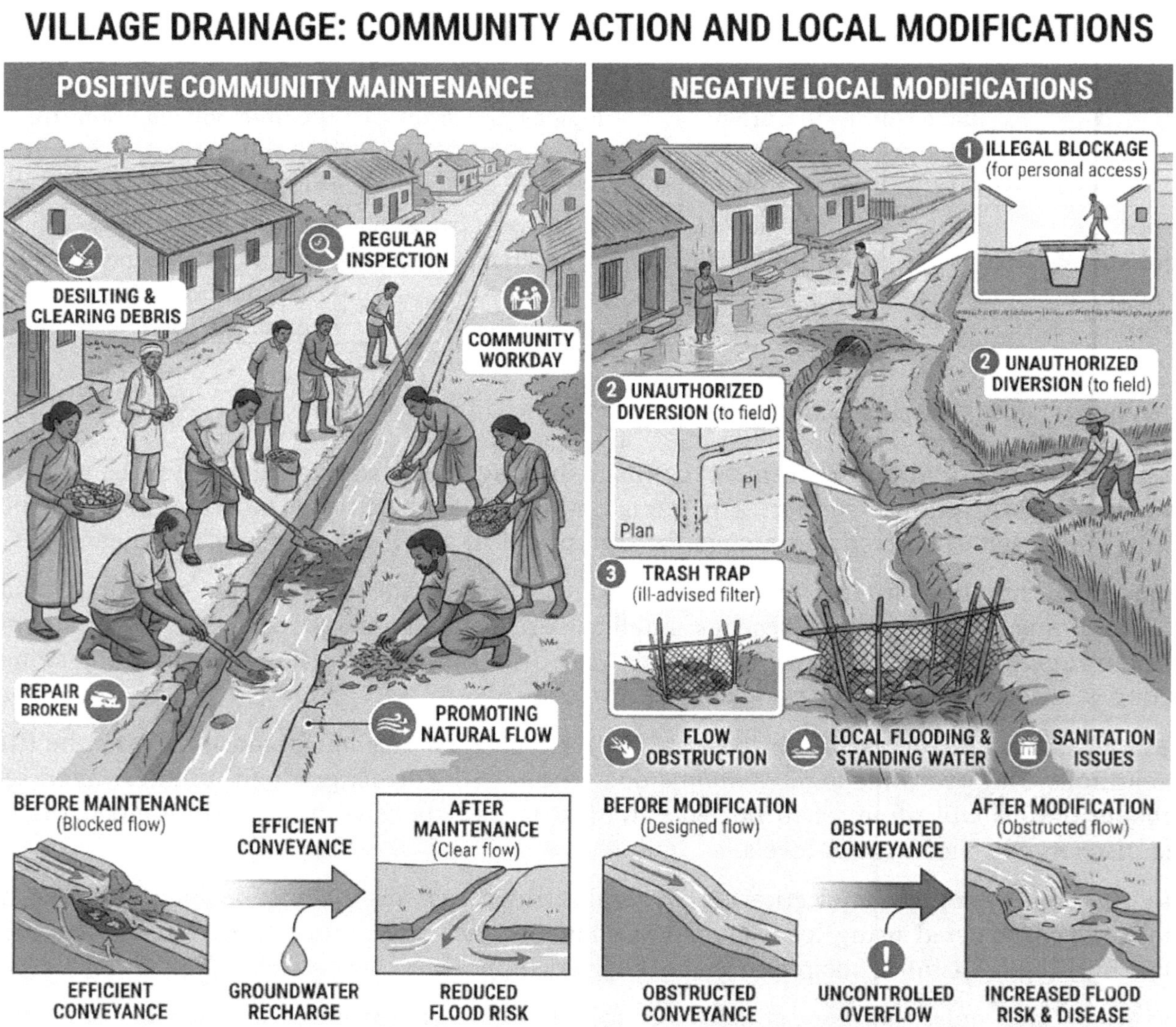

Figure 13.4 Community maintenance and common local modifications affecting village drainage performance.[1][2]

A sound rural drainage strategy therefore combines engineering design with local awareness, routine inspection, shared responsibility, and basic institutional support. Where residents understand the function of the drainage system and participate in its upkeep within a guided framework, long-term performance improves substantially.

Community participation should not replace engineering judgment, but it should reinforce it. The most durable village drainage systems are usually those that are technically simple, locally understood, and institutionally supported.

13.10 Planning Considerations

In village and rural settlement drainage planning, the engineer must work with limited budgets, visible surface flow conditions, and strong dependence on local terrain. For that reason, planning should begin with careful observation of existing runoff paths, pond locations, road levels, field connections, and outfalls rather than with assumptions derived from urban drainage templates.

A practical planning approach should therefore:

1. Identify natural low lines, ponds, depressions, and receiving drains before local modifications obscure them
2. Check continuity of existing roadside drains and open channels, including hidden dead ends and reverse slopes
3. Protect key outfalls from blockage, erosion, and encroachment.
4. Coordinate habitation drainage with adjacent agricultural drainage and road-cross drainage structures.[
5. Give special attention to drainage near homes, schools, markets, and other high-use public spaces.
6. Prefer solutions that can be cleaned, repaired, and understood locally.

The engineer should also recognise that minor works can have major hydraulic consequences in a rural setting. Raising a lane, filling a courtyard edge, narrowing a roadside drain, or blocking a field outlet may shift water problems elsewhere in the settlement. Good planning therefore depends on catchment continuity and local realism rather than isolated structural decisions.

13.12 Summary Points

1. Village drainage systems, though simple, are essential for public health, sanitation, and accessibility.
2. They rely on visible surface flow, local materials, and community-based upkeep rather than buried sewers.
3. Local topography strongly influences performance, as small level differences control flow and ponding.

4. Drainage interacts closely with habitation, roads, ponds, and agricultural fields, creating interdependencies.
5. Systems evolve incrementally, often leading to discontinuities and maintenance challenges.
6. Gravity flow along roads, lanes, courtyards, and depressions defines the basic surface drainage pattern.
7. Roadside drains and open channels—earthen, brick, concrete, or stone-pitched—are the most common engineered elements.
8. Poor grading near homes, schools, and markets causes stagnation, hygiene issues, and unsafe runoff.
9. Outfalls to ponds and natural drains succeed only when capacity, alignment, and erosion protection are ensured.
10. Community participation in cleaning, upkeep, and modification is central to sustainability, supported by low-cost, locally maintainable engineering solutions.

13.13 References

1. United States Department of Agriculture, Natural Resources Conservation Service. (n.d.). *Part 650 engineering field handbook, Chapter 14: Water management (drainage)*.
2. Ritzema, H. P. (Ed.). (1994). *Drainage principles and applications*. International Institute for Land Reclamation and Improvement (ILRI).
3. Food and Agriculture Organization of the United Nations. (n.d.). *Guidelines and computer programs for the planning and design of land drainage systems*.
4. Food and Agriculture Organization of the United Nations. (n.d.). *Drainage of irrigated lands*.
5. United States Department of Agriculture. (n.d.). *National engineering handbook, Section 16, Chapter 1: Principles of drainage*.
6. Drainage Services Department. (n.d.). *Stormwater drainage manual*. Government of Hong Kong.
7. Ministry of Housing and Urban Affairs. (2019). *Manual on storm water drainage systems*. Government of India.
8. Indiana Local Technical Assistance Program. (n.d.). *Stormwater drainage manual*.
9. Federal Highway Administration. (n.d.). *Storm drainage systems design guidance for transportation facilities* (Publication No. FHWA-HIF-24-006).

Chapter 14

Sustainable, Subsurface, and Special Drainage Systems

Learning Objectives

By the end of this chapter, the learner should be able to:

1. Explain the principles of sustainable drainage and their relevance in contemporary drainage engineering.
2. Describe the function and applications of subsurface drainage systems in civil and environmental contexts.
3. Distinguish between conventional surface drainage, subsurface drainage, and special-purpose drainage systems.
4. Assess the engineering role of infiltration, detention, retention, filtration, and controlled discharge methods
5. Identify drainage requirements in special situations such as airports, highways, railways, industrial sites, hill roads, basements, and sports grounds.
6. Develop integrated drainage strategies that combine hydraulic efficiency, environmental responsibility, and long-term maintainability.

14.1 Introduction

Modern drainage engineering is no longer concerned only with the rapid removal of excess water from built environments. Increasing urbanization, reduction of permeable land, groundwater concerns, ecological degradation, climate variability, and infrastructure vulnerability have expanded the scope of drainage design. Systems must now be planned not merely for disposal, but for control, moderation, reuse, infiltration, storage, and environmental compatibility.

In this broader context, sustainable drainage systems, subsurface drainage measures, and special drainage arrangements for unusual sites have become increasingly important. These approaches address conditions that conventional roadside drains and storm sewers alone cannot resolve effectively, especially where runoff quality, groundwater pressure, slope instability, soil saturation, or highly specialized land use must be considered.

This chapter therefore extends drainage study beyond ordinary urban, semi-urban, and rural surface systems. It introduces sustainable drainage concepts, explains subsurface drainage principles, and examines drainage design requirements in selected special environments where site function, safety, and hydraulic control demand more specialized treatment

14.2 Sustainable Drainage Concepts

Sustainable drainage refers to an approach in which stormwater is managed in a way that imitates natural hydrologic behavior as far as practical while reducing flood risk, improving water quality, supporting groundwater recharge where appropriate, and minimizing adverse downstream impacts. Rather than treating runoff only as a nuisance to be removed quickly, sustainable drainage treats it as part of the wider water environment.

This approach becomes necessary because conventional drainage systems often accelerate runoff, increase downstream peak flows, reduce infiltration, and transfer flooding or pollution burdens elsewhere in the catchment. As urban surfaces expand and natural storage areas shrink, drainage systems designed solely for conveyance can become hydraulically efficient in one location while environmentally damaging at a larger scale.

Sustainable drainage seeks to correct this imbalance through distributed measures such as infiltration zones, swales, detention basins, retention ponds, permeable pavements, filter strips, vegetated channels, rainwater storage, and controlled outlets. These features may be used individually or combined into a network that slows, stores, filters, and gradually releases stormwater.

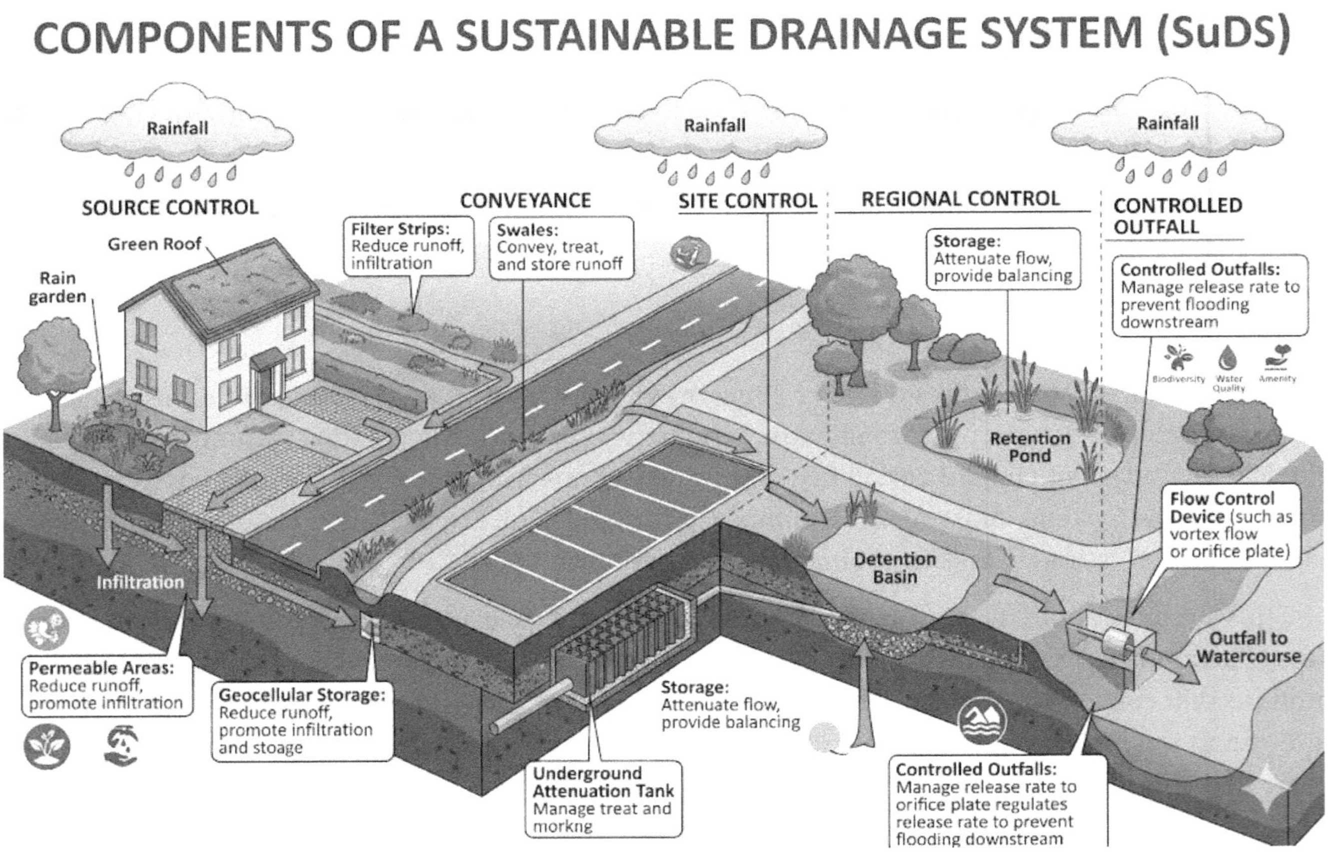

Figure 14.1 Components of a sustainable drainage system including swales, permeable areas, storage, and controlled outfalls.

The engineering objective is not simply ecological improvement; it is also hydraulic resilience. By reducing runoff velocity, delaying peak discharge, and preserving partial infiltration or storage, sustainable systems help urban and developing catchments function more safely under storm conditions.

At the same time, sustainable drainage should be adapted to local soil, rainfall, land use, maintenance capacity, and groundwater conditions. A measure that performs well in one environment may fail in another if infiltration is poor, sediment load is high, or maintenance is neglected.

14.3 Principles of Runoff Control[1]

A central principle of sustainable drainage is runoff control at or near the point where rainfall reaches the ground or built surface. Instead of conveying the entire runoff volume rapidly to a downstream drain, the system attempts to reduce, delay, spread, store, infiltrate, or treat that water progressively across the site and catchment.

Several control mechanisms are commonly used. Infiltration allows part of the runoff to enter the soil. Detention temporarily stores runoff and releases it slowly. Retention holds water for longer durations, sometimes permanently. Filtration improves runoff quality by removing sediment and pollutants. Controlled discharge limits the rate at which water enters downstream drains or natural channels.

These mechanisms may operate together in sequence. For example, runoff from a paved surface may first pass through a filter strip, then enter a swale, then collect in a detention basin, and finally discharge through a controlled outlet to a receiving drain. Such staged control reduces hydraulic shock and improves overall catchment performance.

Table 14.1 Major runoff control principles and their hydraulic and environmental functions.

Runoff Control Principle	Hydraulic Function	Environmental Function
Swales and Vegetated Channels	Convey and temporarily store surface runoff; promote infiltration and reduce peak flow	Filter sediments and pollutants; enhance biodiversity and landscape aesthetics
Permeable Pavements	Allow infiltration through surface layers; reduce surface runoff volume	Recharge groundwater; mitigate heat island effects; improve water quality
Detention and Retention Basins	Store runoff and release it gradually; attenuate flood peaks	Support aquatic habitats; improve sedimentation and nutrient removal
Green Roofs	Intercept rainfall; delay runoff generation	Reduce building energy demand; improve air quality and urban ecology
Infiltration Trenches and Soakaways	Facilitate subsurface infiltration; reduce direct discharge to drains	Recharge aquifers; minimize pollutant transport
Wetlands and Bio-retention Areas	Provide flow regulation and sediment trapping	Enhance ecological diversity; remove nutrients and contaminants biologically
Controlled Outfalls	Regulate discharge rate to receiving water bodies	Prevent downstream erosion; maintain ecological flow balance

From a design standpoint, runoff control requires attention to storage volume, release rate, overflow path, soil permeability, maintenance access, and compatibility with surrounding land use. It also requires recognition that effective control often depends on many small measures working together rather than one large terminal structure.

14.4 Infiltration-Based Drainage Measures

Infiltration-based drainage measures are used where the ground can absorb water at a suitable rate and where infiltration will not create structural or environmental problems. These measures aim to restore part of the natural rainfall-to-soil pathway that is often lost when land is paved or compacted.

Examples include soak pits, infiltration trenches, infiltration basins, recharge pits, porous pavement systems, and landscaped infiltration areas. These measures can reduce runoff volume, delay discharge, and contribute to groundwater recharge under favorable hydrogeologic conditions.

However, infiltration methods are not universally suitable. They may be ineffective in clayey soils, high water-table areas, contaminated ground, or locations close to foundations, retaining walls, basements, or unstable slopes. They may also clog if runoff carries a heavy sediment load and pretreatment is absent.

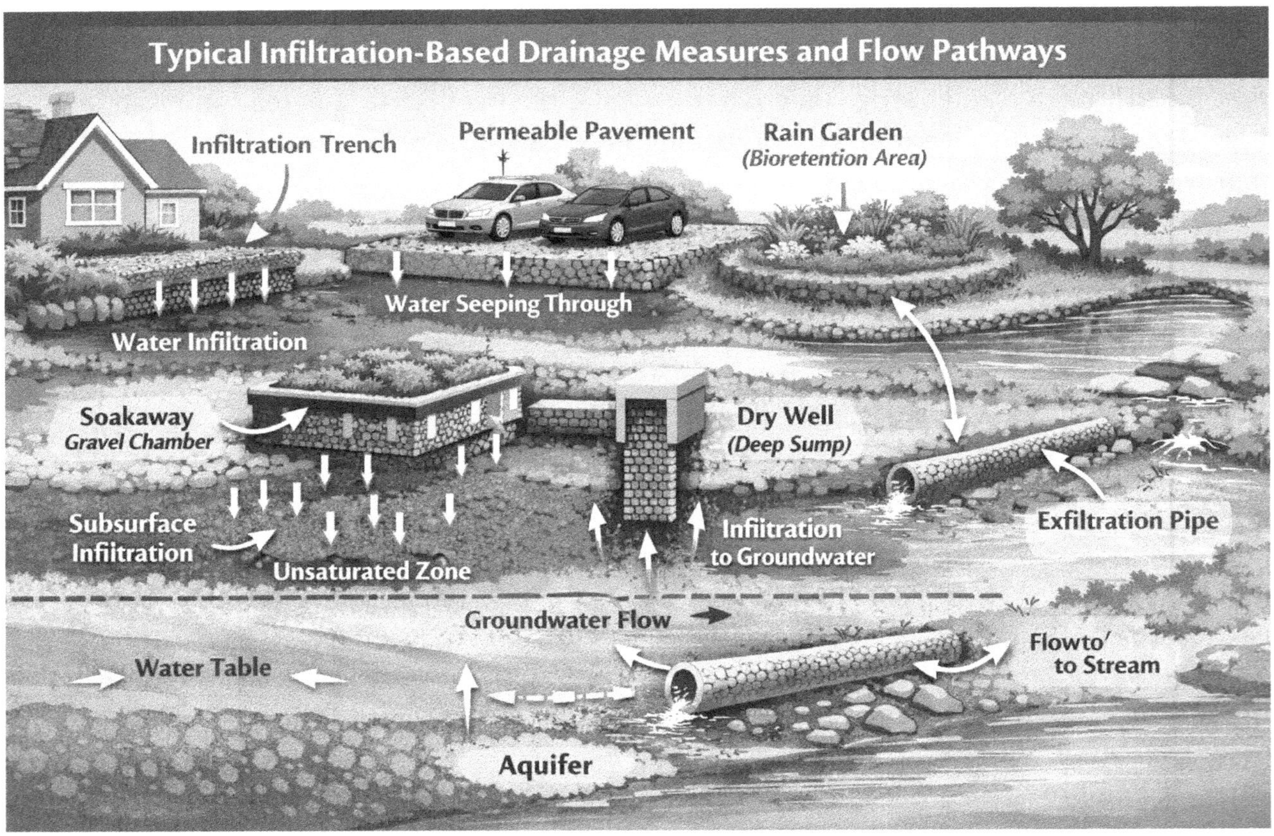

Figure 14.2 Typical infiltration-based drainage measures and flow pathways.

For this reason, infiltration systems should be selected only after evaluating soil permeability, seasonal groundwater variation, expected runoff quality, and maintenance requirements. Pretreatment through silt traps, filter media, or vegetated strips may be necessary to preserve long-term function.

Where conditions are favorable, infiltration measures are especially valuable because they reduce the burden on downstream drains while helping restore a more balanced site hydrology.

14.5 Detention and Retention Systems[1]

Detention and retention systems are designed to moderate stormwater behavior by providing temporary or long-term storage. A detention facility stores runoff during a storm and releases it gradually, while a retention facility stores water for extended periods and may maintain a permanent pool.

These systems are useful in urban developments, institutional campuses, industrial estates, transport corridors, and expanding settlements where peak discharge needs to be reduced before runoff enters public drains or natural receiving channels. They can also provide sediment settling, water quality improvement, and landscape value when designed carefully.

Common examples include detention basins, retention ponds, dry ponds, wet ponds, balancing reservoirs, rooftop storage, and underground storage chambers. Selection depends on land availability, safety, maintenance needs, water quality objectives, and the acceptable rate of discharge downstream.

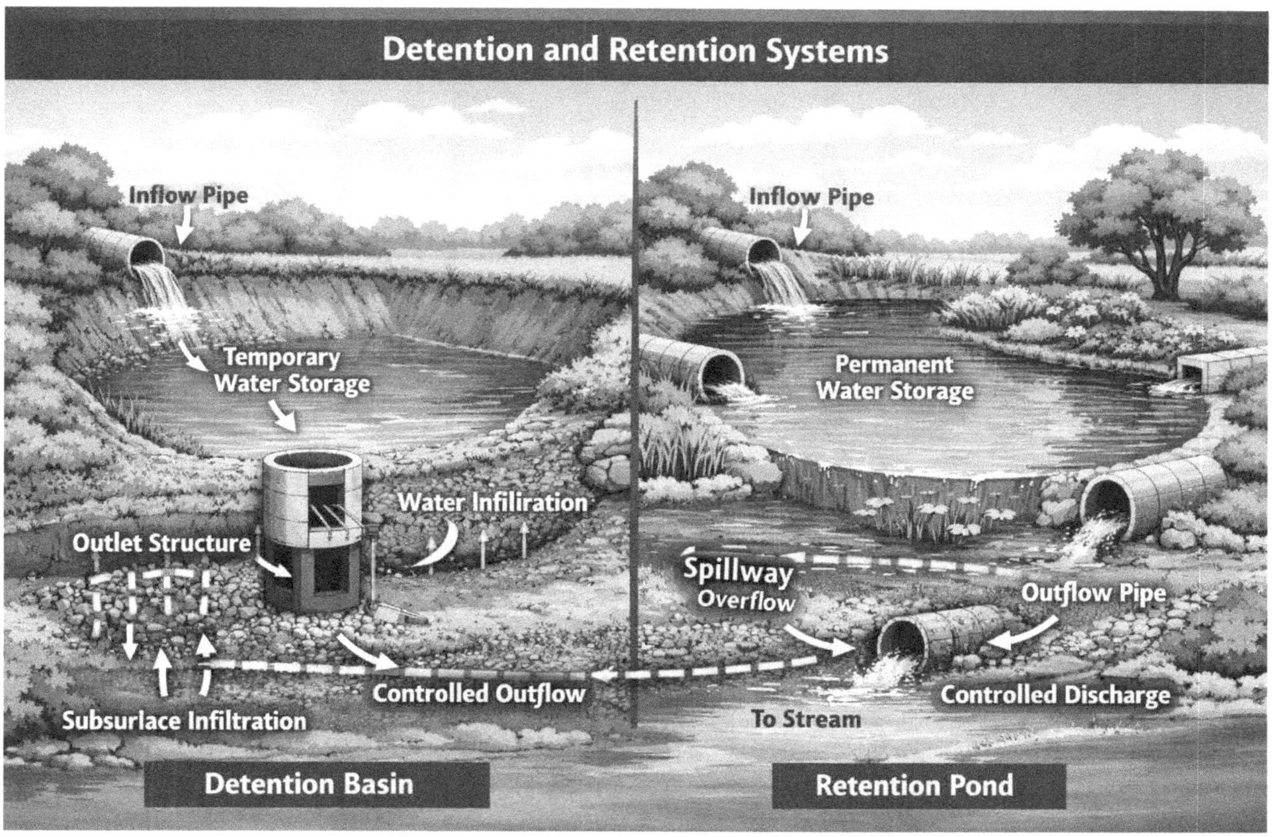

Figure 14.3 Detention and retention systems with inflow, storage, and controlled discharge arrangement.

The main engineering considerations include storage capacity, inlet and outlet control, emergency overflow, embankment safety where applicable, sediment removal, mosquito control, and downstream compatibility. A storage feature that is hydraulically effective but poorly maintained can become unsafe, unsanitary, or nonfunctional.

Therefore, detention and retention should be integrated into a broader drainage plan rather than treated as isolated add-ons. Their value depends on how well they connect with upstream collection, downstream release, and long-term management responsibilities.

14.6 Subsurface Drainage Systems[1]

Subsurface drainage systems are used to control water below the ground surface where excess moisture, seepage, or hydrostatic pressure may damage structures, pavements, slopes, foundations, agricultural land, or underground spaces. Unlike surface drains, these systems work by intercepting and conveying water through buried drains, filter media, underdrains, collector pipes, and outlet arrangements.

Subsurface drainage is important in situations where water cannot be managed adequately by surface grading alone. These include waterlogged soils, retaining structures, basement walls, paved areas with trapped subgrade moisture, sports grounds, embankments, and low-lying ground with poor natural drainage.

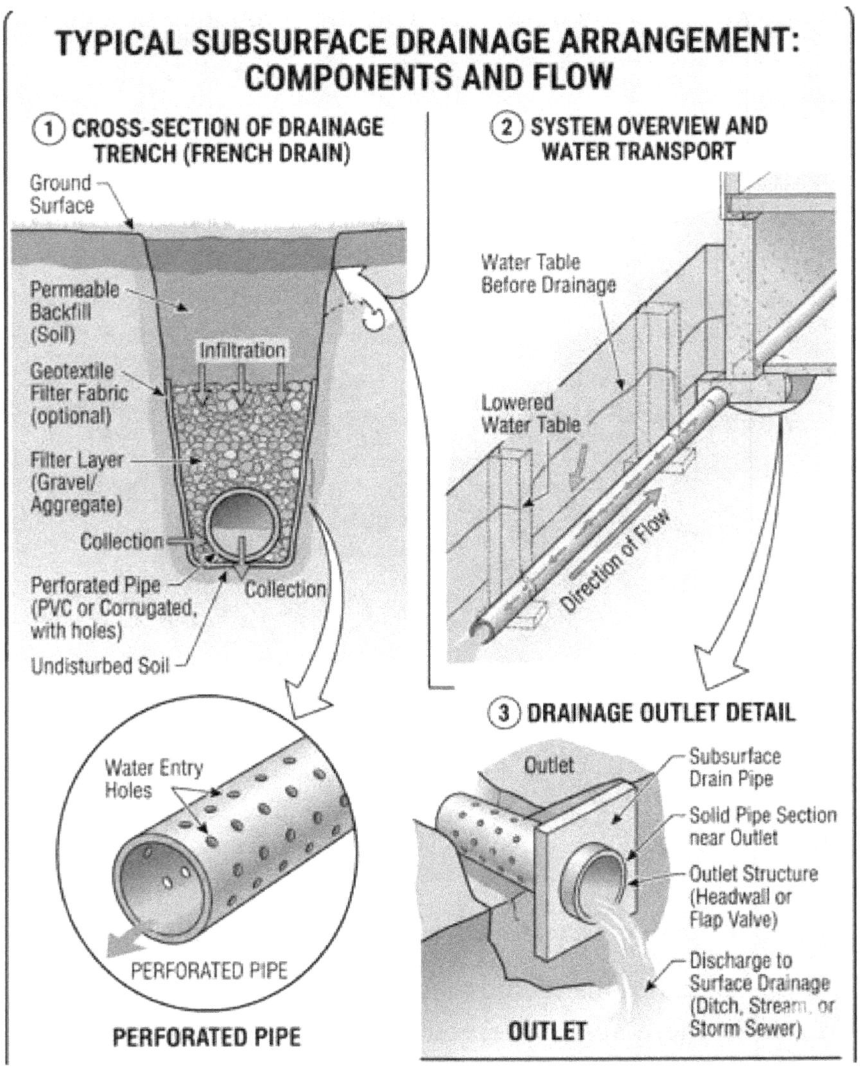

Figure 14.4 Typical subsurface drainage arrangement showing perforated pipe, filter layer, and outlet.

Common forms include perforated pipe drains, French drains, filter drains, edge drains, foundation drains, toe drains, interceptor drains, and underdrain systems beneath pavements or landscaped areas. These installations often require gravel envelopes, geotextile filters, inspection points, and protected outlets to function properly.

The design of subsurface drainage depends on soil characteristics, expected seepage quantity, depth of installation, hydraulic gradient, outlet level, and protection against clogging. A poorly filtered buried drain may fail invisibly over time, making maintenance planning especially important.

Because the system is not visible during normal operation, construction quality becomes critical. Incorrect gradients, damaged geotextiles, blocked perforations, or inaccessible outlets can severely reduce performance without immediate external signs.

Table 14.2 Common types of subsurface drainage systems and their applications.

Type of Subsurface Drainage System	Description / Key Features	Typical Applications
Pipe Drainage (Perforated Pipe System)	Network of perforated pipes laid below ground to collect and convey excess water	Agricultural fields, sports grounds, highways, and landscaped areas
Tile Drainage	Traditional system using clay or concrete tiles joined end-to-end to form conduits	Older agricultural systems; areas with moderate permeability soils
Mole Drainage	Unlined channels formed by pulling a cylindrical mole through clay soils	Heavy clay farmlands; temporary drainage improvement in low-permeability soils
Interceptor Drains	Trenches with perforated pipes installed along slopes to intercept lateral groundwater flow	Hillsides, cut slopes, and embankments to prevent seepage and instability
French Drains	Gravel-filled trenches with or without perforated pipes to promote infiltration and convey water	Gardens, lawns, and small-scale urban drainage applications
Collector Drains	Larger conduits that receive flow from multiple lateral drains and discharge to an outlet	Agricultural drainage networks and urban stormwater systems
Subsurface Infiltration Trenches	Gravel-filled trenches designed to store and infiltrate runoff into the subsoil	Sustainable urban drainage systems (SuDS), parking lots, and green infrastructure
Dry Wells / Soakaways	Vertical or cylindrical chambers that collect runoff and allow infiltration to deeper layers	Residential plots, small catchments, and isolated low-lying areas

14.7 Drainage for Roads, Highways, and Railways

Transportation corridors require specialized drainage because their function depends on structural stability, safe traffic movement, and reliable all-weather performance. Roads, highways, and railways can both generate runoff and obstruct natural flow, so drainage must protect the corridor itself while preserving wider catchment continuity.

For roads and highways, drainage concerns include pavement runoff, shoulder erosion, median drainage, side drains, culverts, embankment protection, slope drainage, and prevention of water entry into the pavement layers. Trapped water in the pavement structure can weaken the subgrade and accelerate deterioration.

Railway drainage has similar concerns but also requires protection of ballast, formation stability, cutting slopes, and cross-drainage beneath track embankments. Water accumulation in railway zones can threaten track geometry, service reliability, and safety.

Drainage design for transport corridors must therefore consider both surface and subsurface components. Open side drains, catch drains on cut slopes, longitudinal edge drains, filter layers, toe drains, and cross-drainage works often need to be combined into a coordinated system.

Failure in one part of the system can affect the whole corridor. A blocked culvert, clogged edge drain, or eroded outfall may produce localized damage that quickly develops into broader structural or operational problems.

14.8 Drainage for Basements and Underground Structures

Basements, underground rooms, retaining walls, tunnels, service corridors, and below-ground utility spaces require drainage measures because they are vulnerable to seepage, hydrostatic pressure, and groundwater entry. Surface runoff may also reach them through ramps, stairwells, depressed access points, or poorly graded surroundings.

Drainage for such structures often combines perimeter foundation drains, filter media, waterproofing systems, sump collection, pump-out arrangements, and surface interception at access points. The objective is to prevent water from accumulating around or within the structure and to relieve pressure that could otherwise cause cracking, leakage, or uplift.

Where the groundwater table is high or the structure lies below surrounding drain level, gravity drainage alone may be insufficient and pumping may be required. This introduces issues of backup power, pump reliability, maintenance, and emergency overflow planning.

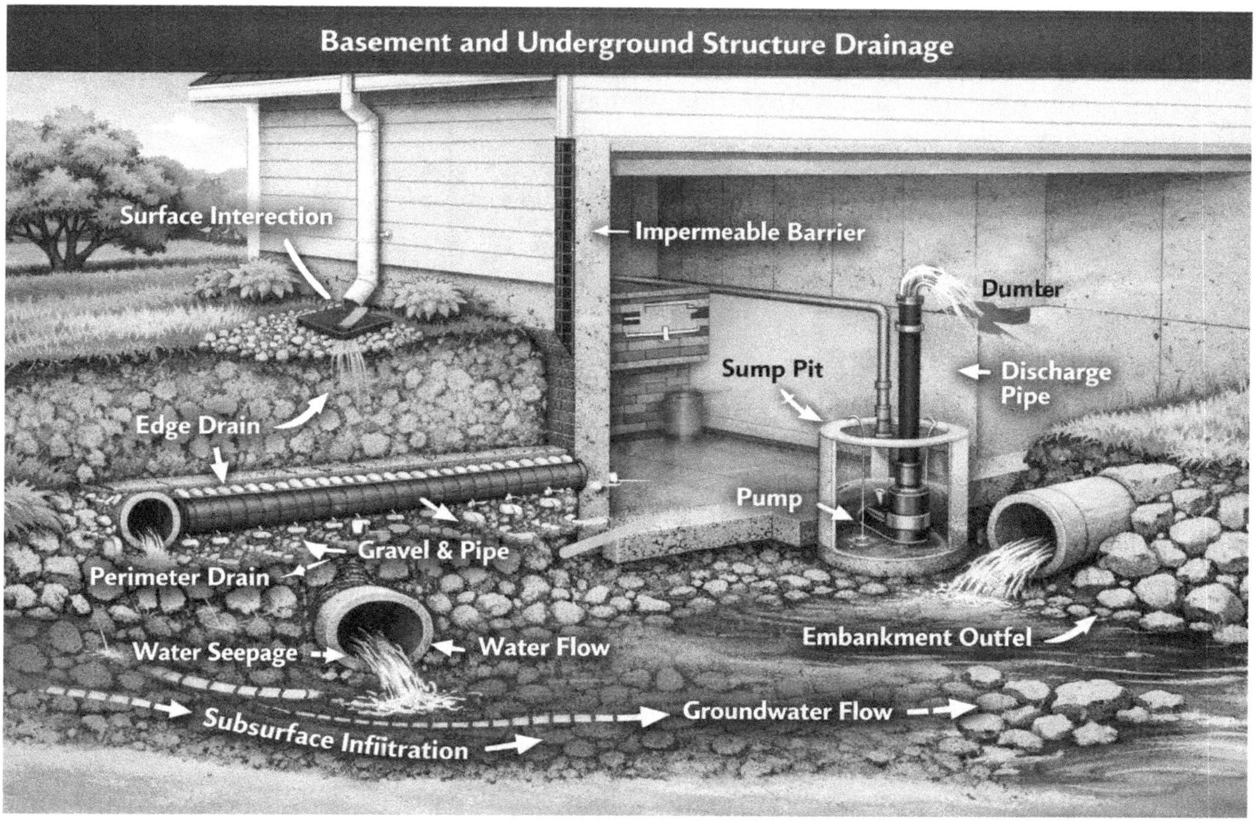

Figure 14.5 Basement and underground structure drainage showing perimeter drain, sump, pump, and surface interception.

Basement drainage design must not be separated from site grading and external drainage layout. A well-drained basement can still fail if adjacent paved surfaces direct runoff toward entry ramps or if roof water is discharged near the foundation.

Accordingly, underground structure drainage should be planned as part of the total site water management system rather than as an isolated protective detail.

14.9 Drainage on Slopes and Hill Roads

Sloping terrain and hill roads create special drainage problems because runoff velocities are high, erosion risk is greater, slope stability may be affected by seepage, and drainage structures are exposed to intense localized flow. In such settings, drainage must control not only water quantity but also the damaging energy of flow.

Surface water on hill roads is commonly managed through side drains, catchwater drains, lined chutes, drop structures, paved gutters, cross drains, and culverts that safely transfer water across or away from the road. On cut slopes, intercepting upslope runoff is often essential to prevent erosion and infiltration into unstable material.

Subsurface drainage may also be needed where seepage weakens slope materials, retaining structures, or road formations. Weep holes, drainage blankets, horizontal drains, and toe drains are used in selected cases to reduce water pressure and improve slope behaviour.

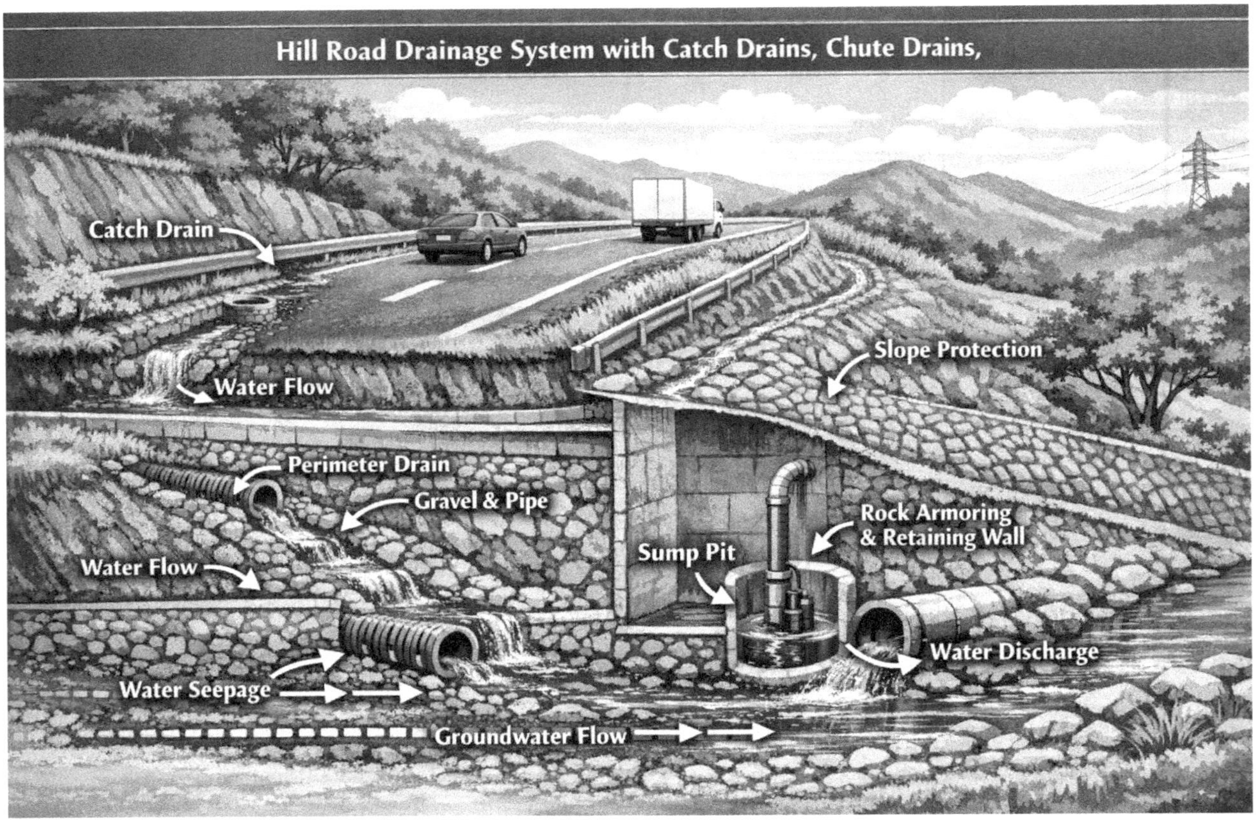

Figure 14.6 Hill road drainage system with catch drains, chute drains, cross-drainage, and slope protection.

The main design challenge is to convey water safely without causing erosion at the outlet or destabilizing the slope below. Structures must therefore be aligned, lined, and protected appropriately, with careful attention to transitions and energy dissipation.

In hill drainage, neglect of small details often leads to major failure. A blocked side drain or eroded chute may rapidly develop into road edge collapse, slope slip, or embankment washout.

14.10 Drainage for Industrial Areas, Airports, and Sports Ground

Certain land uses have special drainage requirements because water affects not only structural safety but also operational performance. Industrial areas, airports, and sports grounds are examples where drainage quality, timing, and surface condition are tightly linked to the purpose of the site.

Industrial areas may require drainage systems that account for contaminated runoff, process water separation, paved yards, loading zones, and large roof catchments. Airports require rapid removal of surface water from runways, taxiways, aprons, and surrounding safety areas in order to maintain operational safety and pavement serviceability. Sports grounds often need carefully controlled surface and subsurface drainage to maintain playable conditions and healthy turf.

Because these sites differ greatly in function, their drainage systems often combine conventional drains with special features such as oil interceptors, slot drains, underdrains, graded surface layers, controlled collection pits, and high-reliability outlet arrangements.

Figure 14.7 Specialised drainage applications for industrial yards, airport pavements, and sports grounds.

The design emphasis in such cases is performance under service conditions. For airports, standing water cannot be tolerated on critical movement surfaces. For sports grounds, even moderate waterlogging may make the facility unusable. For industrial sites, uncontrolled runoff may create environmental or safety hazards.

These examples show that special drainage design must respond directly to the operational demands of the land use rather than relying on generic standards alone.

14.11 Integrated Planning Considerations

In modern drainage engineering, sustainable measures, subsurface systems, and special-purpose drainage should not be treated as separate technical islands. In many projects, they must be combined into one integrated strategy that manages runoff quantity, runoff quality, groundwater interaction, structural protection, and downstream effects at the same time.

A practical planning approach should therefore:

1. Distinguish clearly between water requiring rapid removal and water suitable for infiltration, storage, or treatment
2. Evaluate soil, groundwater, slope, and land-use constraints before selecting sustainable or subsurface measures.

3. Preserve maintenance access to buried drains, storage features, and control structures.
4. Ensure that special-site drainage design is coordinated with the surrounding catchment and outfall system.
5. Consider long-term operation, sediment control, inspection, and institutional responsibility from the beginning.
6. Avoid solutions that appear technically advanced but exceed the site's realistic maintenance capacity.

The engineer must also recognize that drainage decisions increasingly affect environmental performance, asset life, and resilience to extreme rainfall. Good design therefore depends not only on hydraulic calculations, but also on systems thinking across the full site and receiving environment.

14.12 Summary Points

1. Modern drainage engineering focuses not only on removing water but also on managing it sustainably with environmental considerations.
2. Sustainable drainage systems aim to imitate natural hydrological processes to reduce flooding, improve water quality, and support groundwater recharge.
3. Runoff control is achieved through methods such as infiltration, detention, retention, filtration, and controlled discharge.
4. Infiltration-based drainage systems help reduce runoff and recharge groundwater but require suitable soil and proper maintenance.
5. Detention and retention systems store stormwater temporarily or permanently to reduce peak flow and improve water quality.
6. Subsurface drainage systems control underground water and prevent damage to structures, pavements, and soils.
7. Transport infrastructure like roads and railways requires both surface and subsurface drainage to maintain safety and durability.
8. Basements and underground structures need specialized drainage systems to prevent seepage and relieve hydrostatic pressure.
9. Drainage in slopes and hill roads must control high-velocity runoff and prevent erosion and slope instability.
10. Effective drainage design requires an integrated approach that combines sustainability, functionality, and long-term maintenance.

14.13 References

1. CIRIA. (2015). *The SuDS manual (C753).*
2. CIRIA. (2017). *Guidance on the construction of SuDS (C768).*
3. Nijland, H. J., et al. (n.d.). *Subsurface drainage practices: Guidelines for the implementation, operation and maintenance of subsurface pipe drainage systems.*
4. United States Department of Agriculture. (n.d.). *National engineering handbook, Section 16, Chapter 4: Subsurface drainage.*
5. Food and Agriculture Organization of the United Nations. (n.d.). *Materials for subsurface land drainage systems.*
6. Federal Highway Administration. (n.d.). *Storm drainage guidance for transportation facilities* (Publication No. FHWA-HIF-24-006).
7. Defense Technical Information Center. (n.d.). *Airport pavement drainage.*
8. Advanced Drainage Systems, Inc. (2023). *Drainage handbook.*

Chapter 15

Materials and Components of Drainage Systems

Learning Objectives

By the end of this chapter, the learner should be able to:

1. Identify common materials used in drainage construction.
2. Compare material options for different applications.
3. Select components appropriate to hydraulic and environmental conditions.
4. Consider durability and maintenance in material selection.

15.1 Introduction

The performance of a drainage system depends not only on sound planning and hydraulic design but also on the quality of materials and construction. A well-designed drain can still fail if poor masonry is used, concrete is improperly cured, pipe bedding is inadequate, joints leak, slopes are not maintained, or construction tolerances are ignored. For this reason, materials, workmanship, and quality control are central to durable drainage engineering.

This chapter discusses the materials commonly used in settlement drainage systems, the main construction practices for open and closed drains, and the quality control measures needed to ensure safe, stable, and long-lasting performance. The discussion applies to urban, semi-urban, and village drainage works, with attention to both engineered and low-cost contexts.

15.2 Role of Materials and Construction in Drainage Performance

Drainage systems are exposed to flowing water, silt, debris, variable moisture, weathering, traffic loading, ground movement, and in some cases wastewater contamination. These conditions place continuous demands on the structural integrity and durability of the materials used. Therefore, drainage performance is closely linked to material quality and construction practice.

A drain may fail for many construction-related reasons. Poorly compacted bedding can cause pipe settlement, weak mortar may lead to joint opening, inadequate curing can reduce concrete strength, and inaccurate level control can destroy hydraulic continuity. In many cases, recurring field failures are not due to incorrect design calculations, but to avoidable defects introduced during construction.

Construction quality also affects long-term maintenance. Smooth, well-finished channels are easier to clean and less prone to sediment deposition than rough, irregular ones. Proper cover placement, chamber alignment, and access provision improve inspectability and repair. Thus, quality construction improves both immediate functionality and lifecycle performance.

The engineer should therefore view construction as an extension of design. Material choice, workmanship standards, and site supervision are all part of the engineering process, not separate matters left only to contractors.

15.3 Common Materials Used in Drainage Works

A wide range of materials is used in drainage engineering, depending on the type of structure, hydraulic demand, local availability, exposure condition, and cost. Common materials include earth, brick, stone, mortar, concrete, reinforced concrete, precast elements, steel, plastic pipes, and geosynthetics.

Earth is often used for unlined drains, embankments, and backfill. It is economical but vulnerable to erosion and deformation if not properly compacted or protected. Brick masonry is widely used in roadside drains, small chambers, and low-cost lined sections, especially where local skills and materials are available. Stone masonry and pitching are often used for slope protection, aprons, and outlet stabilization.

Concrete is one of the most common drainage materials because of its versatility and durability. It is used for lined drains, slabs, culverts, manholes, channels, and foundations. Reinforced concrete is required where structural strength is more critical, such as in covers, traffic-loaded slabs, culvert sections, and retaining components. Precast concrete units can improve construction speed and dimensional consistency if properly manufactured and installed.

Pipes may be made of reinforced concrete, PVC, HDPE, stoneware, or other materials depending on size, load, cost, corrosion risk, and installation condition. The choice of material should consider hydraulic roughness, joint performance, durability, and maintenance requirements as well as initial cost.

Table 15.1 Common drainage materials and typical uses

Drainage Material	Typical Uses	Key Characteristics / Notes
Earthen (Unlined Soil)	Temporary roadside drains, field channels, small village lanes	Low cost; easy to construct; prone to erosion and silting; requires frequent maintenance
Brick Masonry	Village streets, semi-permanent drains, small culverts	Moderate durability; good for low-flow areas; requires mortar joints and periodic repair
Stone Pitching	Hilly terrain, erosion-prone slopes, outfall protection	High resistance to erosion; blends with rural landscape; labor-intensive construction
Concrete (RCC / Precast)	Main village drains, culverts, high-flow zones	Durable and smooth; high hydraulic efficiency; costly but low maintenance
PVC / HDPE Pipes	Underground drainage, cross-drainage connections	Lightweight, corrosion-resistant; easy installation; suitable for small to medium flows
Cast Iron Pipes	Pumping mains, sump discharge lines	Strong and pressure-resistant; heavy and expensive; used in permanent installations
Gabion Structures	Outfall protection, energy dissipation zones	Flexible and permeable; effective against scour; requires periodic inspection
Vegetated / Grass-Lined Channels	Peripheral areas, low-flow rural drains	Eco-friendly; promotes infiltration; limited capacity during heavy rainfall

15.4 Selection of Materials for Different Settlement Contexts

Material selection should respond to local settlement conditions, not just generic engineering preference. The same material may be suitable in one context and uneconomical or difficult to maintain in another.

In urban areas, reinforced concrete, precast units, pipe systems, and covered drainage elements are often preferred because they can withstand traffic loading, fit constrained corridors, and provide defined geometry. However, such materials require better quality control, accurate installation, and skilled workmanship. In semi-urban areas, mixed solutions may be appropriate, such as brick-lined drains, partial concrete lining, masonry chambers, and pipe culverts integrated with open channels.

In village and rural settings, locally available materials often play a major role. Brick masonry, stone pitching, simple concrete lining, and earth drains with localized protection may offer a practical balance between performance and affordability. Where maintenance depends on local labor, systems should be built with materials that can be repaired without specialized equipment.

Material selection should also consider exposure to erosion, silt, floating debris, wastewater intrusion, and seasonal moisture variation. A low-cost option that fails quickly under local conditions may be less economical than a somewhat stronger and better-finished alternative.

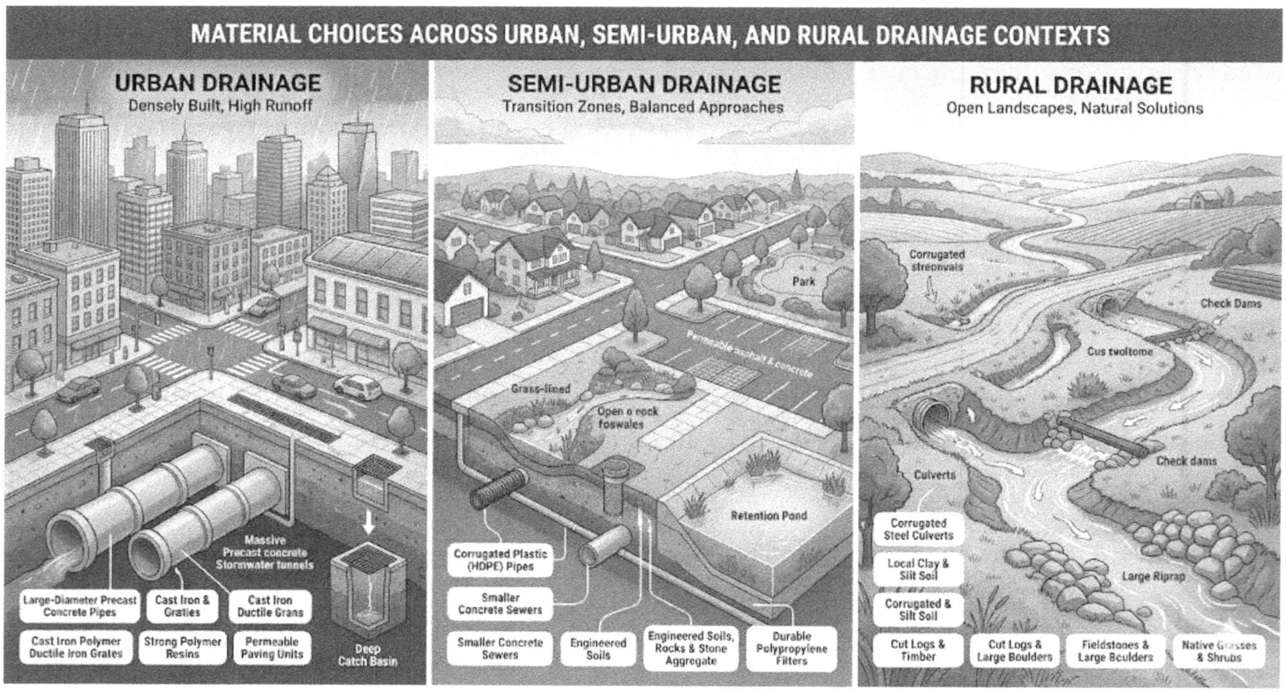

Figure 15.1 Material choices across urban, semi-urban, and rural drainage contexts.

15.5 Earthwork, Excavation, and Foundation Preparation

Earthwork is one of the first and most critical stages in drainage construction. It includes clearing, excavation, trimming of side slopes, foundation preparation, dewatering where needed, and backfilling. Errors at this stage can affect alignment, slope, structural support, and long-term stability.

Excavation for drains, pipes, culverts, and chambers must follow the specified line and level. Over-excavation can weaken support and increase material use, while under-excavation may force incorrect grades or reduce section dimensions. In soft or waterlogged ground, excavation may require temporary support, dewatering, or replacement of weak foundation material.

Foundation preparation is especially important for masonry drains, concrete channels, pipes, and culverts. The base must be stable, level, and compact enough to support the structure without differential settlement. Loose soil, organic material, or saturated pockets should be removed or treated before construction proceeds.

For pipe systems, trench bottom preparation and bedding are crucial. For open drains, the prepared subgrade should maintain the intended slope continuously. If foundation levels are inaccurate, later hydraulic problems are likely even if the visible structure appears satisfactory.

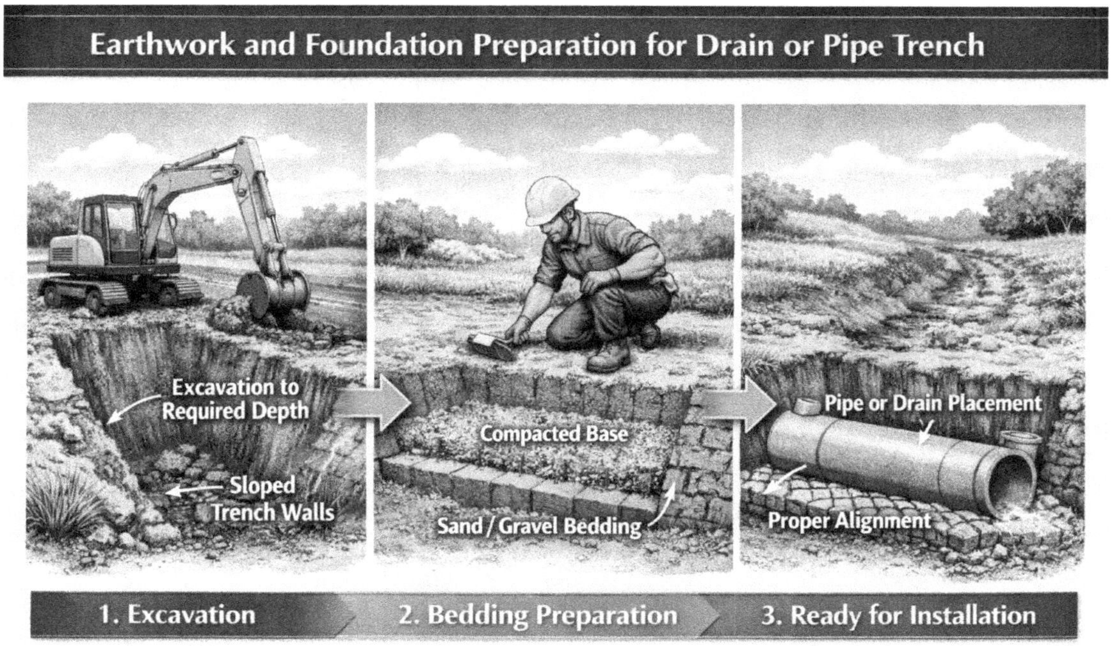

Figure 15.2 Earthwork and foundation preparation for a drain or pipe trench.

15.6 Construction of Open Drains

Open drains may appear simple, but their construction demands careful attention to geometry, slope, lining quality, and outlet continuity. Even minor deviations can create ponding, erosion, or premature structural distress.

Construction begins with accurate setting out of alignment and levels. The bed slope must be maintained continuously, and cross-section dimensions must match the design. For unlined drains, side slopes should be trimmed and protected where necessary to prevent erosion or sloughing. For lined drains, the subgrade should be compacted and prepared before masonry, stone, or concrete lining is placed.

In masonry drains, the quality of brick or stone units, mortar mix, joint thickness, and curing all affect performance. In concrete-lined drains, formwork, mix proportion, compaction, finishing, and curing are critical. The drain bed should be smooth enough to support flow continuity without unnecessary roughness or low pockets.

Transitions, crossings, bends, and outfalls deserve special care because they are common failure points. A drain that is well built in straight reaches may still fail if the outlet is poorly protected or if an access crossing obstructs flow. Therefore, open drain construction should be checked as a complete hydraulic path, not as isolated segments.

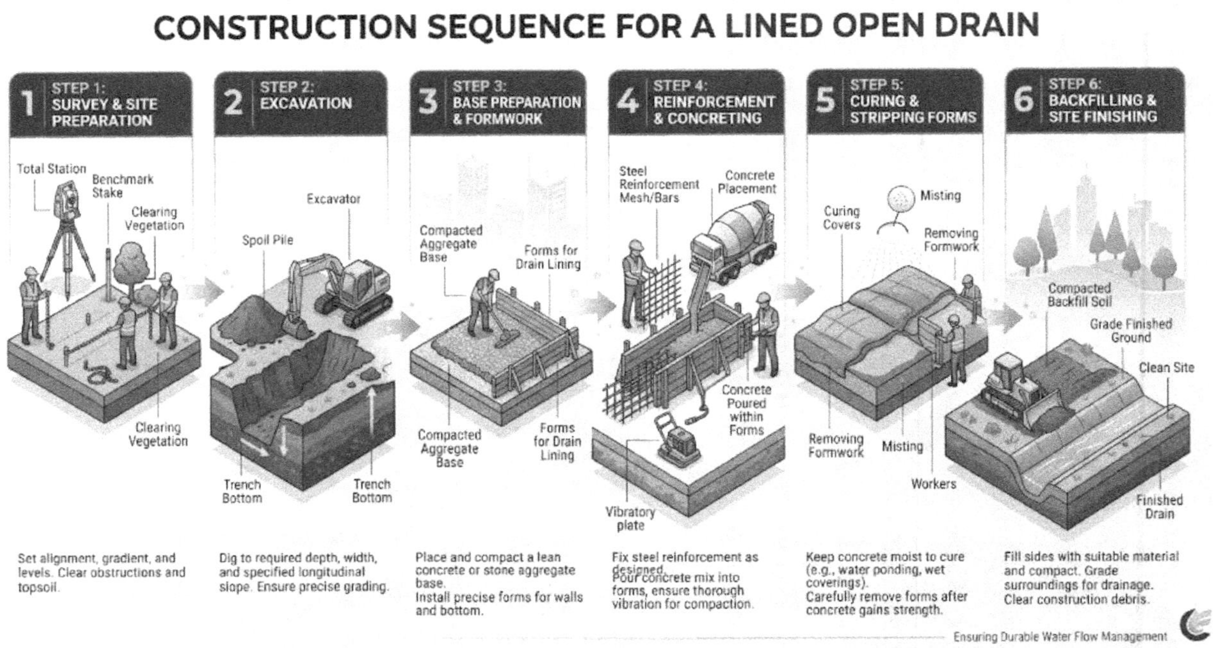

Figure 15.3 Construction sequence for a lined open drain.

15.7 Installation of Pipes, Culverts, and Covered Drains

The installation of buried pipes, culverts, and covered drains requires precision because these structures are less visible after construction and are difficult to correct later. Proper bedding, alignment, jointing, cover, and compaction are essential for both hydraulic and structural performance.

Pipe installation should begin with prepared bedding of the specified grade and thickness. The pipe invert must follow the designed slope, and each pipe length should be aligned so that joints are tight and levels remain continuous. Misalignment or uneven bedding may cause sagging, reverse slope, leakage, or localized sediment deposition.

For culverts, attention must also be given to inlet and outlet levels, embankment integration, headwall construction, apron treatment, and scour protection. A culvert barrel alone does not ensure performance; the surrounding structure and flow approach condition are equally important. Covered drains likewise require sound wall construction, stable cover support, access openings, and continuity with adjoining sections.

Backfilling is a critical part of installation. Improper backfill placement or insufficient compaction can lead to settlement, cracking, pipe displacement, or slab failure under traffic. Therefore, the structural system includes not only the manufactured unit but also the surrounding soil support.

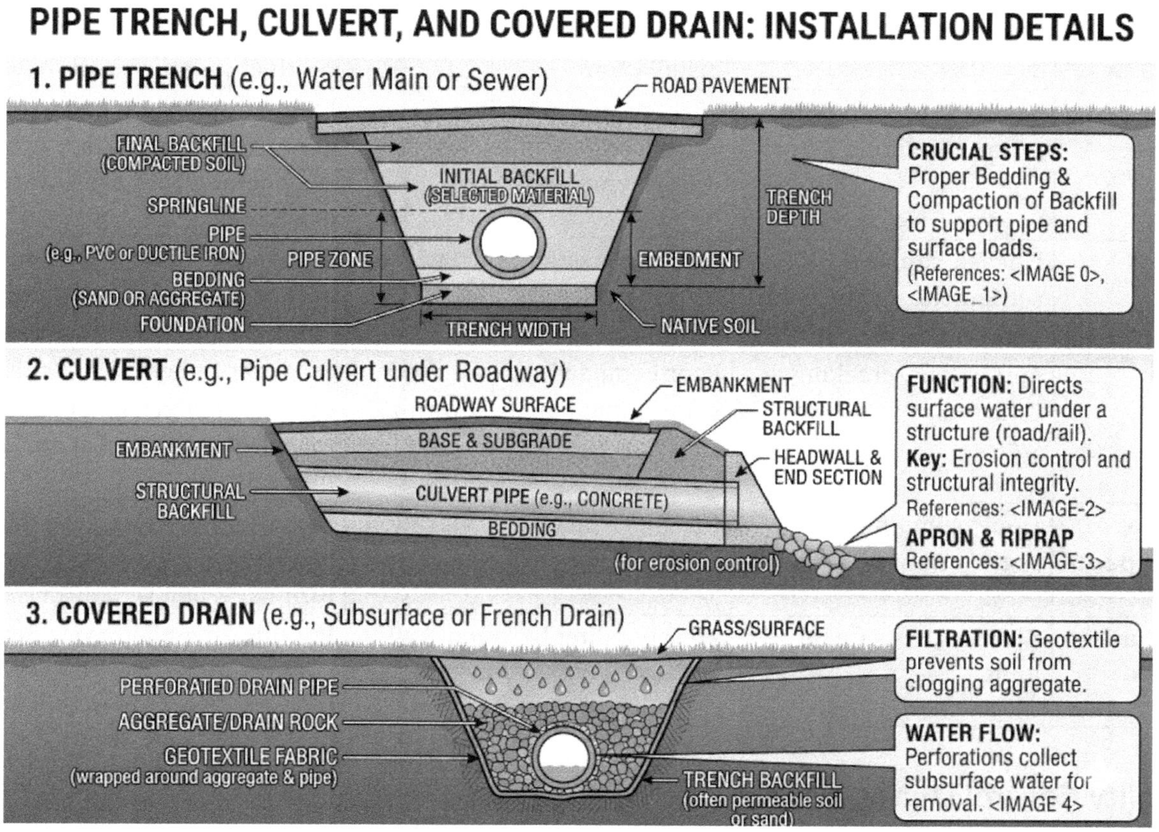

Figure 15.4 Pipe trench, culvert, and covered drain installation details.

15.8 Joints, Bedding, and Backfilling

Joints, bedding, and backfilling are often overlooked in drainage work, yet they are among the most influential aspects of durability. These elements determine whether the structure remains watertight where required, retains its line and level, and resists movement under load and moisture variation.

Joints in masonry, concrete, and pipe systems should be properly formed, sealed where necessary, and finished to avoid leakage and intrusion of soil. Poor joints allow water to escape, fines to enter, and structural weakness to develop over time. In pipe systems, joint failure is a common cause of infiltration, exfiltration, and displacement.

Bedding provides uniform support to pipes and some buried structures. It should be made of suitable material, properly shaped, and compacted to prevent point loading and settlement. Inadequate bedding may cause cracking, misalignment, or loss of hydraulic gradient.

Backfilling should be carried out in layers with appropriate material and compaction. Throwing loose soil into a trench without controlled placement is a frequent cause of future failure. Backfill quality becomes especially important under roads, around chambers, and above pipes carrying traffic loads.

Table 15.2 Common defects in joints, bedding, and backfill with preventive measures.

Component	Common Defects	Causes	Preventive Measures
Pipe Joints	Leakage, misalignment, poor sealing	Inadequate jointing compound, uneven bedding, improper alignment during installation	Use correct jointing materials; ensure uniform bedding; check alignment before backfilling; conduct pressure or flow tests
Brick/ Concrete Joints	Cracked or open joints, mortar washout	Weak mortar mix, insufficient curing, exposure to flowing water before set	Maintain proper mortar ratio; allow adequate curing; protect fresh joints from early water exposure
Bedding Layer	Uneven support, settlement, voids under pipe	Poor compaction, inconsistent bedding thickness, use of unsuitable material	Compact bedding uniformly; use graded sand or gravel; maintain specified thickness and slope
Backfill	Settlement, cracking, infiltration of surface water	Poor compaction, oversized stones, inadequate layering	Compact in layers; use suitable granular material; avoid large stones near pipe; seal surface properly
Concrete Base / Apron	Honeycombing, segregation, weak zones	Improper mixing or vibration, poor formwork	Ensure proper mix design; use mechanical vibrators; inspect formwork before pouring
Drain Wall Interface	Seepage, erosion at joint	Poor bonding between wall and base, lack of keying	Provide keyed joint or dowel connection; seal with waterproof mortar; inspect before plastering

15.9 Quality Control During Construction

Quality control is the organized process of ensuring that materials, workmanship, dimensions, and finished structures conform to design and specification requirements. In drainage works, quality control is needed from the first stage of site preparation to final inspection.

Key quality control activities include checking line, level, and dimensions before and during construction; verifying material quality, mix proportions, and workmanship; inspecting bedding, reinforcement, formwork, and joint treatment; monitoring curing, compaction, and protection of completed work; confirming continuity of flow path and outfall connection; and documenting deviations and corrective measures.

Quality control should be proactive rather than limited to final inspection. Many defects, such as wrong invert level or inadequate bedding, cannot be corrected easily after construction is complete. Therefore, stage-wise checking is more effective than relying only on end-of-work review.

In smaller rural works, formal testing may be limited, but basic quality control is still essential. Even simple drains require correct slope, stable base, proper mortar, and sound workmanship. The level of sophistication may vary, but the principle of quality assurance applies everywhere.

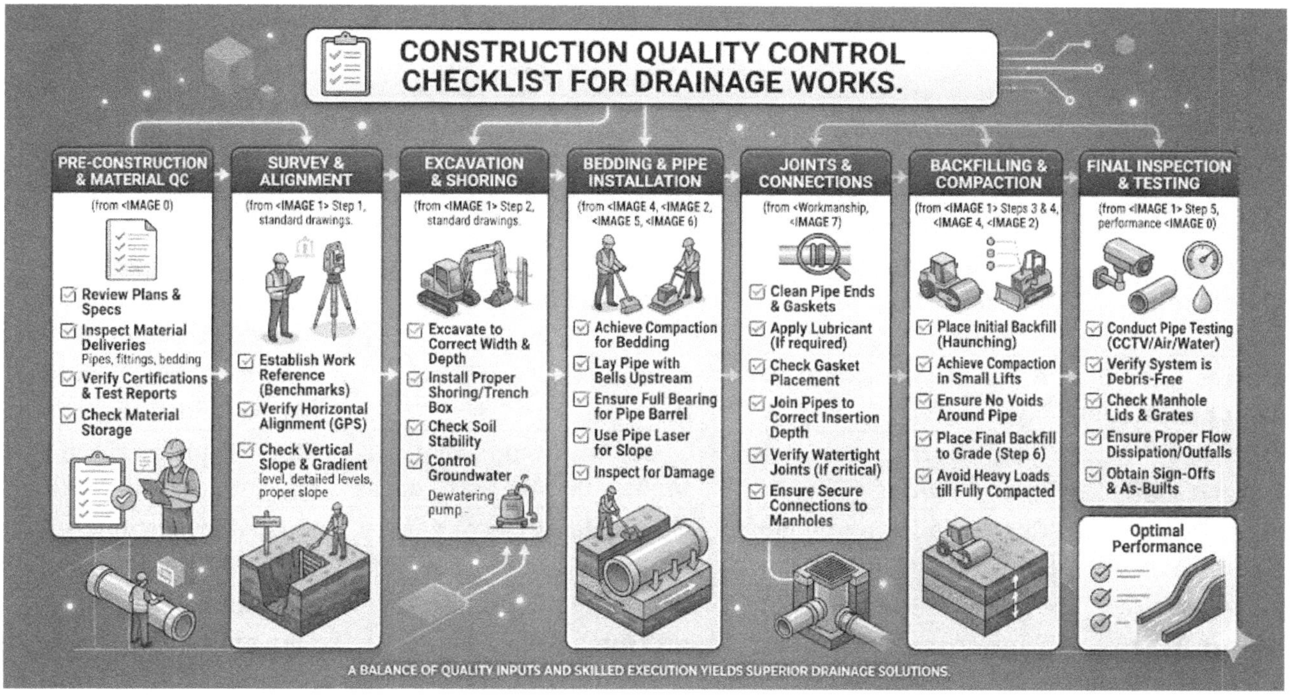

Figure 15.5 Construction quality control checklist for drainage works.

15.10 Field Testing, Tolerances, and Acceptance Criteria

Field testing and dimensional verification help confirm whether constructed drainage works meet the required performance and quality standards. The type and extent of testing depend on the scale and type of work, but basic acceptance should always include checks on geometry, material condition, and hydraulic continuity.

Typical field checks may include line and level verification, section dimension measurement, visual inspection of joints and surface finish, compaction checks for backfill and subgrade, concrete cube or strength records where applicable, water-tightness or flow testing for selected closed systems, and inspection of outlet condition and protective works.

Tolerances define the acceptable range of variation from the specified line, level, dimension, or material property. Some variation is inevitable in construction, but excessive deviation may affect performance. For drainage systems, level tolerance is especially important because small errors can alter hydraulic behavior significantly.

Acceptance criteria should therefore be linked to function as well as form. A drain should not be accepted only because it appears complete; it should also demonstrate continuity, proper gradient, adequate access, and stable construction. Functional inspection after completion is often as important as checking physical measurements.

Table 15.3 Example field tests, tolerances, and acceptance checks for drainage works.

Test / Check	Purpose	Typical Tolerance / Acceptance Criteria	Remarks / Field Procedure
Line and Level Check	Verify alignment and slope of drain or pipe	±10 mm in level; ±25 mm in alignment	Use leveling instrument or laser; check at 10 m intervals
Gradient Verification	Ensure uniform slope for self-cleansing flow	Deviation ≤ 0.05% from design gradient	Measure invert levels at start and end; confirm flow direction
Bedding Compaction Test	Assess firmness of base layer	Minimum 95% of Proctor density	Conduct field density test or plate load test
Concrete Strength Test	Confirm structural integrity of drain lining or base	≥ design strength (e.g., M20 → 20 MPa at 28 days)	Cube samples cured and tested at site or lab
Mortar Joint Inspection	Check sealing and bonding of masonry joints	No visible cracks or gaps; uniform thickness 10–12 mm	Visual inspection; tap test for hollow joints
Pipe Pressure / Leakage Test	Verify watertightness of joints	No leakage under test pressure (typically 1.5 × working pressure)	Fill section with water; maintain pressure for 30 min
Backfill Compaction Test	Ensure stability and prevent settlement	Minimum 90–95% of Proctor density	Layer-wise compaction; field density test using sand cone
Flow Test / Hydraulic Trial	Confirm unobstructed water movement	Continuous flow without ponding or backflow	Run water through completed section; observe discharge
Surface Finish Check	Evaluate smoothness and uniformity of lining	No honeycombing; even plaster finish	Visual inspection and straightedge test
Final Acceptance Inspection	Overall compliance with design and workmanship	All parameters within tolerance; documentation complete	Joint inspection by engineer and contractor before handover

15.11 Common Construction Defects and Preventive Measures

Many drainage failures can be traced to recurring construction defects. Understanding these typical defects helps engineers, supervisors, and students recognize what must be prevented in practice.

Common defects include incorrect slope or reverse gradient, uneven bed levels causing local ponding, weak mortar or poor concrete compaction, honeycombed concrete surfaces, inadequate curing, poor joint sealing, pipe misalignment, insufficient bedding thickness, poorly compacted backfill, blocked or unfinished outfalls, missing access openings in covered drains, and absence of erosion protection at outlets.

Preventive measures are usually straightforward but require discipline. Accurate setting out, staged inspection, use of approved materials, proper supervision, controlled compaction, and protection of completed work can prevent many failures. Equally important is the need to keep drainage paths open during construction so that temporary obstruction does not itself create flooding damage.

The best quality control system is one that anticipates common errors before they become embedded in the finished work. Prevention is almost always less costly than post-construction repair.

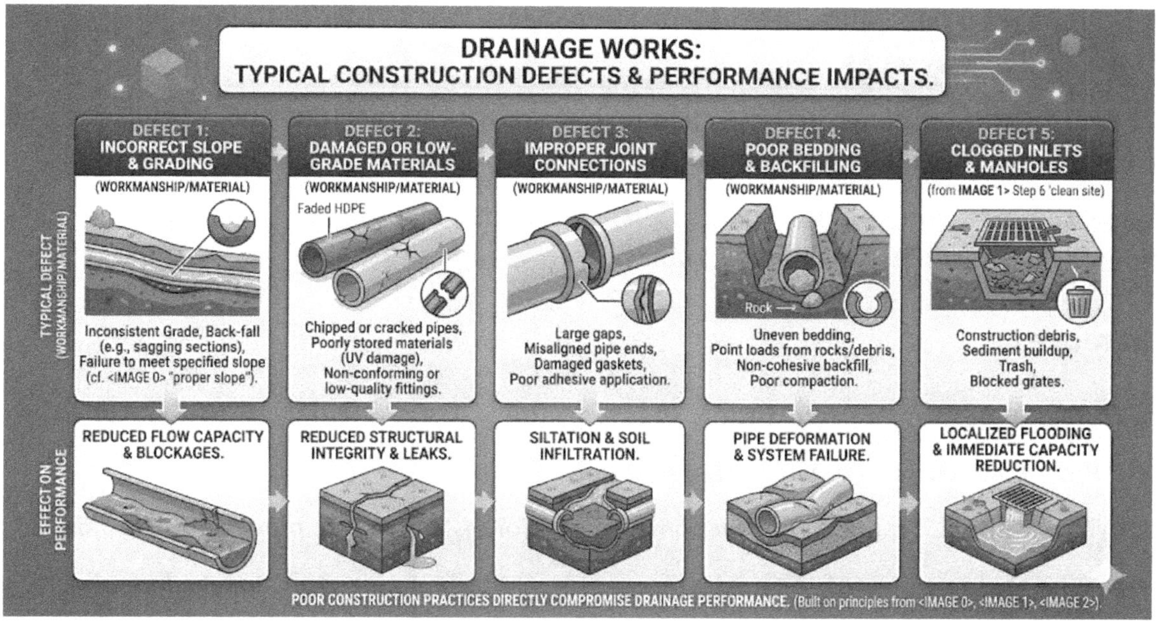

Figure 15.6 Typical construction defects and their effect on drainage performance.

15.12 Learning Interpretation

Chapter 15 should be taught as the bridge between design documents and field reality. Students should appreciate that many drainage problems originate not in theory but in ordinary site mistakes such as poor levels, weak joints, inadequate curing, and insufficient compaction.

It also reinforces the practical idea that a drainage system is only as reliable as its materials, workmanship, and inspection regime.

15.13 Summary Points

1. The performance of a drainage system depends not only on design but also on the quality of materials, construction, and workmanship.
2. Poor construction practices such as weak joints, improper curing, and incorrect levels can lead to drainage failure even with good design.
3. Common drainage materials include earth, brick, stone, concrete, reinforced concrete, pipes (PVC/HDPE), steel, and geosynthetics.
4. Material selection should be based on local conditions, cost, durability, and maintenance requirements.
5. Proper earthwork, excavation, and foundation preparation are essential to ensure stability and correct hydraulic performance.
6. Construction of open drains requires accurate alignment, proper slope, and good-quality lining to prevent erosion and water stagnation.

7. Installation of pipes, culverts, and covered drains must ensure correct bedding, alignment, jointing, and backfilling.
8. Joints, bedding, and backfilling are critical components that influence the durability and structural integrity of drainage systems.
9. Quality control during construction, including stage-wise inspection and testing, is necessary to ensure compliance with design standards.
10. Preventing common construction defects such as reverse slopes, poor compaction, and leakage is essential for long-term drainage performance.

15.14 References

1. Federal Highway Administration. (n.d.). *Urban drainage design manual (HEC-22, 4th ed.).*
2. Concrete Pipe Association. (n.d.). *Design manual.*
3. Illinois Urban Manual. (n.d.). *Construction specification 42: Concrete pipe conduits and drains.*
4. Concrete Pipe Association of Australasia. (n.d.). *Selecting materials for bedding steel reinforced concrete pipe.*
5. Central Public Health and Environmental Engineering Organisation. (n.d.). *Manual on sewerage and sewage treatment.*
6. U.S. Bureau of Reclamation. (n.d.). *Pipe bedding and backfill.*
7. American Concrete Pipe Association. (n.d.). *Post installation inspection for all storm drain pipes.*

Chapter 16

Rehabilitation, Retrofitting, and Asset Management

Learning Objectives

By the end of this chapter, the learner should be able to:

1. Assess when rehabilitation or retrofitting is needed.
2. Compare major techniques for upgrading existing systems.
3. Apply asset management concepts to drainage infrastructure.
4. Prioritize interventions using condition and risk considerations.

16.1 Introduction

Drainage systems rarely remain adequate throughout their full service life without intervention. Even when originally well planned and constructed, they are affected over time by sedimentation, structural deterioration, land use change, hydraulic overload, encroachment, utility conflicts, and inadequate maintenance. As settlements expand and service expectations rise, many existing drainage systems must be repaired, upgraded, adapted, and managed more systematically.

This chapter focuses on the improvement of existing drainage infrastructure through rehabilitation and retrofitting, and on the long-term management of drainage assets through an asset management approach. It explains why drainage systems deteriorate, how improvement needs are identified, how rehabilitation differs from retrofitting, and how lifecycle-based planning supports more reliable and cost-effective drainage services.

16.2 Need for Rehabilitation and Retrofitting

Rehabilitation and retrofitting become necessary when an existing drainage system no longer performs adequately under current conditions. The problem may arise from structural aging, poor original construction, inadequate hydraulic capacity, increased runoff from urbanization, or changing outfall conditions. In many settlements, drainage problems are caused not by absence of infrastructure, but by infrastructure that has become outdated, damaged, or mismatched with present demand.

Rehabilitation is especially important where deterioration has reduced the original function of the system. Retrofitting becomes important where the system must be modified to meet new conditions, such as denser development, increased road width, climate stress, or more stringent service expectations. In practice, many projects involve both, because an old drain may need structural repair and hydraulic upgrading at the same time.

The need for such intervention is strongest in older urban networks, expanding peri-urban corridors, low-lying settlements, and village systems that have evolved incrementally without coordinated planning. In all these settings, existing drains remain valuable assets, but they must be improved rather than simply left to deteriorate.

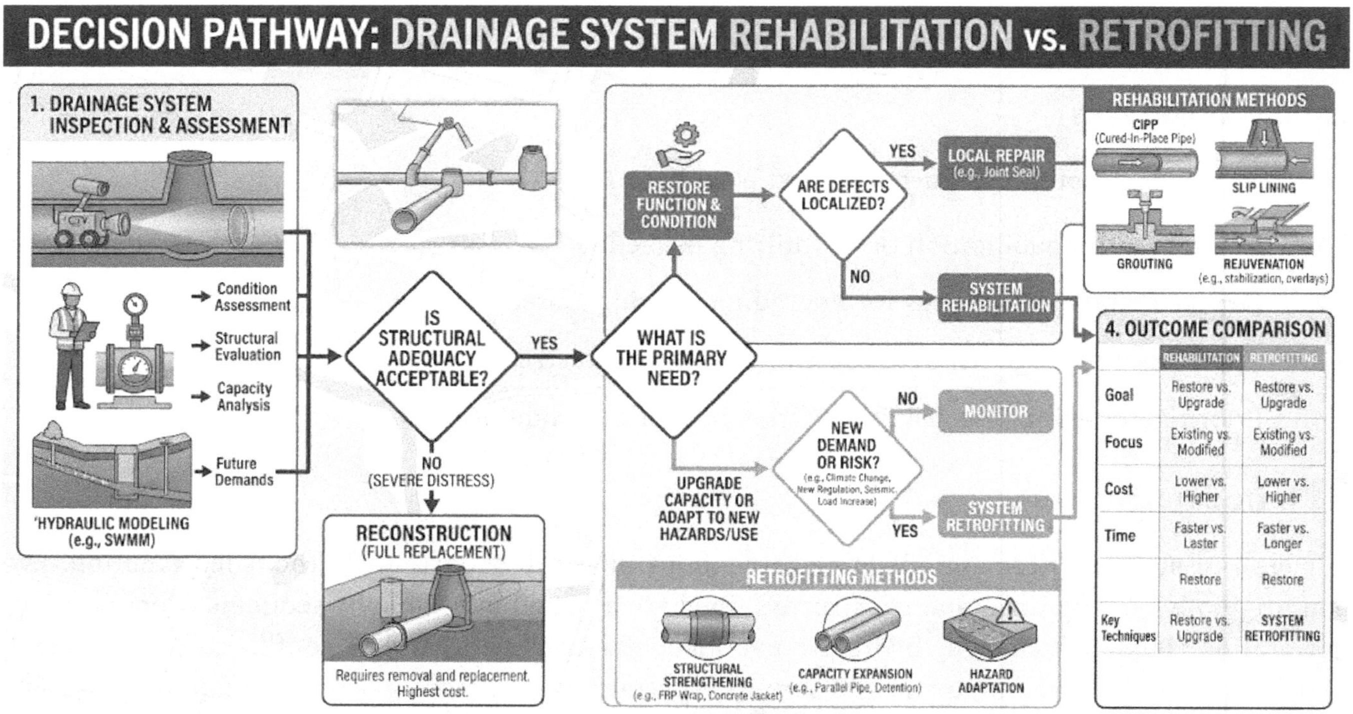

Figure 16.1 Decision pathway for identifying rehabilitation versus retrofitting needs in drainage systems.

16.3 Causes of Functional Obsolescence

A drainage system may remain physically present but become functionally obsolete. This means that it no longer serves current hydraulic, environmental, operational, or social needs, even if parts of it are still structurally intact.

Functional obsolescence often results from increased impervious area and runoff, road widening and raised embankments, plot filling and altered surface grading, blocked or submerged outfalls, sediment accumulation and loss of section, growth of built development over former drainage paths, utility conflicts preventing improvement, and higher expectations for safety, access, or urban service quality.

For example, an open roadside drain built decades earlier may have been adequate when runoff was low and surrounding land was permeable. The same drain may now be undersized because the area has become densely paved and built up. Similarly, a culvert that once passed local flow safely may become a bottleneck after upstream development accelerates runoff.

Recognizing functional obsolescence is important because many systems fail gradually. The infrastructure still exists, but it no longer matches the needs of the settlement.

Table 16.1 Indicators of functional obsolescence and typical physical or operational symptoms.

Indicators of Functional Obsolescence	Typical Physical or Operational Symptoms
Design Inefficiency	Layouts that hinder workflow, excessive travel distances, poor space utilization
Technological Outdating	Equipment incompatible with modern systems, inability to integrate automation or digital controls
Excessive Operating Costs	High energy consumption, frequent breakdowns, costly maintenance
Regulatory Non-Compliance	Failure to meet safety, environmental, or accessibility standards
Capacity Limitations	Inability to handle required volume, undersized storage or processing units
Poor Adaptability	Difficulty in modifying or expanding facilities, rigid layouts
Inadequate Comfort/ Usability	Poor lighting, ventilation, or ergonomics; user dissatisfaction
Safety Hazards	Outdated fire protection, unsafe wiring, structural weaknesses
Environmental Inefficiency	Excessive emissions, poor insulation, water wastage
Market Mismatch	Facilities or equipment not aligned with current demand or industry practices

16.4 Assessment of Existing Drainage Assets

Before rehabilitation or retrofitting can be planned, the existing drainage assets must be assessed systematically. An asset in this context includes any physical component of the drainage system, such as drains, pipes, culverts, manholes, chambers, inlets, outfalls, detention structures, embankment drains, or pump installations.

Assessment should examine location and connectivity, structural condition, hydraulic adequacy, maintenance status, accessibility, exposure to encroachment or utility conflict, and the role of the asset in overall system performance.

A complete assessment usually combines field inspection, level survey, drain inventory, historical flood information, maintenance records, and user complaints. The engineer must not only record whether an asset is damaged, but also understand its significance within the network. A modestly damaged trunk outlet may be more urgent than a severely damaged but hydraulically minor side drain.

This stage is essential because rehabilitation decisions should be evidence-based. Without proper assessment, agencies may spend money on visible but less important defects while leaving major system bottlenecks unresolved.

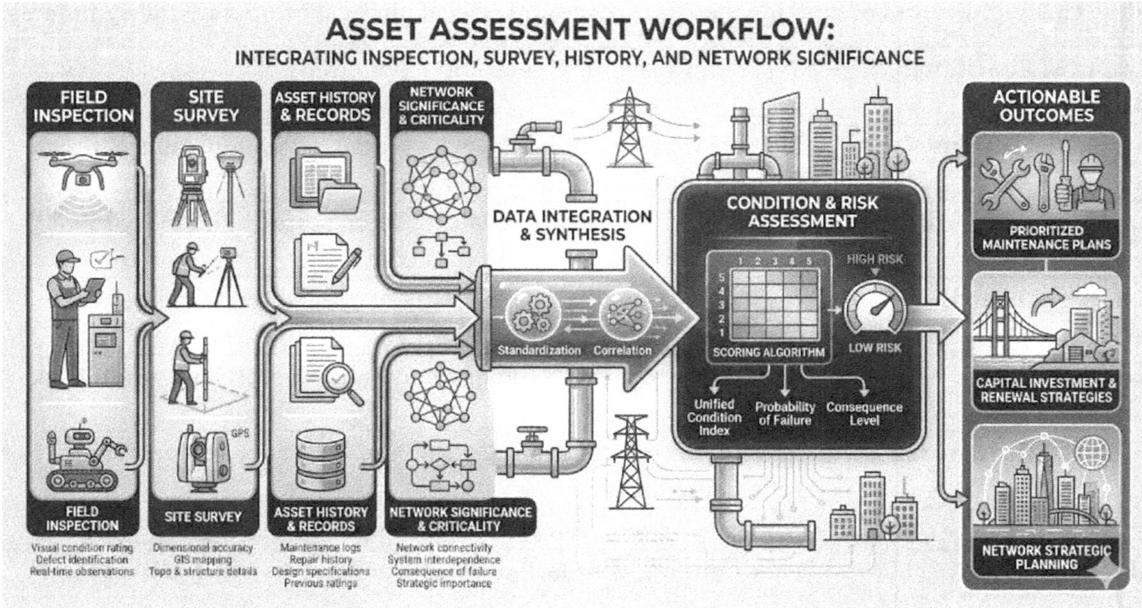

Figure 16.2 Asset assessment workflow combining inspection, survey, history, and network significance.

16.5 Condition Rating and Performance Evaluation

Once drainage assets are identified, they should be evaluated using both condition rating and performance assessment. These are related but distinct concepts. Condition refers to physical state, while performance refers to how well the asset functions within the drainage system.

A condition rating may classify an asset as good, fair, poor, or critical based on observed deterioration such as cracking, collapse, corrosion, joint failure, settlement, erosion, blockage, or material loss. Performance evaluation considers whether the asset still conveys flow adequately, supports maintenance access, avoids flooding, and connects properly to the larger system.

This distinction matters because a structurally poor drain may still perform hydraulically for a time, while a physically intact drain may fail functionally because it is undersized or disconnected from an effective outfall. Therefore, rehabilitation priorities should not be based only on visual damage.

A combined rating system helps decision-making. Assets can be prioritized based on both their physical condition and their service importance. This creates a more balanced basis for repair, replacement, or upgrading.

16.6 Rehabilitation of Deteriorated Structures

Rehabilitation focuses on restoring the performance of drainage assets that have deteriorated physically or operationally. The purpose is to bring the asset back to an acceptable level of service without necessarily changing its original role in the system.

Typical rehabilitation activities include repair of cracked masonry or concrete drains, relining of channels or covered drains, replacement of damaged slabs, covers, or frames, restoration of benching and invert channels in manholes, repair of headwalls, aprons, and culvert barrels, stabilization of eroded outfalls and drain banks, and removal of severe sediment deposits and accumulated obstructions.

Rehabilitation is often appropriate where the system layout is still broadly valid but the structure has deteriorated through age, poor maintenance, weathering, or repeated hydraulic stress. In rural and semi-urban systems, rehabilitation may also include reprofiling of open drains, re-establishment of blocked sections, and protection of vulnerable crossings.

The main aim is not merely cosmetic repair, but restoration of hydraulic and structural function. Therefore, repairs should always be linked to level continuity, flow condition, and future maintainability.

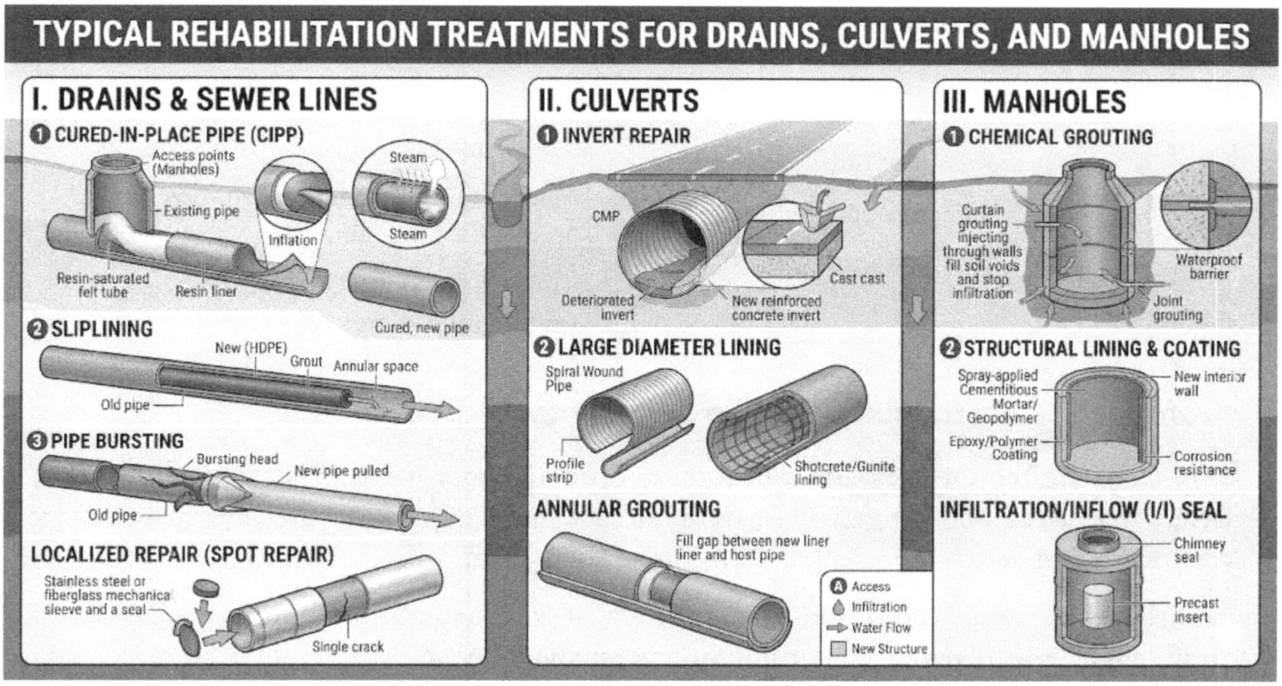

Figure 16.3 Typical rehabilitation treatments for drains, culverts, and manholes.

16.7 Retrofitting for Increased Capacity and New Functions

Retrofitting differs from rehabilitation in that it modifies the existing system to serve a changed or expanded function. It is used when the old drainage infrastructure is no longer adequate because the settlement, catchment, or required service level has changed.

Common retrofitting measures include widening or deepening existing drains, converting open drains into covered drains where appropriate, replacing undersized pipes or culverts, adding new branch drains, inlets, or relief channels, introducing detention, storage, or pumping elements, separating stormwater from inappropriate wastewater inflow, and creating overflow pathways for exceedance conditions.

Retrofitting may also involve adding features that were not part of the original design, such as silt traps, inspection openings, flap gates, smart sensors, or sustainable drainage elements. In urban areas, retrofitting is often constrained by utilities, road traffic, property boundaries, and construction disruption. In peri-urban areas, it may need to anticipate continued future growth.

Good retrofitting does not simply enlarge infrastructure blindly. It should solve the relevant performance problem while fitting within the physical and financial realities of the site.

Table 16.2 Comparison of rehabilitation and retrofitting in purpose, scope, and typical interventions.

Aspect	Rehabilitation	Retrofitting
Purpose	Restore existing facility or system to its original or near-original condition	Upgrade or strengthen existing facility/ system to meet new standards, hazards, or performance requirements
Scope	Focuses on repairing, renewing, or replacing deteriorated components	Focuses on enhancing resilience, safety, and functionality beyond original design
Typical Interventions	- Structural repairs (cracks, leaks, corrosion) - Replacement of worn-out parts - Restoration of finishes and utilities - Improving usability without major design changes	- Seismic strengthening (bracing, base isolation) - Energy efficiency upgrades (insulation, HVAC modernization) - Safety enhancements (fire protection, accessibility compliance) - Integration of modern technologies (automation, smart systems)

16.8 Retrofitting in Constrained Urban Environments

Urban retrofitting is often the most difficult form of drainage improvement because it must be carried out in already developed and congested corridors. Roads are occupied, utilities are crowded, frontage access is sensitive, and construction disruption can be socially and economically costly.

Typical urban constraints include insufficient right-of-way, underground utility conflicts, buildings close to drain edges, traffic management limitations, unknown conditions in older buried systems, and restricted access for machinery and maintenance.

As a result, urban retrofitting may require staged construction, temporary bypass drainage, night work, modular replacements, or selective intervention at bottlenecks rather than full corridor reconstruction. In some cases, even minor grade correction or inlet improvement can produce meaningful benefit where large-scale replacement is impractical.

The engineer must therefore balance ideal design against feasible intervention. In dense urban areas, the best retrofit is often the one that delivers reliable improvement with manageable disruption and realistic maintenance requirements.

16.9 Retrofitting in Semi-Urban and Rural Contexts

Retrofitting in semi-urban and rural settlements presents a different set of challenges and opportunities. There may be more space available than in a city, but the infrastructure is often more fragmented, less documented, and less formally managed.

In semi-urban areas, retrofitting often involves reconnecting discontinuous drains, enlarging culverts affected by land conversion, protecting natural drainage corridors, and improving outfalls that now receive higher runoff than before. In rural settlements, retrofitting may focus on restoring open drain continuity, replacing inadequate road crossings, rehabilitating pond outfalls, and improving slope and lining only where necessary.

Compared with dense urban conditions, such contexts may allow more flexible and lower-cost interventions. However, they also require careful attention to future change, community practices, agricultural drainage interaction, and institutional responsibility for maintenance.

A good retrofit in these settings should remain simple enough to be sustained locally, while still correcting the key hydraulic weakness of the older system.

16.10 Drainage Asset Management Concepts

Asset management is the organized, long-term management of infrastructure so that it continues to provide the required service at acceptable cost and risk. In drainage engineering, this means treating drains, culverts, pipes, chambers, inlets, outfalls, and related structures as assets with a service life, maintenance need, deterioration pattern, and replacement cycle.

The asset management approach includes identifying and inventorying assets, recording condition and performance, estimating remaining service life, planning maintenance and renewal, prioritizing expenditures, and reducing failure risk over time.

This approach differs from purely reactive management, where repairs are made only after visible failure or public complaint. A reactive system often leads to higher long-term costs, repeated emergencies, and inefficient spending. Asset management aims instead for planned, evidence-based, and lifecycle-conscious intervention.

The concept is important because drainage networks are extensive and interconnected. A small local failure can have large service impacts, and delayed maintenance can accelerate deterioration. Asset management helps organise this complexity.

16.11 Asset Inventory, Data Systems, and Lifecycle Planning

A functioning asset management system begins with a reliable asset inventory. This is a structured record of drainage components and their key attributes, such as location, dimensions, material, age, invert levels, condition, inspection history, and maintenance needs.

An effective inventory may include asset identification number, geographic location, type of structure, structural details and dimensions, year of construction or approximate age, condition rating, hydraulic importance, maintenance history, and rehabilitation or replacement priority.

When linked to GIS or digital databases, this information becomes a practical decision-support tool. It helps agencies schedule inspections, estimate budget needs, track deterioration, and coordinate drainage works with road or utility projects. In smaller systems, even a spreadsheet-based inventory can greatly improve continuity of management.

Lifecycle planning builds on this inventory by considering how assets deteriorate over time and when maintenance, rehabilitation, or replacement should occur. The goal is not to replace everything at once, but to intervene at the most appropriate time to preserve service and avoid sudden failure.

Table 16.3 Suggested fields in a drainage asset inventory and digital database.

Category	Suggested Fields	Purpose/Notes
Asset Identification	Asset ID, Name/Label, Location (GPS coordinates, address), Drainage Type (surface drain, culvert, storm sewer, etc.)	Ensures unique identification and geospatial referencing
Physical Attributes	Dimensions (length, width, depth), Material (concrete, brick, PVC, etc.), Age/Year of Construction, Condition Rating	Captures physical characteristics and baseline condition
Operational Attributes	Capacity (flow rate, volume), Connectivity (linked drains, outfalls), Function (primary/secondary), Usage Frequency	Defines operational role and performance
Maintenance Records	Inspection Dates, Maintenance History, Repairs/Upgrades, Responsible Agency	Tracks upkeep and accountability
Environmental/ Regulatory	Compliance Status, Flood Risk Zone, Environmental Sensitivity (wetlands, coastal, etc.), Permits/Approvals	Ensures alignment with regulations and environmental safeguards
Performance Indicators	Hydraulic Efficiency, Blockage Frequency, Overflow Incidents, Service Reliability	Provides measurable indicators of functionality
Digital Integration	GIS Layer Reference, Photographic Records, Sensor Data (IoT monitoring), Linked Documents (design drawings, reports)	Supports digital database integration and smart monitoring
Financial/ Administrative	Asset Value, Replacement Cost, Funding Source, Ownership/Stakeholder	Facilitates budgeting and asset management

16.12 Prioritisation of Renewal and Replacement

Not all drainage assets can be renewed or replaced immediately, so prioritisation is essential. Renewal planning should consider both the physical condition of the asset and the consequences of its failure.

Priority criteria may include severity and frequency of flooding linked to the asset, role of the asset in the broader drainage network, condition rating and risk of collapse, effect on critical roads, buildings, or public services, public health and sanitation implications, cost of continued patch repair versus replacement, and feasibility of intervention under current site constraints.

An old culvert on a major access road may deserve earlier replacement than several deteriorated but low-impact side drains. Similarly, a trunk outfall with repeated backwater failure may justify urgent investment even if its visible damage appears moderate.

Prioritisation should therefore be transparent and technically reasoned. This is especially important for public agencies managing large networks under limited budgets, where competition among drainage needs can be intense.

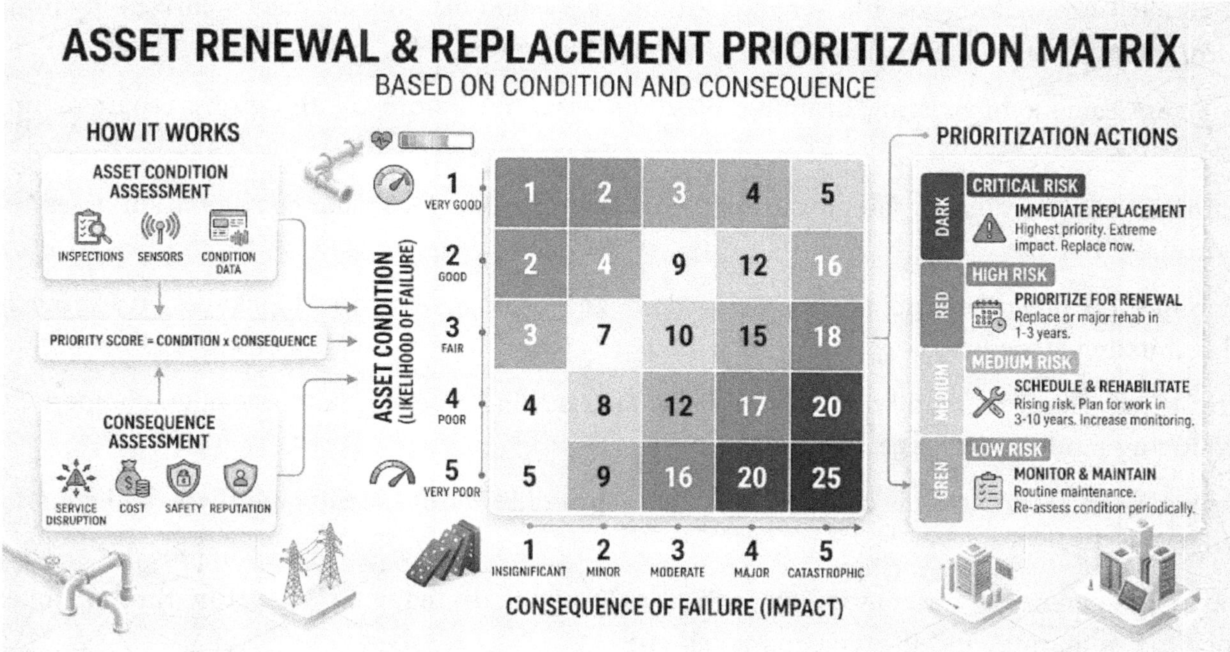

Figure 16.4 Renewal and replacement prioritisation matrix based on condition and consequence.

16.13 Financial and Institutional Aspects of Asset Management

Rehabilitation and asset management are not only technical tasks; they also depend on financial planning, institutional responsibility, and administrative continuity. A technically sound asset strategy cannot function if maintenance budgets are irregular, ownership is unclear, or drainage responsibilities are split among uncoordinated agencies.

Financial planning should distinguish among routine maintenance costs, periodic rehabilitation costs, emergency response costs, and long-term renewal and replacement needs. Without such distinction, agencies often overspend on emergency works while underfunding planned renewal. This leads to a cycle of repeated breakdown and reactive repair.

Institutionally, drainage asset management requires clear assignment of responsibility for inspection, maintenance, data updating, design approval, and capital improvement planning. Coordination is especially important where roads, drainage, sanitation, and urban development are managed by different departments. The better the institutional alignment, the more effective the asset management system becomes.

16.14 Summary Points

1. Drainage systems deteriorate over time due to factors such as aging, sedimentation, urbanization, and inadequate maintenance.
2. Rehabilitation is required to restore the original performance of damaged drainage systems, while retrofitting is needed to upgrade systems for new conditions.

3. Functional obsolescence occurs when a drainage system no longer meets current hydraulic or environmental requirements despite being physically present.
4. Proper assessment of existing drainage assets is essential before planning any repair or upgrade work.
5. Condition rating evaluates the physical state of assets, while performance evaluation assesses their functional efficiency.
6. Rehabilitation measures include repairing damaged drains, removing blockages, restoring levels, and stabilizing structures.
7. Retrofitting involves modifying systems by increasing capacity, adding new components, or introducing modern drainage features.
8. Retrofitting in urban areas is challenging due to space limitations, utility conflicts, and construction disruptions.
9. Asset management is a systematic approach to maintaining drainage infrastructure through inventory, inspection, planning, and lifecycle management.
10. Prioritization of repair and replacement should be based on condition, risk, importance, and available resources for effective long-term performance.

16.15 References

1. Ministry of Housing and Urban Affairs. (n.d.). *Manual on storm water drainage systems (Vol. II: Operation and maintenance)*. Government of India.
2. Ministry of Housing and Urban Affairs. (n.d.). *Manual on storm water drainage systems (Vol. I: Engineering)*. Government of India.
3. Engineers Australia, Institute of Public Works Engineering Australia, & Stormwater Western Australia. (n.d.). *Code of practice: Urban and peri-urban drainage modelling*.
4. Wisconsin Department of Transportation, & Midwest Regional University Transportation Center. (n.d.). *An asset management approach for drainage infrastructure and culverts*.
5. American Association of State Highway and Transportation Officials. (n.d.). *Culvert and storm drain system inspection guide* (1st ed.).
6. Yass Valley Council. (n.d.). *Stormwater drainage asset management plan*.
7. Transportation Asset Management Council. (n.d.). *Culvert asset management best practices*.
8. Bahrami, M. (n.d.). *Asset management of urban drainage systems*.
9. Center for Watershed Protection. (n.d.). *Urban stormwater retrofit manual*.

Chapter 17

Environmental, Health, and Social Aspects of Drainage

Learning Objectives

By the end of this chapter, the learner should be able to:

1. Explain why drainage should be understood as environmental infrastructure rather than only as a water-conveyance system.
2. Identify major pollution pathways associated with drainage flows in urban, semi-urban, and village settings.
3. Describe the public health impacts of stagnant water, contamination, and poor drainage conditions.
4. Discuss the social consequences of flooding, waterlogging, inequity, and vulnerability in drainage planning.
5. Integrate environmental, health, and social considerations into drainage design, maintenance, and management.

17.1 Introduction

Drainage engineering is not only a matter of hydraulics and construction. It also has major environmental, public health, and social consequences. A drainage system influences water quality, vector breeding, sanitation conditions, land use, road accessibility, neighborhood safety, and the overall quality of life in a settlement. For this reason, drainage planning and design must be understood as part of environmental infrastructure and community well-being, not merely as a technical water-conveyance function.

This chapter examines the broader impacts of drainage systems in urban, semi-urban, and village settings. It discusses environmental effects, health risks associated with poor drainage, social implications of flooding and waterlogging, community behavior, equity concerns, and the need for socially acceptable and environmentally responsible drainage planning.

17.2 Drainage as Environmental Infrastructure

Drainage is a form of environmental infrastructure because it directly affects how water, sediment, waste, and pollutants move through settlements and into the surrounding environment. A properly functioning drainage system helps maintain cleaner living conditions, protects natural water bodies, reduces uncontrolled erosion, and supports healthier urban and rural landscapes.

When drainage is poorly planned or neglected, environmental problems quickly emerge. Runoff may carry sediment, litter, wastewater, oil, nutrients, and other pollutants into ponds, canals, rivers, wetlands, and low-lying areas. Open drains may become stagnant and act as informal waste channels rather than stormwater pathways. This changes drainage from a protective system into a source of environmental degradation.

The environmental role of drainage is therefore twofold. First, it manages excess water. Second, it influences the condition of receiving water bodies and surrounding land. A drainage system should thus be evaluated not only by how fast it removes runoff, but also by how responsibly it interacts with the local environment.

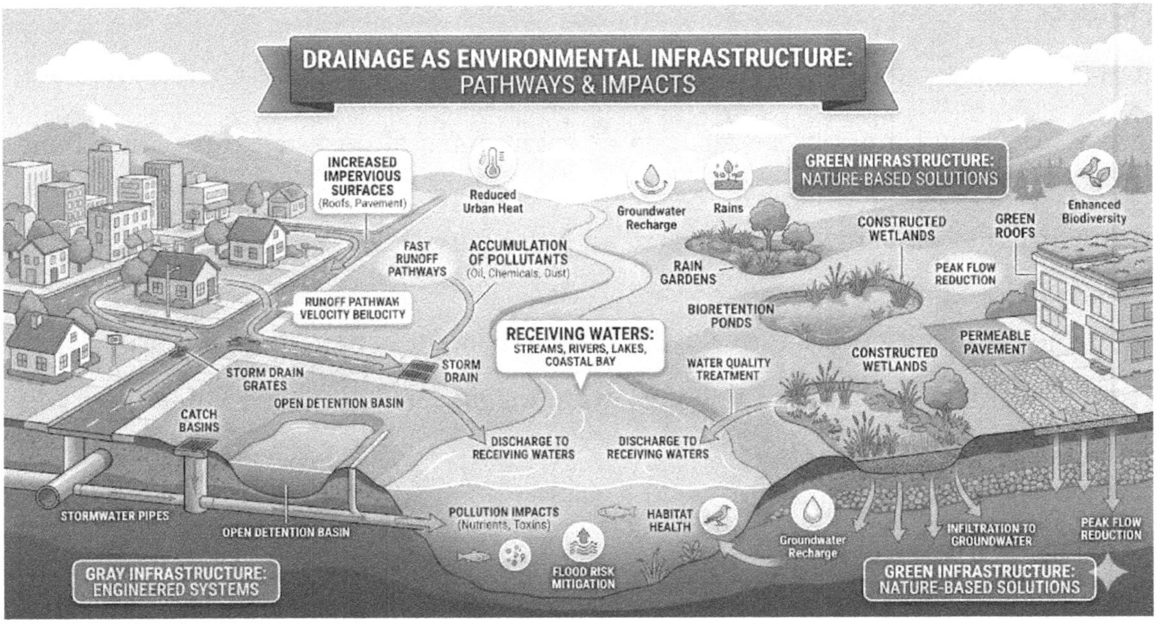

Figure 17.1 Drainage as environmental infrastructure showing runoff pathways, receiving waters, and surrounding settlement impacts.

17.2 Water Quality and Pollution Pathways

Settlement drainage systems often act as transport pathways for pollutants. Rainfall runoff can wash a wide variety of materials from roads, roofs, open yards, markets, construction sites, industrial areas, and waste disposal points into drains and receiving waters.

Common pollutants in drainage flows include suspended sediment, organic matter, household waste, oils and grease from roads, nutrients from wastewater and livestock areas, pathogens from mixed sewage or greywater inflow, and chemicals from workshops, small industries, and urban surfaces.

In many settlements, especially where sanitation systems are incomplete, stormwater drains also receive sullage, greywater, or even direct sewage discharge. This creates a serious water-quality problem because drainage channels become contaminated throughout the year, not only during storms. In village areas, similar issues arise where household wastewater is discharged directly into open roadside drains or where animal waste enters flow paths.

Understanding pollution pathways is important because drainage design decisions affect whether contaminants are dispersed, trapped, diluted, or discharged untreated into sensitive areas. Hydraulic efficiency alone is therefore not a sufficient design criterion; water-quality implications also matter.

Table 17.1 Common pollutants in drainage runoff, likely sources, and environmental consequences.

Pollutant Category	Common Examples	Likely Sources	Environmental Consequences
Sediment	Dirt, sand, silt, clay particles	Construction sites, bare lawns/gardens, eroding streambanks, agricultural fields, road sanding.	Smothers aquatic habitats and fish eggs; clogs fish gills; increases water turbidity (cloudiness), blocking sunlight needed by aquatic plants; transports attached pollutants like phosphorus and metals.
Nutrients	Nitrogen, Phosphorus	Over-applied lawn fertilizers, pet waste, decomposing grass clippings and leaves, failing septic systems, livestock manure.	Causes **eutrophication**: fuels rapid algae blooms that deplete dissolved oxygen when they decompose, leading to fish kills ("dead zones"); some algae blooms can be toxic to humans and animals.
Pathogens	Bacteria (e.g., *E. coli*), Viruses, Protozoa	Pet waste left on ground, urban wildlife (birds, raccoons), leaking sewer lines, failing septic systems, combined sewer overflows (CSOs).	Contaminates water bodies used for recreation (swimming, fishing), leading to beach closures; poses severe health risks to humans if ingested through drinking water or recreational contact.
Heavy Metals	Lead, Copper, Zinc, Cadmium, Nickel	Vehicle brake dust (copper), tire wear (zinc), engine exhaust, weathering of galvanized metal (gutters, roofs), industrial emissions, scrap piles.	Toxic to aquatic life, even at low concentrations; can bioaccumulate in the food chain, posing risks to fish, birds, and humans who consume contaminated seafood; damages the reproductive and nervous systems of aquatic organisms.
Hydrocarbons & Organics	Oil, grease, gasoline, Polycyclic Aromatic Hydrocarbons (PAHs)	Leaking vehicles, fuel spills, engine exhaust, asphalt pavement wear, illegal dumping down storm drains.	Creates oil sheens on water surface that block oxygen transfer and coat aquatic organisms; many components (like PAHs) are toxic and/or carcinogenic to aquatic life and can persist in sediments for long periods.
Pesticides & Herbicides	Insecticides, weed killers	Over-application or improper disposal of lawn and garden chemicals, agricultural runoff, commercial landscaping.	Toxic to non-target aquatic organisms, including beneficial insects, fish, and amphibians; can disrupt aquatic ecosystems and contaminate groundwater aquifers used for drinking water.

(*Contd.*)

Pollutant Category	Common Examples	Likely Sources	Environmental Consequences
Litter & Trash	Plastic bags, bottles, cigarette butts, food wrappers	Careless littering, windblown garbage from containers or landfills, intentional dumping.	Physical hazard to wildlife (entanglement, ingestion); breaks down into microplastics, which enter the food chain; clogs storm drain infrastructure, increasing flooding risk; degrades aesthetics of waterways.
De-icing Salts	Sodium chloride, calcium chloride	Road, sidewalk, and driveway salting during winter months.	High chloride levels are toxic to freshwater aquatic life; contaminates groundwater aquifers; damages road infrastructure, vehicles, and roadside vegetation; can alter soil structure.
Thermal Pollution	Heated water	Runoff flowing over hot impervious surfaces (asphalt, concrete, roofs) on sunny days.	Increases temperature of receiving water bodies, which lowers dissolved oxygen levels; causes thermal shock to cold-water fish species (e.g., trout, salmon), potentially killing them or disrupting their reproductive cycles.

17.3 Stagnant Water and Vector-Borne Disease

One of the most serious public health consequences of poor drainage is the creation of stagnant water. When drains are blocked, slopes are inadequate, or local depressions remain waterlogged, shallow standing water can persist for days or weeks after rainfall.

Stagnant water creates favorable breeding conditions for mosquitoes and other disease vectors. In many regions, this increases the risk of vector-borne illness and contributes to persistent public-health stress in low-lying and poorly serviced neighborhoods. The problem becomes more severe where drainage channels also contain organic waste or wastewater, which further degrades the local environment.

Beyond disease transmission, stagnant drains produce foul odor, unpleasant living conditions, and reduced use of nearby streets and public spaces. Children may play near polluted standing water, and residents may be forced to walk through contaminated wet areas during daily activities. Thus, poor drainage affects both direct health risk and everyday human dignity.

From a preventive perspective, drainage engineering is therefore part of public-health infrastructure. Reducing water stagnation is a practical way to reduce exposure to environmental health hazards.

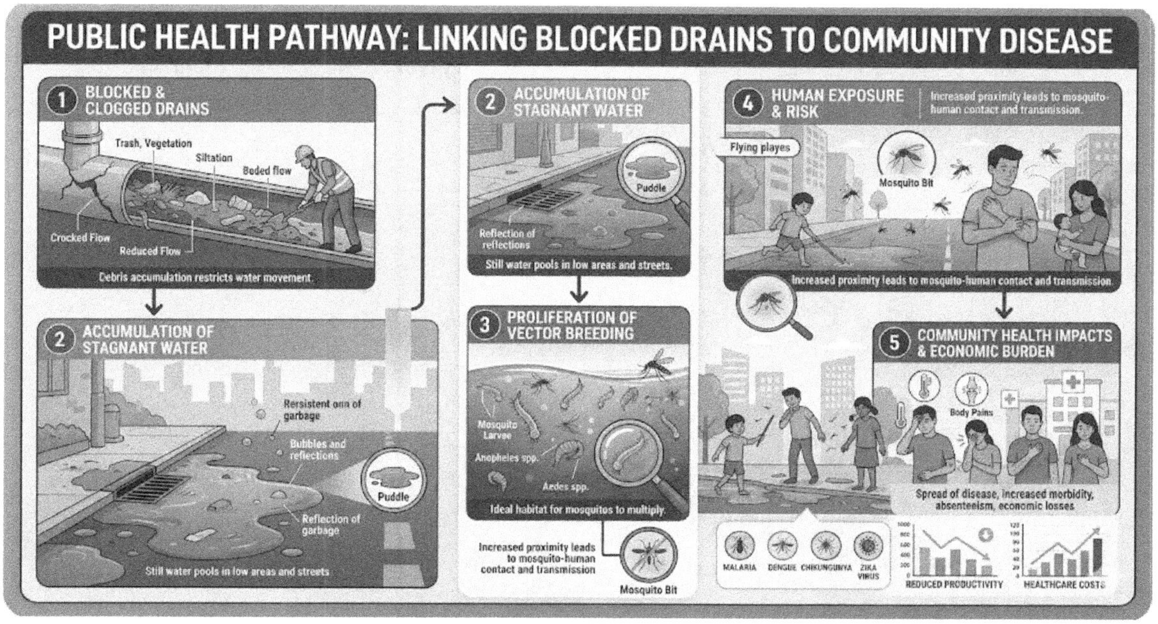

Figure 17.2 about here: Public-health pathway linking blocked drains, stagnant water, vector breeding, exposure, and community health impacts.

17.4 Drainage and Sanitation Linkages

Drainage and sanitation are closely linked, especially in settlements where infrastructure is incomplete or poorly coordinated. Stormwater drains are often misused as channels for household wastewater, sullage, septic overflow, or even direct sewage discharge. This mixing undermines both drainage and sanitation performance.

When wastewater enters storm drains, several problems arise. Drain capacity is reduced by continuous flow and sludge accumulation, foul odor develops, public exposure to pathogens increases, drains remain wet even in dry weather, maintenance becomes more difficult and hazardous, and receiving water bodies become polluted.

In dense urban areas, illegal sewer or household connections to storm drains are common. In villages and peri-urban areas, open drains often serve both stormwater and greywater functions by default because no separate wastewater system exists. While this may appear practical in the short term, it creates long-term environmental and health burdens.

Proper drainage planning should therefore be coordinated with sanitation planning. Even where separate systems are not yet fully available, the engineer should recognize the interaction and design for safe management, maintenance access, and gradual improvement.

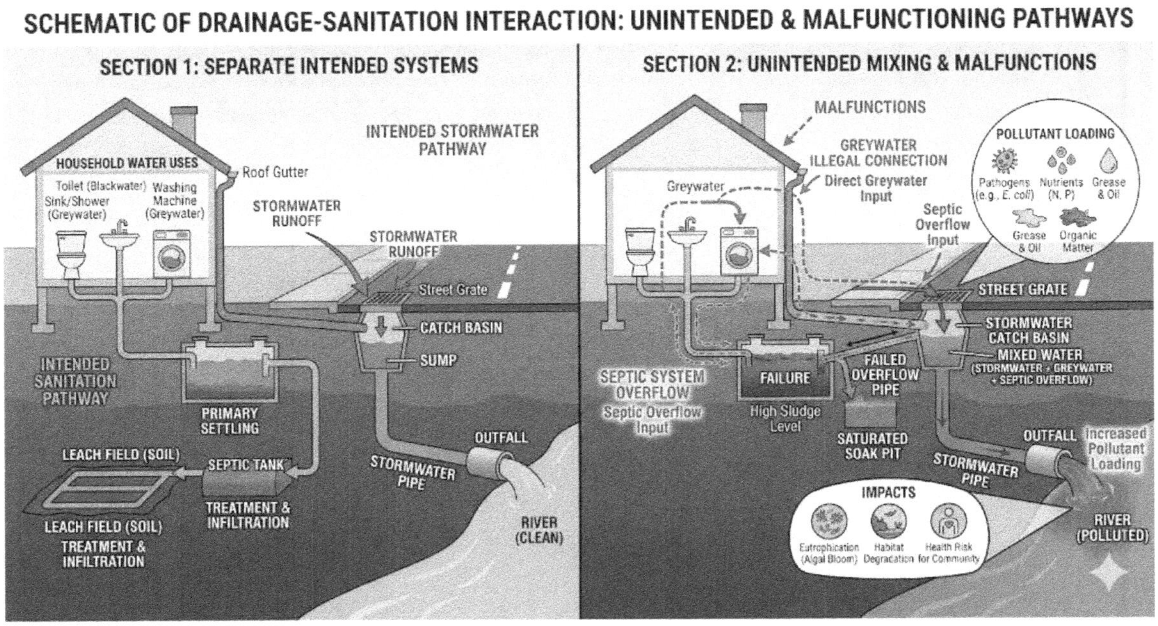

Figure 17.3 Schematic of drainage-sanitation interaction showing stormwater drains receiving greywater, septic overflow, and pollutant loading.

17.5 Impacts on Settlement Livability

Drainage strongly affects how livable a settlement feels and functions. A well-drained area is cleaner, safer, more accessible, and more comfortable during and after rainfall. A poorly drained area is often associated with muddy streets, foul smells, stagnant pools, damaged roads, mosquito nuisance, and repeated inconvenience.

Livability impacts are especially important in dense residential neighborhoods, low-income settlements, village lanes, schools, market areas, and transport corridors. Even moderate ponding near homes or public spaces can discourage movement, damage property access, and reduce the usability of outdoor areas. In flood-prone settlements, these effects become part of everyday life and can shape how people perceive safety and well-being.

Drainage also influences aesthetics and community pride. Open drains carrying waste or stagnant water create a visual and environmental burden, while well-maintained drainage corridors can support cleaner surroundings and better public use of space. Drainage is therefore not only a hidden engineering service; it is also a visible part of settlement quality.

17.6 Social Impacts of Flooding and Waterlogging

Flooding and prolonged waterlogging have significant social consequences beyond physical damage. They disrupt mobility, interrupt schooling, reduce access to work and health services, damage household goods, and place daily stress on affected communities. In low-income and vulnerable settlements, even shallow flooding can have severe social effects because people have fewer resources for recovery.

Waterlogging can isolate homes, block village streets, flood courtyards, and make public spaces unusable. Repeated drainage failure may also influence settlement behavior, with residents raising plinths, building informal barriers, or altering drains in self-protective ways that sometimes worsen conditions for others. Social conflict may emerge where drainage improvements benefit one area while shifting water toward another.

Women, children, elderly residents, and persons with limited mobility often experience drainage problems differently and sometimes more severely. Access to toilets, schools, water points, markets, or health centers may become more difficult during wet periods. The social impact of drainage failure is therefore not evenly distributed across the community.

Engineering planning should recognize these social dimensions. Drainage is not only about water control; it is also about protecting routine life, access, and human well-being.

Table 17.2 Social effects of flooding and waterlogging on mobility, health access, education, livelihoods, and vulnerable groups.

Social Aspect	Key Social Effects of Flooding & Waterlogging	Specific Examples and Details
1. Mobility & Transportation	Severed connectivity and daily disruption.	**Submerged Roads & Bridges:** Routes become impassable for vehicles and pedestrians, paralyzing commute. **Transit Collapse:** Closure of bus, train, or subway stations; cancellation of public transport services. **Travel Delays:** Massive congestion on remaining routes increases commute times. **Isolated Communities:** Areas become effectively cut off ("islands"), preventing the movement of people, goods, and emergency services. **Risk to Travelers:** Drowning risks while walking or driving through floodwater; concealed hazards (open manholes, debris).

(Contd.)

Social Aspect	Key Social Effects of Flooding & Waterlogging	Specific Examples and Details
2. Health Access & Services	Healthcare system strain and compromised well-being.	**Facility Closure/Damage:** Hospitals, clinics, and pharmacies may be flooded, structurally damaged, or inaccessible. **Reduced Access:** Patients are unable to travel for routine care, emergency services, or chronic disease management. **Service Disruption:** Power outages affect refrigeration of medicines (vaccines, insulin) and operation of medical equipment. **Post-Flood Health Surge:** Increased demand for treatment of injuries (cuts, fractures), waterborne diseases (cholera, typhoid), skin infections, and vector-borne diseases (malaria, dengue). **Mental Health Crisis:** Elevated rates of stress, anxiety, depression, grief, and PTSD due to loss, displacement, and fear.
3. Education & Learning	Interruption of human capital development.	**School Closures:** Educational facilities closed due to flooding, damage, or use as emergency shelters. **Attendance Collapse:** Students and teachers are unable to reach school due to transportation failure. **Infrastructure Loss:** Damage to school buildings, classrooms, library books, computers, and supplies. **Disrupted Cycles:** Cancellation of exams; inability to complete the academic year. **Reduced Learning Outcomes:** Decreased enrollment; higher dropout rates, especially among marginalized students; long-term slower grade progression.

Social Aspect	Key Social Effects of Flooding & Waterlogging	Specific Examples and Details
4. Livelihoods & Economics	Erosion of household income and economic stability.	**Workforce Absence:** Employees are unable to reach work due to mobility failure or need to manage home damage. **Business Stoppage:** Flooded storefronts, factories, and warehouses force temporary or permanent closure. **Agricultural Ruin:** Destruction of standing crops; loss of livestock; soil degradation; breakdown of supply chains. **Asset Loss:** Damage to tools, machinery, vehicles, and commercial inventory. **Increased Expenditures:** Spiraling costs for repairs, temporary housing, and basic necessities on reduced income.
5. Vulnerable Groups	Exacerbation of existing inequalities.	**Elderly & Disabled:** Face higher risk of injury/ death due to limited mobility and communication challenges during evacuation; disruption of home care. **Low-Income Households:** Suffer greater relative asset loss; lack financial buffers for recovery or insurance; face housing insecurity (displacement). **Women & Girls:** May face increased domestic workload (clean-up, fetching water); heightened risk of gender-based violence in crowded shelters; disproportionately negative education outcomes for girls. **Children:** High risk of drowning and malnutrition; psychological trauma; increased behavioral problems due to disruption of home and school life.

17.7 Equity and Vulnerability in Drainage Planning

Drainage problems often affect vulnerable populations more severely than others. Low-income neighborhoods, informal settlements, low-lying village margins, roadside habitations, and poorly serviced peri-urban zones frequently experience the worst flooding and waterlogging, even though they often receive the least drainage investment.

This raises an issue of equity in drainage planning. If infrastructure upgrades focus only on high-visibility or high-value areas, existing inequalities may deepen. Floodwater may be diverted away from central roads while still accumulating in marginal settlements. Similarly, lack of maintenance in less powerful neighborhoods may produce chronic health and environmental burdens.

A socially responsible drainage strategy should consider who is exposed to repeated drainage failure, who lacks resources to adapt independently, whose access to services is most affected, and whether proposed interventions shift risk from one group to another.

Vulnerability is shaped not only by topography but also by income, tenure status, housing quality, mobility, and institutional attention. A fair drainage plan should reduce risk in a balanced way and pay particular attention to communities least able to protect themselves.

Figure 17.4 Conceptual vulnerability map showing unequal drainage risk across settlement types and population groups.

17.8 Community Behavior, Participation, and Awareness

Drainage performance is influenced not only by engineering design but also by how people interact with the system. Community behavior affects blockage, maintenance, protection of outfalls, waste disposal, and the long-term success of drainage improvements.

Negative practices may include dumping waste into drains, obstructing channels for private access, connecting wastewater illegally, encroaching on drainage corridors, or neglecting local cleaning responsibilities. Positive practices include reporting blockages, protecting outfall channels, maintaining house-front drains, participating in cleanup campaigns, and supporting proper use of drainage structures.

Community participation is especially important in village, semi-urban, and low-income urban settings where formal maintenance capacity may be limited. Local knowledge can help identify flood-prone areas, recurring bottlenecks, and seasonal problem patterns that may not appear in technical records alone.

Awareness matters because many drainage failures are partly behavioral. Public education on keeping drains clear, protecting natural flow paths, and avoiding misuse of stormwater channels can improve system performance significantly when combined with engineering measures.

17.9 Environmental and Social Considerations in Design

Environmental and social considerations should be integrated into drainage design from the beginning rather than treated as secondary concerns. A technically efficient drain may still be a poor solution if it creates downstream pollution, disrupts community access, becomes unsafe for children, or damages a valued local water body.

Design should therefore consider water-quality protection, prevention of stagnation, safe access across open drains, compatibility with local sanitation conditions, erosion and sediment control, protection of ponds, wetlands, and natural drains, community use of public space, and maintenance feasibility under local conditions.

In some cases, environmental and social improvements can be combined. For example, a well-designed vegetated drainage corridor may improve conveyance, reduce erosion, and create a more acceptable public edge than an unprotected waste-filled ditch. Likewise, covered crossing slabs may improve safety and access without blocking maintenance if detailed properly.

Good drainage design is therefore multidisciplinary in effect. It should serve hydraulic, environmental, and social objectives together as far as possible.

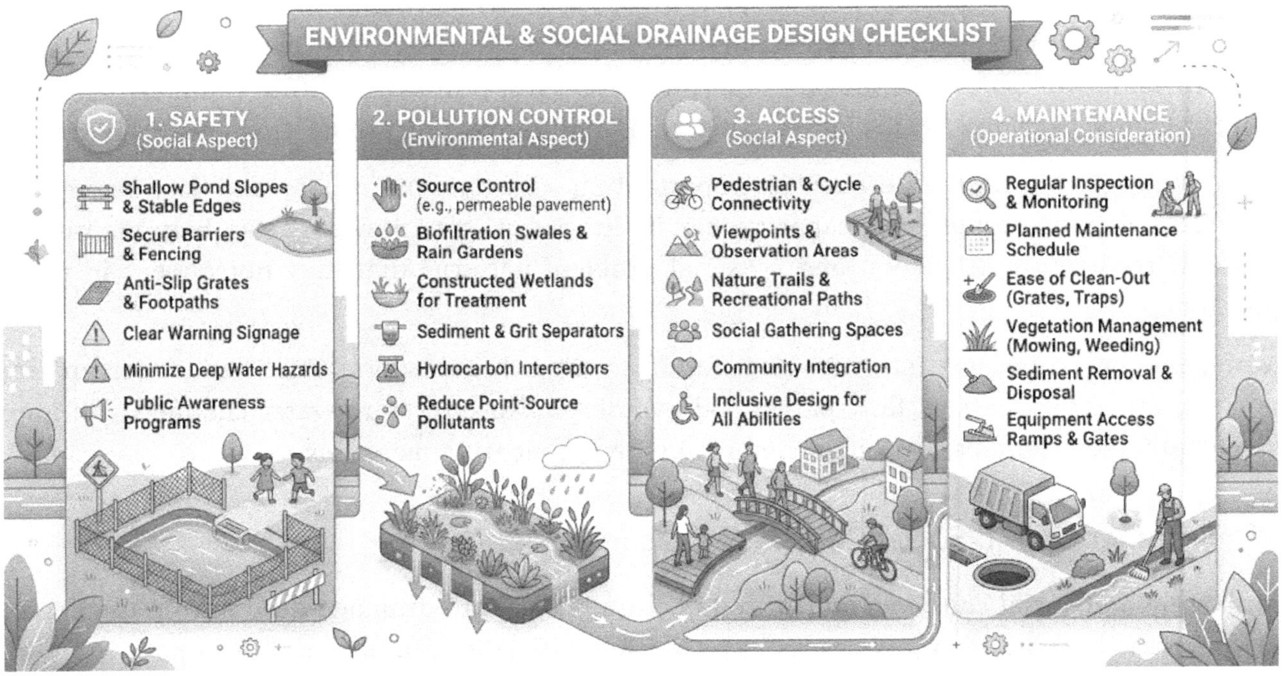

Figure 17.5 Environmental and social drainage design checklist including safety, pollution control, access, and maintenance considerations.

17.10 Summary Points

1. Drainage systems are not only hydraulic structures but also important environmental infrastructure that affect water quality and surrounding ecosystems.
2. Poor drainage can lead to environmental pollution by carrying sediments, waste, chemicals, and pathogens into natural water bodies.
3. Stormwater drains often become polluted due to mixing with wastewater, greywater, and sewage, especially in poorly planned areas.
4. Stagnant water caused by blocked or poorly designed drainage systems creates breeding grounds for mosquitoes and increases disease risk.
5. Drainage and sanitation are closely linked, and improper coordination can lead to serious health and environmental problems.
6. Good drainage improves settlement liveability by keeping areas clean, safe, accessible, and free from waterlogging.
7. Flooding and waterlogging have major social impacts, including disruption of transport, education, health services, and livelihoods.
8. Vulnerable groups such as low-income communities, children, and the elderly are more severely affected by poor drainage conditions.

9. Community behavior, including waste disposal and maintenance practices, plays a significant role in drainage system performance.
10. Effective drainage design must integrate environmental protection, public health, and social equity considerations along with hydraulic efficiency.

17.11 References

1. Arnold, C. L., Jr., & Gibbons, C. J. (n.d.). *Impervious surface coverage: The emergence of a key environmental indicator.*
2. Food and Agriculture Organization of the United Nations. (n.d.). *Health issues related to drainage water management.*
3. Hu, J. (n.d.). *Effects of urban stormwater pollution on watershed diffuse pollution and river water quality.*
4. Modeling the impacts of urban flood risk management on social inequality. (n.d.).
5. Objectives. This study investigated the scale of the public health risk from stormwater runoff caused by urbanization. Public health study available via PubMed Central. (n.d.).
6. Does urban green infrastructure lead to equity issues for flood vulnerable areas? (n.d.).
7. Stormwater strategies: Community responses to runoff pollution. (n.d.).
8. Subramanian, R. (n.d.). *Problems and solutions for managing urban stormwater runoff.*
9. United States Environmental Protection Agency. (n.d.). *Urban stormwater management in the United States.*
10. World Health Organization. (n.d.). *Water and sanitation interventions to prevent and control mosquito-borne diseases.*

Chapter 18

Climate Resilience and Future Directions in Settlement Drainage Engineering

Learning Objectives

By the end of this chapter, the learner should be able to:

1. Explain the need for climate-resilient drainage systems.
2. Describe how climate variability and rainfall extremes affect drainage design and performance.
3. Distinguish between conventional capacity-based design and resilience-based drainage thinking.
4. Explain the concept of adaptive drainage planning.
5. Identify the role of blue-green and nature-based infrastructure in future drainage systems.
6. Discuss the applications of smart drainage and digital monitoring.
7. Explain the idea of integrated urban water thinking in drainage engineering.
8. Compare resilience strategies for urban, semi-urban, and village drainage systems.
9. Identify major future research and professional directions in settlement drainage engineering.

18.1 Introduction

Settlement drainage engineering is entering a period of major change. Traditional drainage systems were often designed on the assumption that past rainfall patterns, land use conditions, and infrastructure demands would remain broadly stable. Today, that assumption is increasingly unreliable. Climate variability, urban expansion, land transformation, environmental stress, digital technology, and rising service expectations are reshaping how drainage systems must be planned, designed, operated, and upgraded.

This chapter examines the need for climate-resilient drainage systems and explores the major future directions of settlement drainage engineering. It focuses on uncertainty in rainfall and runoff, resilience-based planning, adaptive design, smart monitoring, nature-based solutions, integrated infrastructure, and emerging professional and research trends.

18.2 Need for Climate-Resilient Drainage

Drainage systems are directly exposed to climate-related changes because rainfall is their primary loading condition. If rainfall intensity, seasonal distribution, storm sequencing, or flood frequency changes, the performance of drains, culverts, inlets, and outfalls may change as well. Systems designed only for historical averages may therefore become less reliable over time.

Climate resilience in drainage means the ability of the system to continue functioning acceptably under changing, uncertain, and potentially more extreme conditions. This does not imply that all flooding can be eliminated. Rather, it means designing and managing drainage so that failure is less frequent, consequences are reduced, and recovery is faster when extreme events occur.

The need for climate resilience is especially strong in urban, peri-urban, and low-lying settlements where drainage systems already operate near capacity. In such areas, even modest increases in peak rainfall intensity or downstream water level can turn recurring nuisance flooding into serious infrastructure and public-health problems.

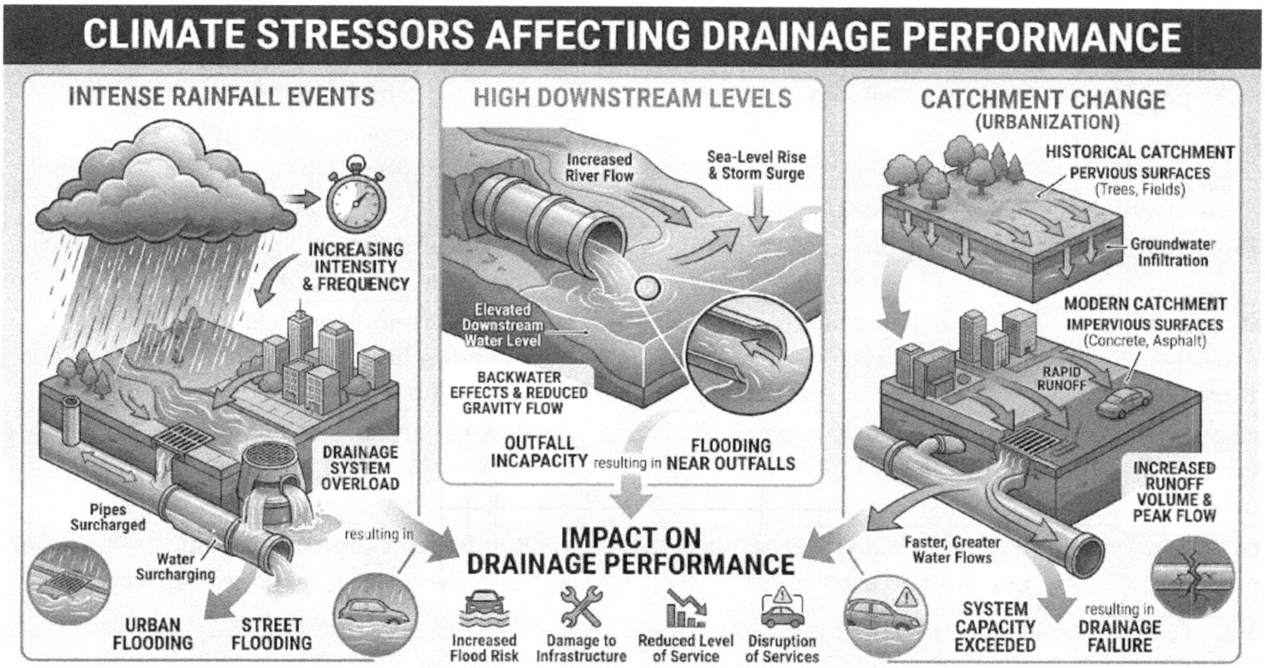

Figure 18.1 Climate stressors affecting drainage performance, including intense rainfall, high downstream levels, and catchment change.

18.3 Climate Variability and Rainfall Extremes

Climate variability affects drainage most directly through changes in rainfall behavior. These changes may include more intense short-duration storms, shifts in seasonal rainfall concentration, longer dry periods followed by heavy rainfall, increased variability between years, and more uncertain design conditions based on historical records.

For drainage engineering, short-duration extreme rainfall is often the most critical factor because local drains and urban systems respond rapidly to intense storms. A drainage network that previously handled seasonal rainfall adequately may begin to fail more often if rainfall peaks become sharper, even if annual rainfall does not increase substantially.

Another issue is uncertainty. Engineers increasingly face conditions in which the past is no longer a fully reliable guide to the future. This does not make hydrologic design impossible, but it means that design must allow for a wider range of plausible conditions and avoid excessive dependence on a single historical assumption.

Table 18.1 Examples of climate-related changes and their likely implications for drainage design.

Climate-Related Change	Description / Trend	Likely Implications for Drainage Design
Increased Rainfall Intensity	More frequent short-duration, high-intensity storms	Larger pipe diameters, enhanced inlet capacity, increased detention storage, and improved overflow routing
Extended Dry Periods	Longer intervals between rainfall events	Need for sediment flushing systems, maintenance of dry-weather flow, prevention of odor and stagnation
Sea-Level Rise	Rising coastal and estuarine water levels	Elevated outfall structures, backflow prevention valves, tidal gates, and coastal pump stations
Urban Heat Island Effect	Higher surface temperatures and evapotranspiration	Integration of green infrastructure (permeable pavements, vegetated swales) to reduce runoff and heat stress
Catchment Land-Use Change	Urban expansion, deforestation, and impervious surface increase	Revised runoff coefficients, expanded retention basins, and adaptive zoning for floodplain management
Extreme Weather Variability	Unpredictable rainfall patterns and storm surges	Flexible design standards, modular drainage components, and real-time monitoring systems
Groundwater Level Fluctuations	Rising or falling water tables due to climate and extraction	Reinforced sub-surface drainage, improved infiltration design, and groundwater recharge integration
Temperature Rise & Evaporation	Increased evaporation rates and soil moisture loss	Enhanced infiltration zones, water reuse systems, and drought-resilient landscaping around drains

18.4 Resilience Versus Conventional Capacity Design

Traditional drainage design often emphasizes capacity, meaning the ability to convey a selected design discharge without overtopping or surcharge. Capacity remains important, but resilience introduces a broader perspective. A resilient drainage system is not judged only by whether it avoids failure under one design event, but also by how it behaves when exceeded, how quickly it recovers, and how well it limits damage.

The difference can be understood in simple terms: capacity-based design asks, "How much flow can the system carry?" Resilience-based design asks, "How will the system perform under stress, exceedance, disruption, and recovery?"

A system with large capacity but poor recovery options, no overflow routing, and weak maintenance may still be fragile. By contrast, a system with moderate capacity but protected flood pathways, storage, redundancy, and adaptable components may perform better under uncertain future conditions.

This does not mean abandoning conventional hydraulic design. Rather, it means extending it. Resilience complements capacity by considering uncertainty, exceedance behavior, operational reliability, and long-term adaptability.

18.5 Adaptive Drainage Planning

Adaptive drainage planning is an approach in which drainage systems are designed and managed so they can be adjusted over time as conditions change. Instead of assuming that one final design will remain adequate for decades without modification, adaptive planning recognizes that rainfall, land use, and service requirements may evolve.

An adaptive strategy may include phased development of drainage infrastructure, protection of future drainage corridors, provision for later widening or deepening of major drains, modular storage systems, reserved sites for future detention or pumping, and updating of design standards as new data emerge.

This approach is particularly useful in expanding peri-urban areas and rapidly developing settlements, where present conditions may differ greatly from future ones. It is also useful where climate uncertainty makes it difficult to justify a single fixed design assumption with complete confidence.

Adaptive planning requires institutions to think beyond one-time construction. It links present-day design choices with future decision points, making drainage more flexible and less vulnerable to becoming obsolete too quickly.

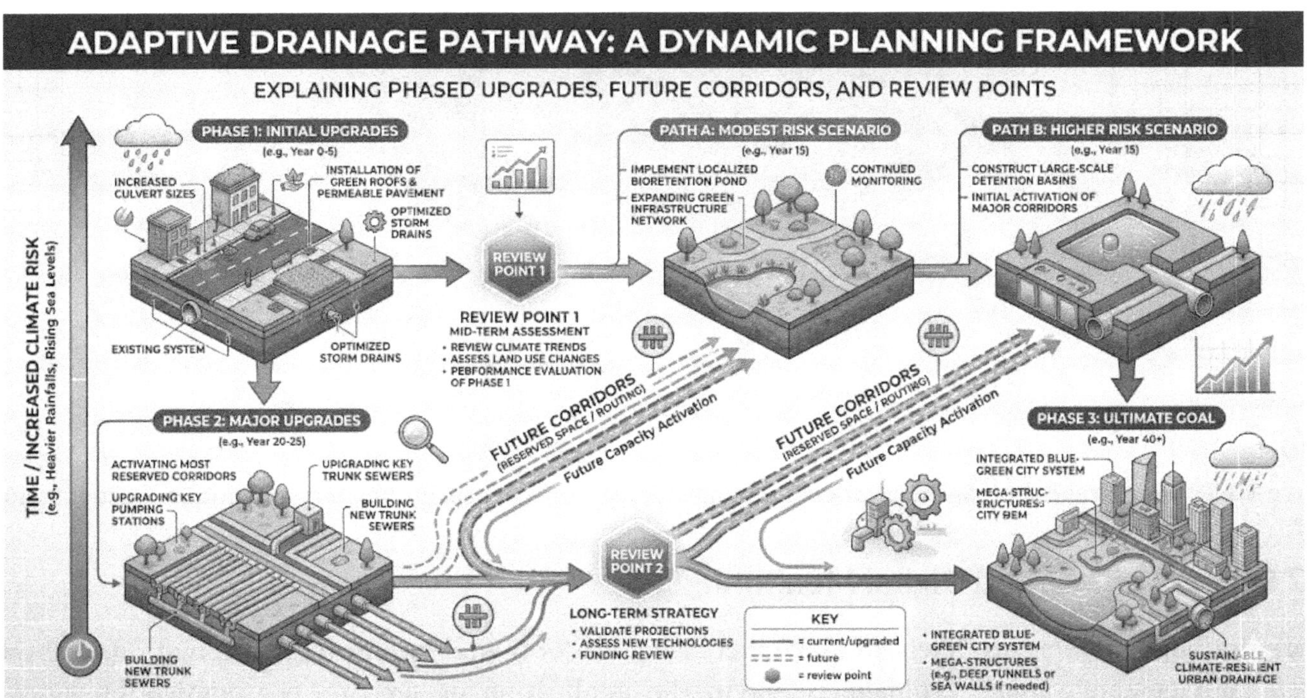

Figure 18.2 Adaptive drainage pathway showing phased upgrades, future corridors, and review points.

18.6 Blue-Green and Nature-Based Drainage Futures

One of the most important future directions in drainage engineering is the growing role of blue-green infrastructure and other nature-based approaches. These systems integrate water management with vegetation, landscape, open space, and ecological processes rather than relying entirely on buried or heavily engineered conveyance.

Examples include bioswales and vegetated channels, retention ponds and restored urban water bodies, rain gardens and bioretention cells, green streets and permeable surfaces, floodable parks and open detention spaces, and wetland-based drainage buffers.

The value of blue-green approaches lies in their multiple benefits. They can reduce peak runoff, improve infiltration, support water quality management, lower heat stress, enhance public space, and increase visual and ecological value in settlements. In climate-resilient design, these systems are attractive because they provide distributed storage and reduce dependence on a single closed network.

However, nature-based drainage is not automatically superior in every setting. It requires space, maintenance, local acceptance, and correct hydraulic integration. Its future importance lies not in replacing all conventional drains, but in complementing them more intelligently.

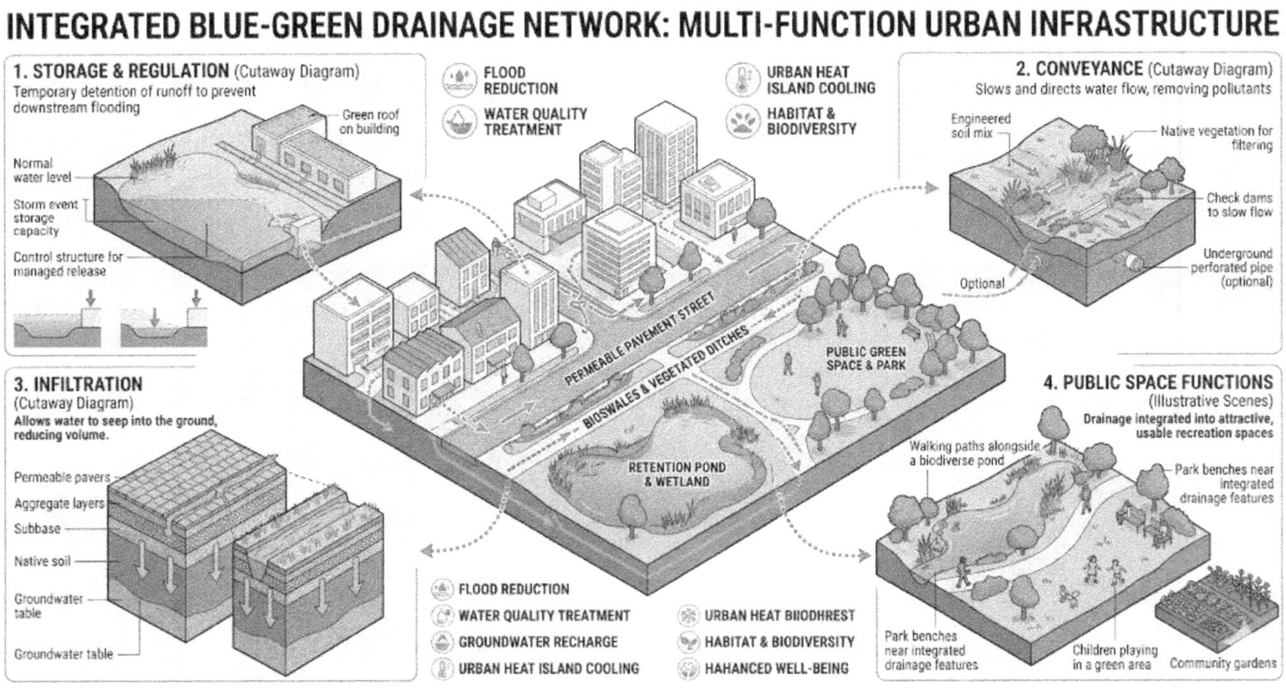

Figure 18.3 Blue-green drainage network combining storage, conveyance, infiltration, and public space functions.

18.7 Smart Drainage and Digital Monitoring

Future drainage systems are likely to make increasing use of digital monitoring, automation, and data-driven management. Smart drainage refers to the application of sensors, communication systems, forecasting tools, and digital decision support to improve real-time and long-term drainage performance.

Potential applications include rainfall and water-level sensors at critical nodes, real-time flood alerts, automated pump and gate control, digital asset inventories, performance dashboards, integration of field data with GIS and hydraulic models, and predictive maintenance based on condition trends.

Smart systems can improve responsiveness during storms and support more efficient maintenance between events. For example, a monitored low point can be watched during intense rainfall, or a pump station can be operated more effectively if water levels are tracked automatically. Over time, digital data can also improve understanding of system bottlenecks and support better rehabilitation planning.

Even so, smart drainage should not be seen as a substitute for basic engineering or maintenance. Technology adds value when it supports well-planned systems, but it cannot compensate for fundamentally inadequate layout, poor construction, or neglected field upkeep.

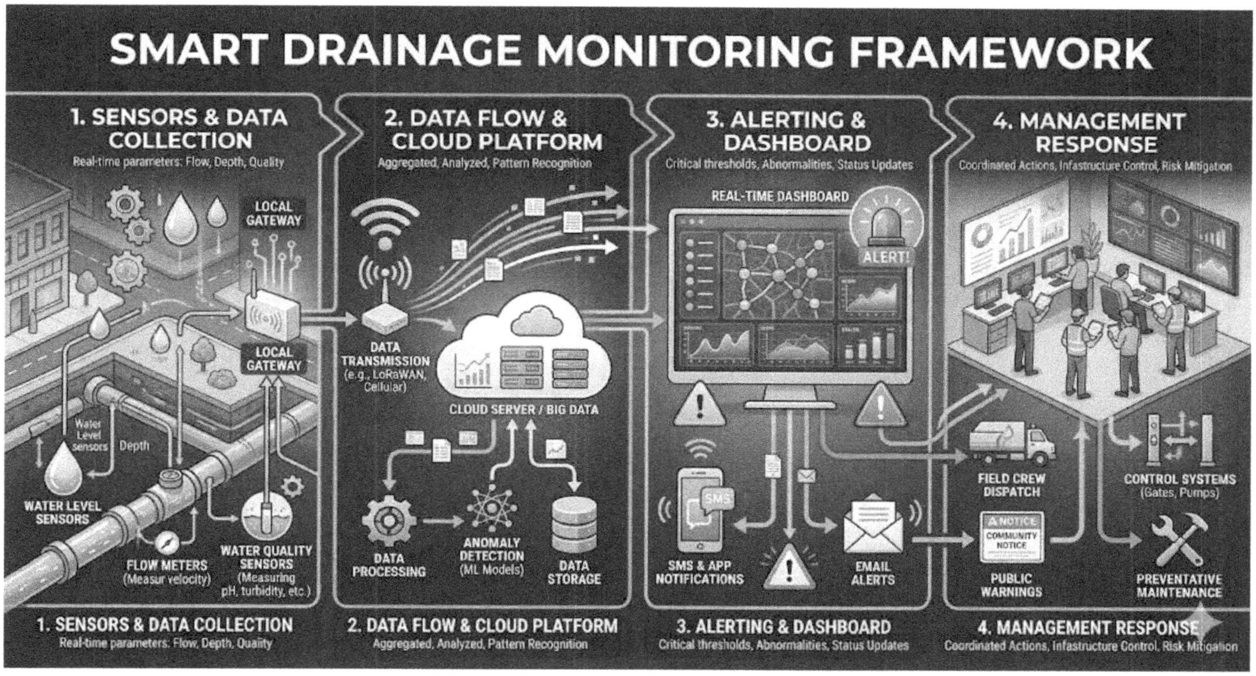

Figure 18.4 Smart drainage monitoring framework showing sensors, data flow, alerts, and management response.

18.8 Integrated Urban Water Thinking

A major future trend in drainage engineering is the movement toward integrated urban water thinking. This approach recognizes that drainage, water supply, wastewater, groundwater, public space, ecology, and land development are interconnected parts of a larger settlement water system.

Under this view, drainage is no longer treated only as a disposal function. It becomes part of a broader strategy that may include stormwater reuse, groundwater recharge, flood-compatible land use planning, combined management of roads and drainage corridors, pollution reduction at source, coordination between sanitation and drainage infrastructure, and protection of water-sensitive urban landscapes.

Integrated thinking is especially important in climate-stressed settlements where separate infrastructure decisions can create unintended consequences. For example, sealing large surfaces without recharge measures can increase flooding and reduce groundwater replenishment at the same time. Similarly, drainage improvements that ignore sanitation may simply transfer polluted water faster downstream.

The future of drainage engineering therefore lies increasingly in integration across sectors, rather than isolated hydraulic design alone.

18.9 Resilient Infrastructure for Different Settlement Types

Climate resilience does not look the same in every settlement. Urban, semi-urban, and village systems face different exposure conditions, constraints, and adaptation options.

In urban areas, resilience may depend on upgrading overloaded trunk systems, protecting flood corridors, improving inlet efficiency, adding storage, using smart monitoring, and coordinating with transport and utility networks. In semi-urban and peri-urban areas, resilience may rely more on preserving natural drainage paths, controlling land conversion, planning phased systems, and avoiding premature encroachment on future drainage corridors. In village settings, resilience often depends on maintaining open drainage continuity, protecting ponds and natural outfalls, strengthening road cross-drainage, and supporting community-based maintenance.

This distinction matters because resilience is not a fixed design template. It must be tailored to settlement form, governance capacity, land availability, and maintenance realities. A high-technology response may work in one context, while a simpler but better-protected open drainage network may be more resilient in another.

Table 18.2 Settlement-type resilience strategies for urban, peri-urban, and village drainage systems.

Settlement Type	**Characteristics & Key Drainage Challenges**	**Resilience Strategies: Structural / Technical (Engineering)**	**Resilience Strategies: Non-Structural / Policy / Social**
1. Urban	Characteristics: High density, highly impervious surfaces, complex existing infrastructure, high land value. Challenges: Rapid peak runoff, aging pipes, limited space for expansion, combined sewer overflows (CSOs).	• Smart Drainage & Monitoring: Deploy IoT sensors, flow meters, and AI to monitor real-time capacity and predict blockages. • Nature-Based Solutions (NBS) & SuDS: Integrate green roofs, permeable pavement, bioretention cells, and bioswales to reduce and filter runoff at the source.	• Adaptive Planning & Zoning: Mandate post-development runoff rates, enforce floor-elevation standards, and protect vital future corridors. • Real-Time Alert Systems: Implement early-warning systems and communication protocols for rapid public alert and evacuation.

Settlement Type	Characteristics & Key Drainage Challenges	Resilience Strategies: Structural / Technical (Engineering)	Resilience Strategies: Non-Structural / Policy / Social
		• Pipe Upsizing & Retrofitting: Strategically enlarge key trunk sewers and replace deteriorating pipes. • Localized Storage: Construct underground stormwater detention tanks or deep tunnel storage systems. • Flood Barriers: Install permanent or deployable flood gates and barriers at critical points.	• Stormwater Utility Fees: Introduce "pay-as-you-dump" fees based on impervious surface area to fund resilient infrastructure. • Inter-Agency Collaboration: Create coordinated task forces for watershed-wide management.
2. Peri-Urban	Characteristics: Transitional zone (mix of land use), rapid (often informal) development, developing infrastructure. Challenges: Unplanned conversion of natural retention areas, inadequate funding, fragmented systems, conflict with agricultural water needs.	• Blue-Green Corridors: Preserve and formalize natural waterways and floodplains to serve as resilient conveyance paths. • Detention Basins & Wetlands: Construct large-scale basins and restore wetlands at watershed connection points to slow and treat water. • Flexible/Modular Systems: Design adaptable drainage systems (e.g., open swales transitioning to pipes) that can easily expand with development. • Community-Maintained NBS: Implement simpler NBS (e.g., unlined swales, sand filters) that can be managed by local groups.	• Master Planning & Right-of-Way Protection: Enforce clear land-use zoning and protect drainage right-of-ways *before* formal development. • Community-Based Organization (CBO): Empower local communities to manage and maintain localized drainage assets. • Formalizing Informal Drainage: Identify and upgrade critical existing informal channels to integrated standards. • Capacity Building: Train local developers and officials on sustainable drainage practices and climate risk assessment.

Settlement Type	Characteristics & Key Drainage Challenges	Resilience Strategies: Structural / Technical (Engineering)	Resilience Strategies: Non-Structural / Policy / Social
3. Village	Characteristics: Low density, natural/agricultural land dominance, limited engineered infrastructure. Challenges: Over-reliance on natural channels (vulnerable to erosion/blockage), lack of maintenance funds, exposure to agricultural runoff, potential erosion.	• Watershed & Riverbank Stabilization: Plant native, deep-rooted vegetation (e.g., vetiver grass) along natural channels and riverbanks to prevent erosion. • Contour Farming & Terrace Drainage: Implement farming practices (e.g., contour plowing, terracing with diversion ditches) to manage agricultural runoff. • Small-Scale Water Harvesting: Build check dams, farm ponds, and bioswales to capture and store water for irrigation or slow release. • Natural Channel Dredging & Clearing: Regularly dredge natural streams and remove vegetation to maintain hydraulic capacity. • Low-Cost, Decentralized Systems: Use locally available materials and techniques to build simple, effective drains.	• Watershed-Level Management: Coordinate management efforts at the community level to protect the larger natural watershed. • Local Knowledge Systems: Integrate traditional knowledge of local hydrology into drainage planning. • Participatory Maintenance: Establish community-wide maintenance days and clear "who does what" protocols for upkeep of simple drains. • Erosion Control Policy: Implement community agreements or rules to reduce land clearance and protect steep slopes.

18.10 Emerging Research and Professional Directions

The future of settlement drainage engineering is also shaped by emerging research and professional practice. Several directions are becoming more important: better representation of nonstationary rainfall and climate uncertainty in design; integration of GIS, remote sensing, and digital terrain analysis with drainage planning; use of real-time monitoring and digital twins for urban drainage management; greater focus on lifecycle performance, resilience metrics, and service reliability; wider adoption of nature-based and hybrid drainage systems; improved understanding of social vulnerability, equity, and community-based adaptation; stronger linkage between drainage, public health, and environmental quality; and application of data science and machine learning in flood prediction, maintenance planning, and performance analysis.

Professionally, this means drainage engineers increasingly need interdisciplinary skills. Hydraulic knowledge remains essential, but so do geospatial analysis, environmental understanding, infrastructure management, and awareness of social risk and climate adaptation.

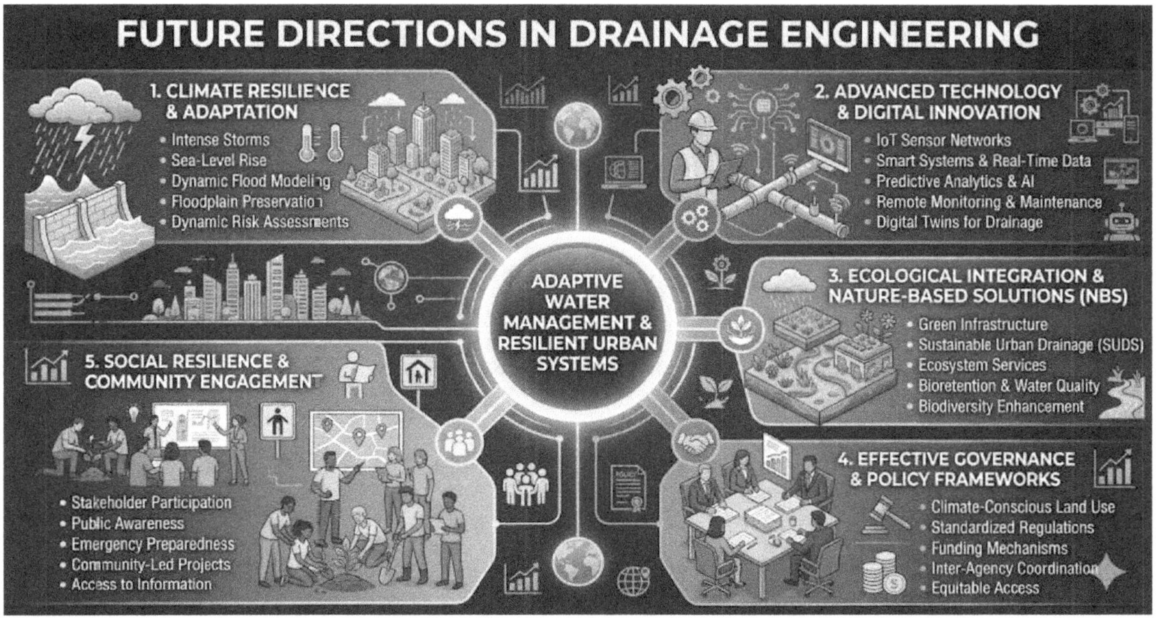

Figure 18.5 Future directions in drainage engineering showing climate, technology, ecology, governance, and social resilience

18.11 Summary Points

1. Climate change and variability are making traditional drainage design based on past rainfall data less reliable.
2. Climate-resilient drainage systems are designed to function effectively under uncertain and extreme weather conditions.
3. Increasing rainfall intensity and variability can lead to more frequent flooding and failure of existing drainage systems.
4. Resilience-based design focuses not only on system capacity but also on its ability to adapt, recover, and reduce damage during extreme events.
5. Adaptive drainage planning allows systems to be modified over time to meet changing environmental and urban conditions.
6. Blue-green and nature-based infrastructure, such as swales and wetlands, play an important role in sustainable and resilient drainage systems.

7. Smart drainage systems use sensors, monitoring, and digital tools to improve real-time management and maintenance.
8. Integrated urban water management considers drainage as part of a larger system including water supply, sanitation, and environmental protection.
9. Different settlement types (urban, peri-urban, and rural) require different strategies for achieving drainage resilience.
10. Future drainage engineering will increasingly rely on interdisciplinary approaches, combining technology, environmental science, and planning.

18.12 References

1. Arnold, C. L., Jr., & Gibbons, C. J. (n.d.). Impervious surface coverage: The emergence of a key environmental indicator.
2. Food and Agriculture Organization of the United Nations. (n.d.). Health issues related to drainage water management.
3. Hu, J. (n.d.). Effects of urban stormwater pollution on watershed diffuse pollution and river water quality.
4. Modeling the impacts of urban flood risk management on social inequality. (n.d.).
5. Objectives. This study investigated the scale of the public health risk from stormwater runoff caused by urbanization. Public health study available via PubMed Central. (n.d.).
6. Does urban green infrastructure lead to equity issues for flood vulnerable areas? (n.d.).
7. Stormwater strategies: Community responses to runoff pollution. (n.d.).
8. Subramanian, R. (n.d.). Problems and solutions for managing urban stormwater runoff.
9. United States Environmental Protection Agency. (n.d.). Urban stormwater management in the United States.
10. World Health Organization. (n.d.). Water and sanitation interventions to prevent and control mosquito-borne diseases.

APPENDIX A

SCS Curve Number Method for Estimation of Runoff

A1. Introduction

Runoff estimation is a fundamental task in hydrology and drainage engineering because it supports the design of stormwater drains, culverts, detention basins, flood-control structures, and watershed management systems. Among the available rainfall-runoff methods, the Soil Conservation Service Curve Number method, now commonly referred to as the NRCS Curve Number method, is one of the most widely used empirical approaches for estimating direct runoff from storm rainfall. It relates runoff generation to rainfall depth, watershed storage, land use, soil type, treatment practice, hydrologic condition, and antecedent moisture condition.

The popularity of the method arises from its simplicity, modest data requirements, and suitability for ungauged or partially gauged catchments. It is especially useful in small watershed hydrology, urban drainage planning, agricultural watershed studies, and GIS-based runoff mapping.

A2. Basic Concept of the SCS Curve Number Method

The SCS-CN method estimates direct runoff depth from a storm event by assuming that a watershed has a finite storage capacity before runoff begins. Part of the rainfall is lost through interception, depression storage, infiltration, and other abstractions, while the remaining rainfall becomes direct runoff.

The runoff potential of the watershed is represented by the **curve number (CN)**. A low CN indicates low runoff potential and high infiltration, while a high CN indicates high runoff potential and low retention. In general, permeable soils under dense vegetation have low CN values, whereas impervious or compacted urban surfaces have high CN values.

A3. Fundamental Assumptions

The SCS-CN method is based on the following assumptions:

a) Runoff is estimated for a single storm event rather than continuous long-term simulation.

b) Initial abstraction is approximated as a fixed fraction of potential retention, commonly

c) Watershed response can be represented adequately by a curve number or a composite curve number.

d) Standard curve numbers are generally associated with antecedent moisture condition II unless otherwise adjusted.

These assumptions make the method easy to use, but they also introduce limitations when watershed conditions depart strongly from standard empirical behavior.

A4. Governing Equations

The general runoff equation of the SCS-CN method is:

$$Q = \frac{(P - I_a)^2}{(P - I_a) + S}$$

for $P > I_a$, and $Q = 0$ for $P \leq I_a$.

Under the standard TR-55 assumption $I_a = 2S$, the equation becomes:

$$Q = \frac{(P - 0.2S)^2}{P + 0.8S}$$

for $P > 0.2S$.

Where:

- Q = direct runoff depth (mm).
- P = total rainfall depth (mm).
- I_a = initial abstraction (mm).
- S = potential maximum retention (mm). [3]
- CN = curve number.

The storage parameter is related to the curve number by:

$$S = \frac{25400}{CN} - 254$$

when SI units are used.

A5. Curve Number and Its Significance

The curve number is an empirical index representing the runoff response of a watershed. It integrates the combined effects of land use, soil infiltration characteristics, surface condition, and antecedent wetness into one parameter.

The CN scale generally ranges from about 30 to 100. Lower values correspond to forests, open space in good condition, and highly permeable soils, while higher values correspond to paved surfaces, commercial areas, and poorly drained soils.

Table A(1) Curve Numbers for Different Land Use/Cover and Hydrological Soil Groups

Land use / cover	**Hydrological Soil Groups**			
	A	**B**	**C**	**D**
Open space, good	39	61	74	80
Open space, fair	49	69	79	84
Open space, poor	68	79	86	89
Impervious surface	98	98	98	98
Commercial district	89	92	94	95
Industrial district	81	88	91	93
Residential 1/4 acre	61	75	83	87
Residential 1/2 acre	54	70	80	85
Row crops, straight row, good	67	78	85	89
Row crops, straight row, poor	72	81	88	91
Pasture, good	39	61	74	80
Pasture, fair	49	69	79	84
Woods, good	30	55	70	77
Woods, fair	36	60	73	79

A6. Factors Affecting Curve Number

6.1 Land Use and Land Cover

Land use is one of the most important controls on runoff generation. Forests, pasture in good condition, and open space with dense cover usually have lower CN values than residential, industrial, or commercial land uses.

6.2 Hydrologic Soil Group

The NRCS classifies soils into four hydrologic soil groups: A, B, C, and D. Group A soils have high infiltration and low runoff potential, while Group D soils have low infiltration and high runoff potential.

6.3 Treatment and Hydrologic Condition

Agricultural practices such as contouring, terracing, and good residue management can reduce curve number, while poor surface cover or degraded hydrologic condition increases it. The same land-use category may therefore have different CN values depending on management and condition.

6.4 Antecedent Moisture Condition

Antecedent moisture influences the amount of infiltration that can occur before runoff begins. Standard tabulated CN values usually correspond to average antecedent moisture condition (AMC II), while dry condition (AMC I) lowers CN and wet condition (AMC III) raises CN.

A7. Hydrologic Soil Groups

Hydrologic soil groups are used to classify soils according to minimum infiltration rate when thoroughly wetted. Their hydrologic interpretation is essential in CN selection.

a) **Group A:** High infiltration rates and very low runoff potential.

b) **Group B:** Moderate infiltration rates.

c) **Group C:** Low infiltration rates.

d) **Group D:** Very low infiltration rates and high runoff potential.

A8. Antecedent Moisture Condition Adjustment

Curve numbers are commonly tabulated for AMC II. When watershed conditions are unusually dry or wet, CN may be adjusted using standard relations.

$$CN_1 = \frac{CN_{II}}{2.281 - 0.01281CN_{II}}$$

$$CN_{III} = \frac{CN_{II}}{0.427 + 0.00573CN_{II}}$$

These equations provide dry and wet condition curve numbers from the AMC II value.

Table A(2): Antecedent moisture adjustment of curve number from AMC II.

CN (AMC II)	CN (AMC I)	CN (AMC III)
30	15.8	50.5
40	22.2	63.1
50	30.5	73.8
60	40.5	82.3
70	51.4	88.6
80	63.7	93.0
90	79.8	95.5

A 9. Composite Curve Number

Natural watersheds are rarely homogeneous, and a single watershed may contain cultivated land, settlements, grass cover, bare soil, and road surfaces. In such situations, a composite curve number is determined as an area-weighted average of all land-use-soil units.

$$CN_c = \frac{\Sigma A_i CN_i}{\Sigma A_i}$$

where A_i is the area of the -th land unit and CN_i is the corresponding curve number.

This approach is commonly used in GIS-based hydrologic analysis where land-use and soil maps are overlaid to generate spatial CN classes.

A10. Stepwise Procedure for Runoff Estimation

The standard procedure for estimating runoff by the SCS-CN method is as follows:

1. Delineate the watershed boundary and subareas.
2. Identify land use, treatment, and hydrologic condition.
3. Determine hydrologic soil group for each mapping unit.
4. Select the appropriate CN values from standard tables for AMC II.
5. Adjust CN if moisture condition differs from AMC II.
6. Compute composite CN if more than one land unit exists.
7. Calculate potential retention S.
8. Calculate initial abstraction $I_a = 0.2S$.
9. Use rainfall depth to estimate runoff depth Q.
10. Convert runoff depth to runoff volume if required.

A11. Worked Example

Consider a watershed with curve number $CN = 78$ under AMC II and a storm rainfall depth of 100 mm. Estimate the direct runoff depth using the SCS Curve Number method.

11.1 Given Data

- Rainfall depth, $P = 100$ mm.
- Curve number, $CN = 78$.

11.2 Calculation of Potential Retention

$$S = \frac{25400}{CN} - 254$$

$$S = \frac{25400}{78} - 254 = 71.64\, mm$$

Thus, the potential maximum retention of the watershed is 71.64 mm.

11.3 Calculation of Initial Abstraction

$$I_a = 0.2S$$
$$I_a = 0.2 \times 71.64 = 14.33\,mm$$

This means the first 14.33 mm of rainfall is abstracted before direct runoff begins.

11.4 Calculation of Direct Runoff

Since $P > I_a$, runoff occurs. The runoff equation is:

$$Q = \frac{(P - 0.2s)^2}{P + 0.8S}$$

Substituting the given values:

$$Q = \frac{(100 - 14.33)^2}{100 + 0.8(71.64)}$$
$$Q = \frac{(85.67)^2}{100 + 57.31}$$
$$Q = \frac{7339.35}{157.31}$$
$$Q = 46.66\,mm$$

Therefore, the direct runoff depth is **46.66 mm**.

11.5 Interpretation

This result indicates that out of 100 mm of rainfall, approximately 46.66 mm appears as direct runoff, while the remainder is retained or abstracted within the watershed. The example demonstrates the strong role of watershed storage and curve number in determining runoff yield.

Table A(3) Computed runoff depth for selected rainfall and curve number combinations.

Rainfall (mm)	CN 40	CN 50	CN 60	CN 70	CN 80	CN 90
25	0.00	0.00	0.00	0.09	1.67	9.09
50	0.00	0.00	1.40	5.81	13.80	27.11
75	0.00	0.54	7.92	17.49	28.99	45.79
100	0.00	4.84	18.57	32.71	50.54	67.27
125	0.00	11.73	30.84	48.43	70.90	89.77
150	0.31	20.20	43.87	64.87	91.74	112.12
200	7.88	40.50	72.63	98.07	130.38	156.45
250	19.48	62.29	101.91	130.59	166.16	196.08

A12. Supplementary charts

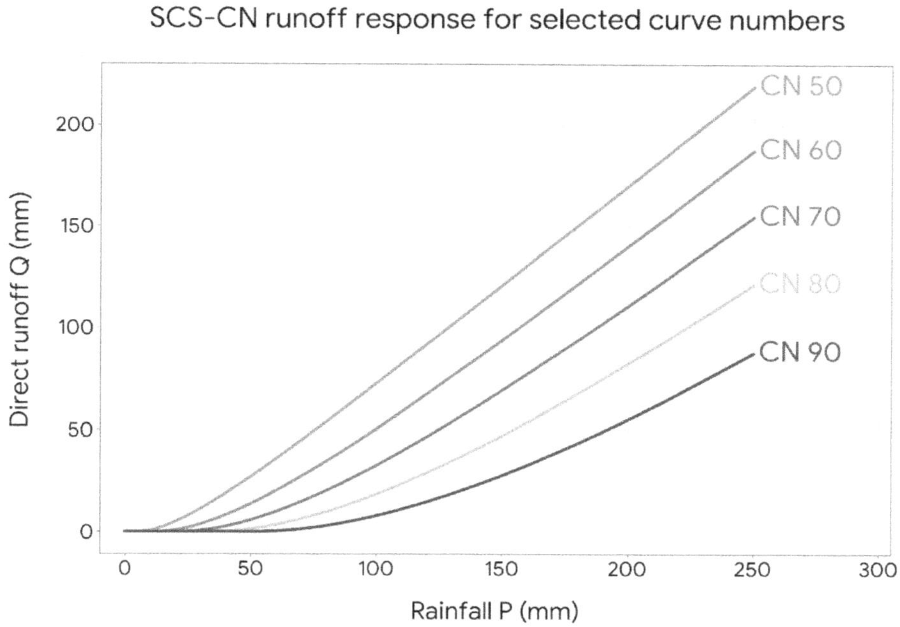

Figure A(1). SCS-CN runoff response curves showing direct runoff depth as a function of rainfall for selected curve numbers.

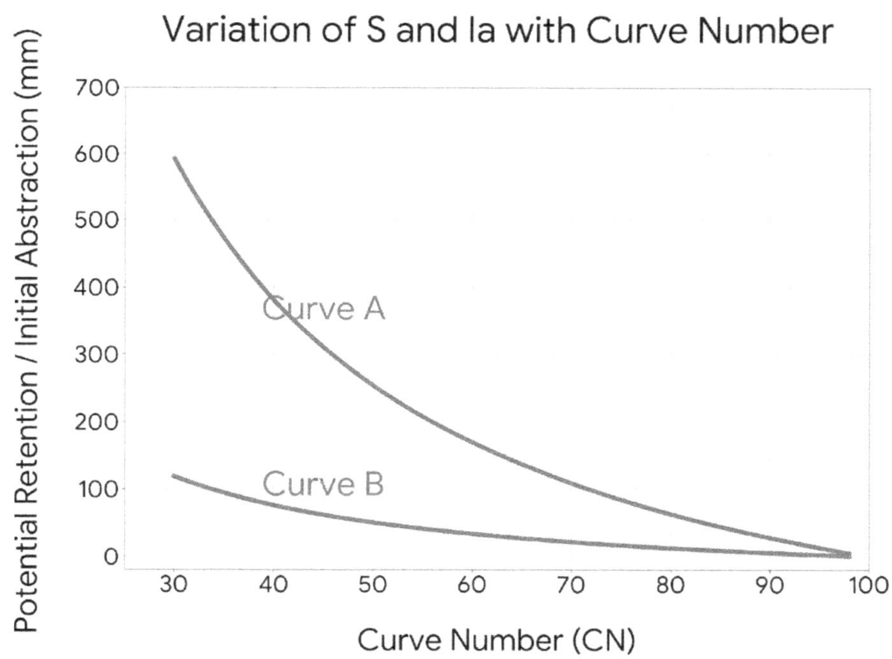

Figure A(2): Variation of (A)potential retention and (B) initial abstraction with the curve number.

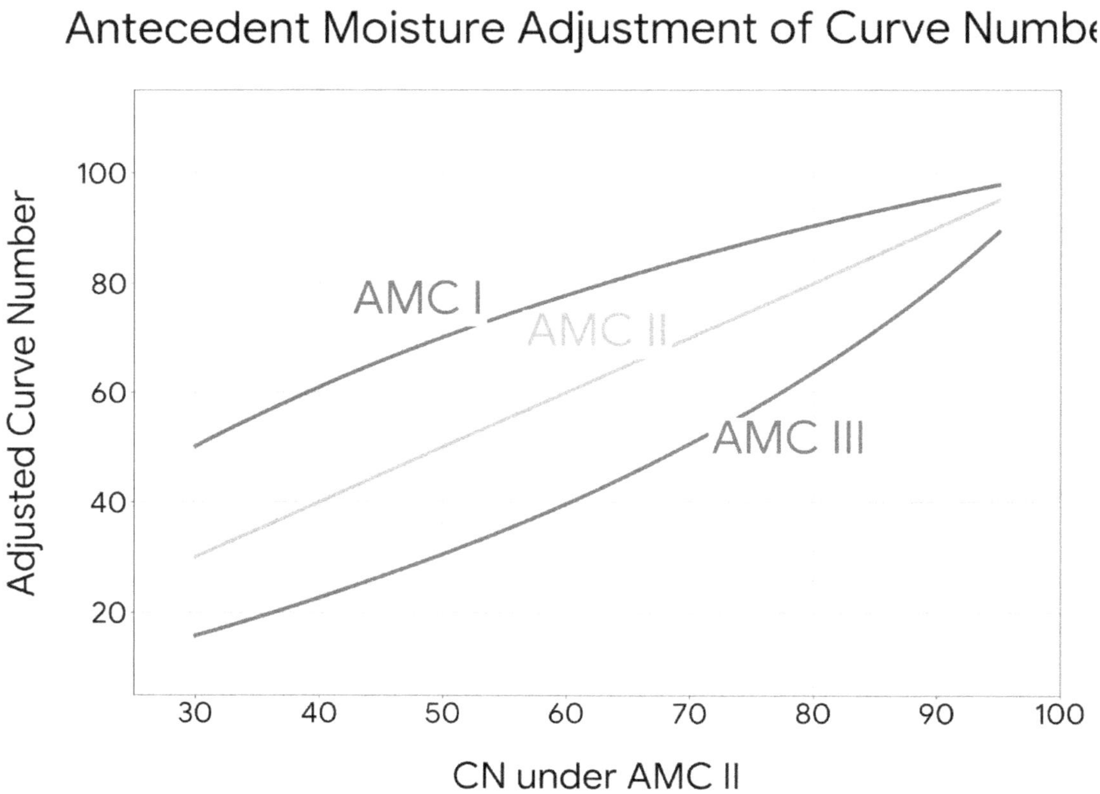

Figure A(3) Antecedent moisture adjustment curves for CN under AMC I, AMC II, and AMC III.

A13. Applications of the SCS-CN Method

The SCS-CN method is widely used in urban stormwater planning, agricultural watershed studies, highway drainage design, flood estimation in ungauged basins, and GIS-based runoff assessment. It is especially useful in preliminary design and planning studies where observed runoff records are limited or absent.

The method is also implemented in hydrologic models and engineering software for event-based watershed analysis. In practical drainage engineering, it often serves as the first stage in deriving runoff hydrographs and peak flow estimates.

A14. Advantages

The major advantages of the SCS-CN method are its simplicity, broad standardization, low data requirement, and adaptability to both manual and GIS-based analysis. These features make it especially attractive for engineering design, teaching, and regional hydrologic assessment.

The method also provides an intuitive connection between watershed surface condition and hydrologic response. This helps students and practitioners understand how land-use change affects runoff generation.

A15. Limitations

Despite its wide use, the method has several limitations. It is empirical, and the assumption may not always represent local field behavior accurately.

Its performance may be less reliable when applied to highly heterogeneous watersheds, unusual rainfall patterns, or regions where the original empirical basis differs substantially from local conditions. For accurate design, calibration against observed rainfall-runoff data is desirable whenever possible.

APPENDIX B

The Rational Method for Peak Discharge Estimation (Q)

B1. Overview and Assumptions

The Rational Method is a widely used hydrological model for estimating peak surface runoff rate from small urban and rural watersheds. It assumes that the maximum rate of runoff occurs when the entire watershed area is contributing to the outlet.

Key Assumptions:

- Rainfall intensity is uniform over the entire watershed.
- Rainfall intensity is uniform over the duration of the storm.
- The peak runoff rate occurs when the storm duration equals the Time of Concentration (Tc).
- The Runoff Coefficient (C) remains constant throughout the storm.

B2. The Rational Formula

The core of the Rational Method is expressed as:

$$Q = C \times I \times A$$

US Customary Units

- **Q** = Peak discharge rate (cfs - cubic feet per second)
- **C** = Runoff coefficient (dimensionless, ranging from 0.0 to 1.0)
- **I** = Rainfall intensity (in/hr)
- **A** = Drainage area (acres)

(Note: 1 acre-inch per hour is approximately 1.008 cfs. For practical engineering purposes, the conversion factor of 1.0 is used).

SI (Metric) Units

To account for unit conversions in the metric system, the formula includes a divisor:

$$Q = (C \times I \times A) / 360$$

- **Q** = Peak discharge rate (m^3/s)
- **C** = Runoff coefficient (dimensionless)

- **I** = Rainfall intensity (mm/hr)
- **A** = Drainage area (hectares)

B3. The Runoff Coefficient (C)

The coefficient C represents the fraction of rainfall that becomes surface runoff. It depends on surface type, soil, and slope.

Table B(1) Typical Runoff Coefficients Values

Surface Type	Description / Condition	C Value Range
Pavement	Asphalt or Concrete	0.70 – 0.95
Roofs	Standard pitched or flat	0.75 – 0.95
Lawns (Sandy Soil)	Flat (2%)	0.05 – 0.10
Lawns (Sandy Soil)	Steep (>7%)	0.15 – 0.20
Lawns (Heavy Soil)	Flat (2%)	0.13 – 0.17
Lawns (Heavy Soil)	Steep (>7%)	0.25 – 0.35
Woods/Forest	Dependent on ground cover	0.10 – 0.25
Industrial	Heavy industrial areas	0.60 – 0.90

(Field Tip: For watersheds with mixed land uses, calculate a composite C value using an area-weighted average: $C_{composite} = \Sigma(C_i \times A_i) / \Sigma A_i$*)*

B4. Time of Concentration (Tc)

Tc is the time required for a drop of water to travel from the most hydrologically remote point in the watershed to the point of analysis. You must determine Tc to find the correct Rainfall Intensity (I).

Method A: Kirpich Formula

Best suited for natural basins with well-defined channels and steep slopes.

$$Tc = 0.0078 \times (L^{0.77} / S^{0.385})$$

- **Tc** = Time of concentration (minutes)
- **L** = Longest flow path/channel length (feet)
- **S** = Average basin slope (ft/ft)

(Note: Multiply Tc by 0.2 for concrete channels, or 0.4 for earth channels).

Method B: FAA Formula

Best suited for urban areas, airfields, or areas with overland sheet flow.

$$Tc = [1.8 \times (1.1 - C) \times L^{0.5}] / S\ 0.333$$

- **Tc** = Time of concentration (minutes)

- **C** = Rational runoff coefficient
- **L** = Longest flow path (feet)
- **S** = Surface slope (percent, e.g., use 5.0 for 5%)

B5. Rainfall Intensity (I) Derivation from IDF Curves

Intensity-Duration-Frequency (IDF) curves are regional graphical or tabular datasets that relate rainfall intensity to storm duration and return period.

Figure IDF Curves for Gurugram (Haryana)

Step-by-Step Derivation:

- **Determine the Return Period:** Select the design storm frequency based on local codes (e.g., 10-year, 25-year, or 100-year storm).
- **Calculate Tc:** Use the Kirpich, FAA, or TR-55 methods to find your Time of Concentration.
- **Set Duration:** In the Rational Method, set the storm duration *equal* to the calculated Tc. (If Tc is calculated as less than 5 minutes, engineers typically default to a 5-minute minimum duration to avoid overestimating intensity).
- **Read the Curve:** Locate the Tc duration on the x-axis of the IDF chart. Move vertically to intersect the curve for your chosen Return Period.
- **Extract Intensity:** Move horizontally from that intersection to the y-axis to read the Rainfall Intensity (I) in in/hr or mm/hr.

APPENDIX C

Gumbel Distribution for Flood Frequency Analysis

C1 Gumbel Distribution

The Gumbel Extreme Value Type-I distribution is widely applied to annual maximum floods. Parameters are estimated from sample statistics: scale parameter and location parameter , where is sample mean and s is sample standard deviation.

The reduced variate for return period T is computed as:

$$y_T = \ln\left[-\ln\left(1 - \frac{1}{7}\right)\right]$$

and the design quantile is . $X_T = u + a_{yT'}$ [

C2 Worked Example: Flood Frequency Analysis

Objective: Estimate the 100-year peak discharge from a 20-year annual peak series (m^3/s):

320, 450, 290, 510, 380, 425, 360, 480, 305, 395, 410, 370, 520, 340, 385, 465, 330, 440, 355, 475.

Step 1 (sample statistics):

- Sum = 8,265 m^3/s, Mean m^3/s
- Sum of squared deviations = 82,462.75; Variance = 4,340.14; Standard deviation s = 65.88 m^3/s

Step 2 (Gumbel parameters):

$$\text{Scale } a = \frac{65.88}{1.2825}, \quad m^3/s$$

Location u = 413.25 – 0.5772 (51.35) = 383.62 m^3/s

Step 3 (reduced variate for 100 years):

The example table in the chapter lists y_{100} = 4.6001.

Step 4 (100-year quantile):

Using $X_{100} = u + ay_{100}$ with u = 383.62, a=51.35 and y_{100}= 4.6001: the corrected 100-year estimate corresponds to ~618 m^3/s in the example's corrected table.

Table C(1) Return-period Chart

Return Period	y_t	X_t (m^3/s)
2-year	0.3665	402.46
5-year	1.4999	460.15
10-year	2.2504	499.93
25-year	3.1985	548.79
50-year	3.9019	583.81
100-year	4.6001	619.83
200-year	5.2960	654.92

Interpretation and cautions: Uncertainty is high when estimating long return periods from short records (20 years). The chapter recommends comparison with regional patterns, use of regional frequency analysis where possible, and attention to non-stationarity under climate change (traditional stationarity assumptions may not hold.

www.ingramcontent.com/pod-product-compliance
Lightning Source LLC
LaVergne TN
LVHW060029210926
848481LV00013B/3818
* 9 7 9 8 9 0 4 3 1 7 3 9 3 *

Mastering Denial Management

A Physician Advisor's Playbook

Stuart M. Hoffman, MD, FACS, CHCQM

Mastering Denial Management: A Physician Advisor's Playbook is independently published.

ISBN: 979-8-3493-5508-0

Stuart M. Hoffman, MD, FACS, CHCQM, Author

Advice given is general. Readers should consult professional counsel for specific legal, ethical, or clinical questions.

Foreward

Ronald Hirsch, MD, FACP, ACPA-C, CHRI, CHCQM-PHYADV, FABQAURP

Vice President, Regulations and Education, R1 RCM Inc.

When I started in medical practice in 1991, things were different. I had a full-time office practice as a general internist but also cared for my patients in the hospital and at local nursing facilities. The hospital had a chief medical officer and department chairs but for the most part, physicians were left to practice as they saw fit.

Fast forward a few years and the growth of health maintenance organizations and Medicare Plus Choice program, later to be rebranded Medicare Advantage, along with increasing hurdles to care and payment put in place by most payers, led to the need for more physician involvement in the administrative functions of hospital care. That led to the creation of the physician advisor role to act as an intermediary between the practicing physicians and the regulators and payers.

Initially physician advisors were tasked solely with guiding the attending physician to follow the payer rules to ensure the patient was in the right admission status. But as hospital C-suites saw the value of having a physician advisor providing the practicing physician perspective on other parts of hospital operations such as length of stay oversight and also including areas generally left to the revenue cycle team, such as denial management.

Which leads to this important place this book, Mastering Denial Management A Physician Advisor's Playbook, by Dr. Stuart Hoffman, plays in the armamentarium of the physician advisor. In many hospitals, the role of physician advisor is established and willing physicians appointed to serve in that role, but training and educational resources are often left to the doctor to seek out.

As Dr. Hoffman notes in his book, there are two prominent professional organizations, the American College of Physician Advisors and the American Board of Quality Assurance and Utilization Review Physicians, both of whom provide resources. And I co-authored a book, The Hospital Guide to Contemporary Utilization Review, that many find a valuable resource.

This book, though, serves as the first text dedicated to the knowledge and skills required specifically in managing the ever-growing requirement to actively and aggressively prevent and manage denials. From the often-dreaded peer-to-peer phone calls with a payer medical director to writing denial letters to collecting meaningful performance indicators to collaborating with payers, Dr. Hoffman covers each area in depth, with expertise, and with case examples, checklists, and sample scripting to help guide physician advisors to perform at their best, advocating for their colleagues and their hospital.

Dr. Hoffman also delves into the ever-growing role of Artificial Intelligence in denials management and the role of the physician advisor in another area that is attracting more attention, good and bad, value-based care.

This practical guide is an important addition to the resources available to physician advisors and to physicians considering a role as a physician advisor. I expect before long it will find a place on the bookshelf of many physician advisors with multiple pages soon to be dog eared and repeatedly referenced.

Table of Contents

About the Author

Stuart M. Hoffman, MD, FACS, CHCQM

Stuart M. Hoffman, MD, FACS, CHQM is a board-certified general surgeon and network physician advisor at the University of Vermont Health Network, a six hospital academic health network with three hospitals in Vermont and three hospitals in Northern New York. He completed his residency training in surgery at Brown University School of Medicine. He is a member of the American College of Physician Advisors and a fellow of the American College of Surgeons.

Acknowledgments

I would like to express my sincere gratitude to Steve Grant, MD, FHM, CHCQM. Together, we undertook the challenge of building our health network's first physician advisor program in a new and rapidly evolving field of medicine. His approachable nature and transformational leadership have been inspiring throughout this journey.

INTRODUCTION

The Emergence and Evolut on of the Physician Advisor Role

The physician advisor role has emerged as a vital response to the increasing complexities of the modern healthcare landscape. What began as a loosely defined advisory position in the late 20th century has evolved into a critical function that bridges clinical expertise, administrative oversight, and financial stewardship. The trajectory of this evolution is closely tied to major shifts in healthcare systems, payment models, and regulatory frameworks over the past several decades.[1]

The Early Roots: Addressing Uncontrolled Healthcare Costs

The roots of the physician advisor role are deeply intertwined with the cost-control measures that dominated U.S. healthcare policy in the 1970s and 1980s. During this period, healthcare spending began to rise at an unsustainable rate, driven by factors such as:

- Advances in medical technology that increased the scope of treatment options.
- A lack of standardized criteria for hospital admissions, lengths of stay, and diagnostic testing.
- A fee-for-service payment model that incentivized volume over value.[2]

To address these challenges, managed care organizations (MCOs) and utilization review programs were introduced. Their goal was to align clinical decisions with evidence-based guidelines, ensuring that care was both necessary and cost-effective. However, these early initiatives were often led by non-clinical staff or physicians without formal training in utilization management, leading to variability in outcomes and friction between providers and payers.[3]

For example, a landmark 1984 study by John E. Wennberg revealed significant disparities in hospital admission rates for similar patient populations across different regions. This variability highlighted the need for more systematic oversight, paving the way for the physician advisor role as it exists today.[4]

The Managed Care Era and the Rise of Utilization Review

The 1990s marked a transformative period for healthcare in the United States. The proliferation of health maintenance organizations (HMOs) and other managed care models brought stricter oversight of clinical decisions. Utilization review programs became a cornerstone of managed care, focusing on hospital admissions, lengths of stay, and the appropriateness of procedures.

Hospitals faced mounting pressure from payers to justify clinical decisions, prompting a shift toward more structured utilization management practices.[5]

This period also saw the introduction of key regulatory changes that influenced the development of the physician advisor role. The implementation of Medicare's prospective payment system (PPS) for hospital inpatient services in 1983—and later for outpatient services in 2000—placed greater financial risk on hospitals.[6] Physician advisors emerged as a critical resource for ensuring compliance with these payment models, advocating for appropriate levels of care, and minimizing financial losses from denied claims.

Professionalization in the 21st Century

While the physician advisor role gained traction in the 1990s, it was not until the early 21st century that it began to achieve professional recognition and standardization. The establishment of organizations like the American College of Physician Advisors (ACPA) marked a turning point in the role's development. Through training programs, certification pathways, and networking opportunities, these organizations elevated the physician advisor role from an informal function to a respected profession.

In addition to utilization management, physician advisors increasingly took on responsibilities in clinical documentation integrity (CDI). This shift was driven by growing recognition of the financial and operational implications of accurate documentation. CDI programs, often led by physician advisors, aimed to ensure that medical records fully reflected the complexity of patient care, supporting both appropriate reimbursement and compliance with quality reporting requirements.

The early 2000s also saw a surge in payer audits, including the introduction of Medicare's Recovery Audit Contractor (RAC) program in 2003.[7] Physician advisors played a pivotal role in defending hospitals against audit findings, ensuring that documentation supported the clinical necessity of services provided.

The Expanding Scope of the Modern Physician Advisor

Today, the physician advisor role encompasses a wide range of responsibilities that extend far beyond its original focus on utilization management. These include:

1. **Denial Management and Appeals:** Physician advisors analyze denial trends, prepare for peer-to-peer (P2P) conferences, and advocate for overturned claims. Their expertise is critical for ensuring that hospitals receive appropriate reimbursement while maintaining payer relationships.
2. **Regulatory Compliance**: Physician advisors navigate complex payer policies and regulatory requirements, helping hospitals mitigate risks associated with non-compliance. This includes responding to audits and ensuring adherence to Medicare and Medicaid guidelines.

3. **Quality Improvement Initiatives:** Many physician advisors are involved in hospital-wide efforts to enhance patient outcomes and align care delivery with value-based reimbursement models. Their clinical expertise allows them to identify opportunities for improvement and implement evidence-based strategies.
4. **Data Analytics and Workflow Optimization:** By leveraging analytical tools, physician advisors can identify high-risk claims, optimize documentation practices, and streamline workflows.[8]

Personal Experience and Inspiration for This Book

The expansion of the physician advisor role became particularly evident to me in my experience as a network physician advisor at the University of Vermont Health Network (UVMHN), where I helped establish our first formal physician advisor program. One of the central initiatives of this program was to take ownership of responding to insurance denials through peer-to-peer (P2P) conferences—a responsibility that previously fell to the clinicians directly involved in patient care.

We recognized early that clinicians faced significant barriers in managing P2P conferences. Their unpredictable schedules, coupled with urgent clinical demands, often left little time to thoroughly prepare for or conduct these critical discussions. This lack of dedicated focus resulted in missed opportunities to overturn denials and, ultimately, added administrative burden to already overextended providers.

By building a network-wide team of physician advisors specifically tasked with managing P2P conferences, we were able to refine our expertise in the denial management process. This concentrated effort led to a notable increase in the number of denials we successfully addressed and overturned. Equally important, the initiative improved the quality of life for frontline clinicians, freeing them to focus fully on patient care without the distraction of payer disputes.

This experience underscored for me the transformative potential of dedicated physician advisor programs—not only in financial and operational outcomes but also in supporting provider well-being and enhancing patient care. It is this journey that inspired me to write this book, in the hopes of sharing practical strategies and insights with other physician advisors navigating similar challenges.

The Critical Importance of Peer-to-Peer Conferences

Among the many responsibilities of physician advisors, peer-to-peer (P2P) conferences stand out as one of the most impactful functions. These discussions provide a direct channel for appealing payer denials, ensuring that hospitals receive appropriate reimbursement, and advocating for uninterrupted patient care. The stakes are high: unresolved denials can disrupt treatment plans, delay hospital payments, and place significant administrative burdens on healthcare teams.

The Purpose of P2P Conferences

P2P conferences serve as a forum for resolving disagreements between providers and payers regarding the medical necessity of a service or procedure. Unlike written appeals, which are often reviewed by non-clinical staff, P2P conferences allow physician advisors to engage directly with insurance medical directors. This direct communication creates opportunities to:

- Clarify clinical nuances that may not be apparent in written documentation.
- Address payer-specific criteria or guidelines that may have been misapplied.
- Advocate for medically necessary care based on evidence and clinical judgment.

Case Study: A Denial Overturned for Sepsis Care

Consider the case of a 74-year-old patient admitted with severe sepsis. Despite clear signs of organ dysfunction, including hypotension refractory to fluids and elevated lactate levels, the payer denied inpatient status, citing insufficient evidence of medical necessity. The denial hinged on a narrow interpretation of criteria that failed to account for the patient's clinical complexity.

During the P2P conference, the physician advisor presented a compelling argument, emphasizing:

- Persistent hypotension despite aggressive fluid resuscitation, meeting criteria for septic shock.
- The need for continuous monitoring and IV vasopressors, which could not be safely administered in an outpatient setting.
- Alignment with MCG's inpatient criteria for sepsis with hemodynamic instability.

The insurance medical director agreed to overturn the denial, citing the advisor's clear documentation of organ dysfunction and alignment with established criteria. This case highlights the critical role of physician advisors in bridging clinical and administrative perspectives.

Key Elements of Successful P2P Conferences

To maximize the likelihood of overturning denials, physician advisors must approach P2P conferences with preparation, precision, and professionalism. The following elements are essential for success:

1. **Preparation**

- **Review the Denial Letter:** Understand the specific reasons for the denial and identify any discrepancies or gaps in documentation.
- **Gather Clinical Evidence:** Compile relevant medical records, including diagnostic findings, treatment details, and trends in vital signs or lab values.
- **Prepare a Prep Sheet:** Create a concise document summarizing the strongest arguments for overturning the denial. This document should highlight key clinical data and align it with payer-specific criteria.

2. Framing the Argument

   - **Be Concise and Focused:** Present the most compelling evidence early in the conversation, ensuring that the key points are clear and concise.
   - **Use Clinical Guidelines:** Reference authoritative sources, such as InterQual, MCG, or payer-specific criteria, to support the argument.
   - **Avoid Overloading with Details:** While comprehensive documentation is important, excessive detail can obscure the main argument.

3. Collaboration

   - **Engage Empathetically:** Acknowledge the insurance medical director's perspective and seek common ground.
   - **Respond to Objections:** Be prepared to address counterarguments or requests for additional clarification.
   - **Maintain Professionalism:** Approach the conversation as a collaborative effort to ensure the best outcome for the patient.

Challenges in P2P Conferences

While P2P conferences are a powerful tool for overturning denials, they also present unique challenges that require careful navigation.

1. Time Constraints

   - P2P conferences are often scheduled with limited notice, requiring physician advisors to prepare quickly and efficiently. Developing standardized templates and processes for P2P preparation can mitigate this challenge.

2. Documentation Gaps

   - Incomplete or poorly structured medical records are a common barrier to successful appeals. Physician advisors must proactively address these gaps by working closely with CDI specialists and clinical staff.

3. **Payer-Specific Guidelines**
    - Insurance companies often rely on proprietary criteria that may differ from widely recognized guidelines such as MCG or InterQual. Physician advisors must familiarize themselves with these criteria to craft effective arguments.
4. **Bias Toward Denial**
    - Payers may approach P2P conferences with a bias toward upholding the original denial. Overcoming this bias requires both clinical evidence and persuasive communication.

Case Study: Navigating Proprietary Payer Guidelines

A 60-year-old patient was admitted with acute exacerbation of chronic obstructive pulmonary disease (COPD). The payer denied inpatient status, citing proprietary guidelines that emphasized outpatient management for similar cases. However, the patient's clinical presentation included:

- Severe dyspnea despite maximum bronchodilator therapy.
- Persistent oxygen desaturation requiring non-invasive ventilation (NIV).
- Comorbidities, including heart failure with preserved ejection fraction, complicating management.

During the P2P conference, the physician advisor focused on the following points:

- The patient's failure to respond to outpatient-level interventions, necessitating continuous monitoring and NIV.
- The risk of decompensation in the setting of multimorbidity, which justified inpatient admission.
- Alignment with InterQual criteria for COPD exacerbation requiring NIV.

The denial was overturned after the payer acknowledged that the proprietary guidelines had been applied too rigidly. This case underscores the importance of tailoring arguments to payer-specific criteria while emphasizing clinical judgment.

The Role of P2P Conferences in Driving Systemic Improvement

Beyond individual outcomes, P2P conferences contribute to broader organizational goals, including:

- **Identifying Documentation Gaps:** Trends observed during P2P conferences can inform targeted CDI initiatives to improve documentation quality.

- **Strengthening Payer Relationships:** Constructive dialogues with insurance medical directors can foster mutual understanding and collaboration.
- **Enhancing Physician Advisor Skills:** Each conference provides an opportunity for physician advisors to refine their communication techniques and deepen their understanding of payer dynamics.

Conchohocken, PA-based Mercy Health System built a denials management program composed of a full-time team of physician advisors and realized a reduction in denials by 18 percent annually, on average, amounting to $2.2 million savings for the health system in the first year of the program.[9] This data underscores the strategic value of structured denials management program in achieving both financial and operational excellence.

Recognizing the Broader Landscape of Denials

Although this book concentrates predominantly on level of care denials—by far the most common type encountered by physician advisors—hospitals grapple with a wide array of additional payer denials. These include, but are not limited to, readmission denials, diagnosis-related group (DRG) downgrades or clinical validation denials, skilled nursing facility (SNF) denials, and procedure-specific payment disputes. Each of these denial types presents its own challenges, requiring tailored strategies for prevention and appeal.

- **Readmission Denials** often revolve around the clinical appropriateness of a patient's return to the hospital within a specific time frame. Payers scrutinize these short-interval readmissions, alleging that care could have been better managed on an outpatient basis or prevented altogether.
- **Clinical Validation Denials** typically arise when payers contend that the hospital's documentation does not fully support the severity level assigned to a patient's stay. This scenario underscores the importance of robust clinical documentation integrity (CDI) efforts.
- **SNF Denials** center on whether a patient truly requires skilled nursing care, often questioning the necessity of rehabilitation services or intravenous medications in a post-acute setting.

While these broader denial categories can overlap with the appeal and peer-to-peer processes discussed throughout this text, *Mastering Denial Management: A Physician Advisor's Playbook* specifically focuses on level of care denials and their resolution via peer-to-peer conferences and strong appeal letters. By honing in on this critical aspect of denial management, we delve into the day-to-day realities physician advisors face when they must rapidly determine inpatient versus observation status or outpatient care. The strategies covered—such as compiling compelling clinical evidence, navigating payer-specific guidelines, and sharpening communication tactics—can nonetheless inform how you address other forms of denials. The underlying principles of thorough preparation, effective argumentation, and cohesive documentation transcend any single type of denial.

This decision to center on level of care denials should not diminish the importance of other categories. Instead, the aim is to give readers a deep dive into the complexities of the denial most frequently encountered in practice. With this foundational skill set, readers will be well-equipped to engage in all aspects of denial management, whether they confront a readmission denial, a clinical validation dispute, or a question over SNF eligibility.

Overview of the Book's Goals and Structure

The physician advisor role is at the intersection of clinical care, administrative operations, and financial stewardship. While the importance of the role continues to grow, formalized resources to guide physician advisors remain limited. *Mastering Denial Management: A Physician Advisor's Playbook* addresses this gap by providing actionable strategies, practical tools, and expert insights tailored to the unique challenges physician advisors face.

Goals of the Book

This book is designed to empower physician advisors with the skills and knowledge necessary to excel in their roles. Its goals are threefold:

1. **Empower Physician Advisors**
   - Equip readers with proven techniques for navigating peer-to-peer (P2P) conferences, denial management, and regulatory compliance.
   - Provide practical frameworks for interpreting payer policies, preparing compelling arguments, and improving documentation practices.
2. **Enhance Patient Outcomes**
   - Advocate for medically necessary care, ensuring that patients receive appropriate treatments without unnecessary delays or disruptions.
   - Minimize the administrative burden on clinical teams, allowing them to focus on delivering high-quality care.
3. **Optimize Hospital Operations**
   - Improve denial overturn rates, streamline revenue cycle processes, and reduce financial losses associated with denied claims.
   - Strengthen organizational compliance with payer and regulatory requirements, mitigating audit risks.

These goals align with the broader mission of physician advisors to support patient care while safeguarding the financial and operational health of their organizations.

Structure of the Book

The book is organized into ten chapters, each addressing a critical aspect of the physician advisor role. Together, these chapters provide a comprehensive roadmap for mastering the complexities of denial management and P2P conferences.

Chapter Summaries

Chapter 1: The Role of the Physician Advisor

This chapter traces the history and evolution of the physician advisor role, from its origins in managed care to its current status as an essential hospital function. It explores the key responsibilities of physician advisors and their growing influence in areas such as utilization management, clinical documentation integrity (CDI), and regulatory compliance. Case examples illustrate how the role has adapted to meet the challenges of modern healthcare.

Key Takeaway: Understanding the historical context of the role helps physician advisors appreciate its significance and adapt to emerging demands.

Chapter 2: Understanding Insurance Company Processes

Denial patterns often reflect payer workflows and priorities. This chapter demystifies the internal processes of insurance companies, including the roles of claims adjusters, medical directors, and appeals committees. It also identifies common reasons for denials, such as incomplete documentation or misalignment with payer-specific criteria.

Key Takeaway: Gaining insight into payer workflows enables physician advisors to craft more effective arguments during P2P conferences.

Chapter 3: **Understanding Inpatient Level of Care Across Different Payer Types**

Inpatient admissions are defined differently by Medicare, Medicaid, and Commercial insurers, each with its own requirements and documentation standards. This chapter examines how criteria such as the Two-Midnight Rule, social determinants of health, and standardized guidelines like MCG and InterQual influence the determination of inpatient vs. observation status. By clarifying these variations, physician advisors can ensure that clinical documentation supports the most appropriate level of care and mitigates the risk of preventable denials.

Key Takeaway: A firm grasp of payer-specific definitions and standardized guidelines is crucial for establishing medical necessity and safeguarding proper reimbursement.

Chapter 4: Preparing for Peer-to-Peer Conferences

Preparation is critical to the success of P2P conferences. This chapter provides practical tools and techniques for gathering clinical evidence, interpreting denial letters, and creating concise Prep Sheets. It also explores strategies for addressing common challenges, such as ambiguous documentation or payer-specific guidelines.

Key Takeaway: A well-prepared physician advisor is better equipped to advocate for overturning denials and ensuring appropriate reimbursement.

Chapter 5: Building Persuasive Arguments

Effective communication is the cornerstone of successful denial management. This chapter outlines strategies for framing arguments logically, using clinical guidelines to support claims, and addressing objections from insurance medical directors. It also highlights the importance of tailoring arguments to specific payer criteria.

Key Takeaway: Clear and persuasive arguments increase the likelihood of successful appeals during P2P conferences.

Chapter 6: Communication Techniques

Managing difficult conversations requires both professionalism and empathy. This chapter explores techniques for handling objections, de-escalating tension, and fostering collaboration during P2P conferences. It also includes tips for navigating cultural differences and language barriers.

Key Takeaway: Strong communication skills are essential for building rapport with insurance medical directors and achieving positive outcomes.

Chapter 7: Avoiding Common Pitfalls

Even experienced physician advisors can encounter pitfalls in denial management. This chapter identifies common mistakes, such as misinterpreting payer criteria, overloading arguments with irrelevant details, or failing to address the financial implications of denials. Practical tips and case examples illustrate how to avoid these errors.

Key Takeaway: Avoiding common pitfalls enhances the physician advisor's credibility and effectiveness.

Chapter 8: Case Studies

Real-life examples bring the book's concepts to life. This chapter features detailed case studies of successful P2P conferences, including an analysis of the strategies that led to favorable outcomes. It also includes lessons learned from challenging cases, highlighting the importance of adaptability and persistence.

Key Takeaway: Case studies provide valuable insights into the practical application of denial management strategies.

Chapter 9: Measuring Success and Improving Skills

Continuous improvement is essential for physician advisors. This chapter explores methods for tracking overturn rates, seeking feedback, and mentoring others. It also discusses the importance of staying updated on payer policies, clinical guidelines, and emerging trends in denial management.

Key Takeaway: Measuring success and fostering professional development ensures sustained effectiveness in the physician advisor role.

Chapter 10: Responding to Upheld Denials – Writing Appeal Letters

Even with thorough preparation, some denials will inevitably be upheld. This chapter focuses on crafting effective appeal letters that directly address payer objections, align with clinical guidelines, and adhere to regulatory standards. It also emphasizes the importance of timeliness in the appeal process and explores how emerging technologies like artificial intelligence (AI) can enhance the efficiency and success of appeals.

Key Takeaway: Effective appeals turn setbacks into opportunities, reinforcing advocacy for patients while maintaining organizational integrity.

Chapter 11: The Future of the Physician Advisor Role

The healthcare landscape is constantly evolving, and the physician advisor role must evolve with it. This chapter examines emerging trends, including the integration of artificial intelligence (AI) and predictive analytics, the shift toward value-based care, and the need for standardized training programs. It also explores opportunities for physician advisors to shape the future of healthcare through advocacy and innovation.

Key Takeaway: Embracing change and innovation positions physician advisors for long-term success in an increasingly complex healthcare environment.

Practical Resources

The book concludes with appendices that provide practical tools and resources for physician advisors, including:

- Templates for preparing for P2P conferences.
- A glossary of key terms and concepts.
- References to authoritative guidelines, such as InterQual, MCG, and Medicare regulations.

Reflections on the Broader Impact of the Physician Advisor Role

The physician advisor role extends far beyond the immediate responsibilities of denial management and utilization review. It is a pivotal force in improving patient care, supporting organizational goals, and driving systemic change across the healthcare landscape. By fostering collaboration, advocating for evidence-based care, and streamlining operational processes, physician advisors create value at multiple levels of the healthcare ecosystem.

Enhancing Patient Care

At its core, the physician advisor role is designed to improve patient outcomes. By advocating during peer-to-peer (P2P) conferences, physician advisors ensure that medically necessary care is delivered without disruption. Additionally, they contribute to:

Reduced Care Fragmentation

Physician advisors help minimize disruptions caused by denied claims by advocating for uninterrupted treatment. For example, they often highlight critical patient-specific factors during P2P conferences to justify the necessity of certain admissions or treatments.

Empowering Clinical Teams

Physician advisors act as mentors and resources for clinicians, guiding them through payer requirements and improving documentation practices. This support reduces administrative burdens and allows clinical staff to focus on delivering high-quality care.

Aligning Care Delivery with Guidelines

By integrating authoritative guidelines like InterQual and MCG into their advocacy, physician advisors ensure that clinical decisions are both evidence-based and defensible, enhancing care quality and avoiding unnecessary denials.

Supporting Organizational Goals

In addition to their impact on patient care, physician advisors are instrumental in supporting the financial and operational health of healthcare organizations. Key contributions include:

Improved Financial Performance

Strong physician advisor programs can reduce denial rates and improve resolution times, contributing to better financial outcomes. Streamlined processes and effective advocacy can support cash flow improvements and organizational sustainability.

Regulatory Compliance

Physician advisors ensure adherence to complex regulatory frameworks, minimizing risks associated with audits and penalties. Their expertise in interpreting guidelines helps hospitals remain compliant while optimizing reimbursement.

Strengthening Payer Relationships

Constructive dialogue during P2P conferences fosters trust between providers and payers. Physician advisors act as mediators, reducing friction and building collaborative relationships.

Addressing Emerging Challenges

As healthcare systems evolve, physician advisors must navigate new challenges that require adaptability and innovation.

Administrative Overload

High denial volumes and payer scrutiny contribute to increasing workloads for physician advisors. Streamlining processes through standardized templates and AI-driven tools can alleviate these burdens.

Workforce Shortages

The ongoing shortage of clinical staff places additional pressure on physician advisors to allocate resources efficiently. This includes advocating for inpatient admissions when outpatient care could compromise patient safety.

Regulatory and Policy Changes

Evolving payer policies and legislative reforms require physician advisors to stay informed and agile. Adjustments to procedures and policies often demand immediate adaptation to ensure compliance.

Opportunities for Innovation and Growth

Despite these challenges, the future of the physician advisor role is rich with opportunities for growth and innovation.

Harnessing Artificial Intelligence

AI has the potential to transform denial management by identifying high-risk claims, automating repetitive tasks, and generating actionable insights. Physician advisors who adopt AI tools can improve efficiency and focus on complex cases.

Expanding Influence Across Care Settings

As healthcare systems integrate outpatient services, accountable care organizations (ACOs), and population health initiatives, physician advisors will play a larger role in these settings. Their expertise in compliance and denial management is invaluable for aligning care delivery with value-based reimbursement models.

Driving Systemic Reforms

Physician advisors can advocate for changes that reduce administrative burdens and enhance transparency in medical necessity determinations. Collaborative efforts between physician advisors and payers could lead to more streamlined prior authorization processes and clearer admission criteria.

Mentorship and Education

Sharing knowledge through workshops, publications, and mentorship ensures a steady pipeline of skilled physician advisors. Experienced advisors can guide new colleagues, fostering a culture of excellence and continuous improvement.

A Vision for the Future

Looking ahead, the physician advisor role will continue to evolve in response to the changing healthcare environment. Key opportunities include:

- **Leveraging Emerging Technologies:** AI and predictive analytics will enhance decision-making and streamline workflows.
- **Advocating for Policy Reforms:** Physician advisors will play a critical role in shaping payer and regulatory policies that benefit patients and providers.
- **Expanding the Role Beyond Hospitals:** Physician advisors will increasingly contribute to care delivery in outpatient and population health settings.

By embracing innovation, fostering collaboration, and advocating for systemic improvements, physician advisors will remain at the forefront of efforts to deliver high-quality, efficient, and equitable healthcare.

References

[1] Quinn, E (2022). Evolution of the Physician Advisor: A Current Perspective - RACMonitor, https://racmonitor.medlearn.com/evolution-of-the-physician-advisor-a-current-perspective/

[2] Bovbjerg, R. R., Griffin, C. C., & Carroll, C. E. (1993). U.S. Health Care Coverage and Costs: Historical Development and Choices for the 1990s. The Journal of Law, Medicine & Ethics, 21(2), 141–162. doi:10.1111/j.1748-720x.1993.tb01239.x

[3] Kaufman, B. G., Spivack, B. S., Stearns, S. C., Song, P. H., & O'Brien, E. C. (2017). Impact of Accountable Care Organizations on Utilization, Care, and Outcomes: A Systematic Review. Medical Care Research and Review, 107755871774591. doi:10.1177/1077558717745916

[4] Wennberg, J. E. (1984). Dealing With Medical Practice Variations: A Proposal for Action. Health Affairs, 3(2), 6–33. doi:10.1377/hlthaff.3.2.6

[5] Harrington, C., Lynch, M., & Newcomer, R. J. (1993). Medical Services in Social Health Maintenance Organizations. The Gerontologist, 33(6), 790–800. doi:10.1093/geront/33.6.790

[6] Mayes, R. (2006). The Origins, Development, and Passage of Medicare's Revolutionary Prospective Payment System. Journal of the History of Medicine and Allied Sciences, 62(1), 21–55. doi:10.1093/jhmas/jrj038

[7] Terra, S. M. (2009). Regulatory Issues. Professional Case Management, 14(5), 217–223. doi:10.1097/ncm.0b013e3181b094fe

[8] Zander, Karen. "Getting more from physician advisors: building a formal job description to improve effectiveness." Case Management Monthly, vol. 4, no. 12, Dec. 2007, pp. 4+. Gale OneFile: Health and Medicine, link.gale.com/apps/doc/A171623700/HRCA?u=nysl_oweb&sid=googleScholar&xid=a49f3714. Accessed 2 Dec. 2024.

[9] Baumel, Mark J., and Robert R. Corrato. "Dedicating the right resources to denials management: full-time physician attention helped reduce one health system's denied claims and length of stay." *Healthcare Financial Management*, vol. 57, no. 12, Dec. 2003, pp. 66+. *Gale Academic OneFile*, link.gale.com/apps/doc/A111696061/AONE?u=nysl_oweb&sid=googleScholar&xid=83446def. Accessed 28 Dec. 2024

CHAPTER 1

The Role of the Physician Advisor

Introduction

The physician advisor role has become a cornerstone of hospital operations, evolving in response to growing regulatory demands, payer scrutiny, and financial pressures. Initially focused on utilization review, the role has expanded to encompass responsibilities such as clinical documentation integrity (CDI), regulatory compliance, denial management, and peer-to-peer (P2P) conferences. Today, physician advisors are essential to aligning clinical care with operational goals, safeguarding hospital revenue, and ensuring compliance with ever-changing healthcare regulations.[10]

Hospitals increasingly depend on physician advisors to navigate these complexities, bridging the gap between administrative and clinical priorities. This chapter explores the history and development of the physician advisor role, its key responsibilities, and its growing importance in hospital networks.

History and Development of the Field

The physician advisor role has its roots in the 1980s, when managed care organizations (MCOs) began implementing utilization review programs to control rising healthcare costs. Hospitals faced increasing pressure to justify admissions, lengths of stay, and procedural appropriateness, often resulting in inconsistencies and inefficiencies when managed without clinical expertise.[11]

The Managed Care Era: Laying the Groundwork

The rise of health maintenance organizations (HMOs) in the 1990s introduced stricter oversight of hospital practices through standardized criteria for medical necessity, such as early versions of InterQual and MCG guidelines.[12] However, the lack of clinical expertise in applying these criteria often led to inappropriate denials and strained relationships between providers and payers.[13]

Hospitals began appointing part-time physician advisors to address these challenges. These early advisors focused on reviewing medical necessity and resolving disputes with payers, laying the foundation for a more structured approach to utilization management.[14]

Professionalization of the Role: The Early 2000s

The physician advisor role became more formalized in the 2000s with the development of training programs and the establishment of organizations like the American College of Physician Advisors (ACPA). This evolution elevated the role from an informal function to a recognized profession, with expanded responsibilities such as CDI, denial management, and regulatory compliance.[15]

Modern Evolution: Beyond Utilization Review

Today, physician advisors have moved beyond their origins in utilization management to serve as strategic partners in hospital operations. They influence areas such as:

1. **Technology Adoption:** Physician advisors leverage technology, including AI, to streamline workflows and improve documentation accuracy.
2. **Policy Advocacy:** They collaborate with professional organizations to influence payer policies and regulatory frameworks.
3. **Systemic Improvements:** Physician advisors identify inefficiencies and implement initiatives to enhance care delivery and financial performance.

Key Responsibilities of Physician Advisors

Physician advisors play a multifaceted role within healthcare organizations, bridging clinical expertise, administrative acumen, and strategic vision. Their responsibilities are dynamic, reflecting the evolving challenges of modern healthcare. This section expands on their key responsibilities to provide a comprehensive understanding of their impact.

---

1. Utilization Management

Utilization management (UM) ensures that hospital resources are used efficiently while maintaining high standards of patient care. Physician advisors are central to this process, helping hospitals align admissions, continued stays, and discharge plans with payer-specific criteria for medical necessity.

Responsibilities:

- **Case Reviews:** Collaborating with case management teams to evaluate flagged cases for appropriate level of care.
- **Payer Communication:** Acting as intermediaries between hospitals and payers to resolve discrepancies in medical necessity determinations.

- **Developing Protocols:** Creating or refining UM protocols to reduce denials and improve workflow efficiency.

Challenges:

- Navigating the diverse criteria and policies of different payers.
- Balancing clinical judgment with administrative requirements, particularly in borderline cases.
- Managing patients who do not meet strict inpatient criteria but still require significant resources (e.g., extended monitoring or specialized care).

Impact:
Effective utilization management improves patient flow, reduces unnecessary costs, and minimizes disruptions to care delivery caused by inappropriate denials or delays.

---

2. Clinical Documentation Integrity (CDI)

Accurate clinical documentation is critical for reflecting the complexity of patient care, ensuring appropriate reimbursement, and maintaining compliance with regulatory standards. Physician advisors are instrumental in driving CDI initiatives.

Responsibilities:

- **Education and Training:** Teaching clinicians to document with specificity, emphasizing terminology that aligns with coding and payer requirements.
- **Collaboration:** Partnering with CDI specialists to identify and address documentation gaps.
- **Quality Improvement:** Leading initiatives to enhance documentation processes, such as implementing standardized templates or checklists.

Example:
Documenting "acute hypoxemic respiratory failure" instead of "shortness of breath" ensures that the severity of the condition is accurately captured, leading to better alignment with payer terminology and reimbursement guidelines.

Impact:
Improved documentation reduces denial rates, enhances risk adjustment for value-based payment models, and bolsters organizational quality metrics.

---

3. Denial Management

Denial management is one of the most financially impactful responsibilities of physician advisors. They analyze denied claims, identify trends, and work to overturn inappropriate denials through appeals and peer-to-peer (P2P) conferences.

Responsibilities:

- **Trend Analysis:** Identifying patterns in denials to address systemic issues.
- **P2P Advocacy:** Engaging in discussions with payer medical directors to provide clinical justifications for care.
- **Developing Prevention Strategies:** Establishing proactive measures to reduce the likelihood of denials, such as pre-admission reviews or enhanced documentation protocols.

Example:
During a P2P conference, a physician advisor might overturn a denial by presenting clinical evidence that a patient required inpatient care due to refractory hyperglycemia and multiple electrolyte abnormalities, aligning the case with established guidelines.

Impact:
Effective denial management protects hospital revenue, reduces administrative burdens, and fosters trust between healthcare providers and payers.

---

4. Regulatory Compliance

Navigating the complex and ever-changing landscape of healthcare regulations is a critical aspect of the physician advisor's role. By ensuring compliance, physician advisors mitigate risks associated with audits, penalties, and accreditation deficiencies.

Responsibilities:

- **Staff Education:** Providing training on regulatory requirements, such as CMS guidelines and The Joint Commission standards.
- **Audit Preparedness:** Identifying documentation or billing vulnerabilities to reduce risks during audits.
- **Policy Development:** Collaborating on policies that align clinical and operational practices with regulatory standards.

Impact:
Proactive compliance efforts safeguard hospital operations, avoid costly penalties, and enhance organizational credibility.

---

5. Leadership and Education

Physician advisors often serve as leaders and mentors, fostering a culture of accountability, continuous learning, and clinical excellence.

Responsibilities:

- **Staff Training:** Educating clinicians on documentation best practices, denial prevention strategies, and payer-specific criteria.
- **Mentorship:** Supporting new physician advisors and guiding clinical teams in navigating administrative challenges.
- **Leadership Roles:** Participating in hospital committees and strategic planning efforts to align clinical care with organizational goals.

Example:
At a large hospital, a physician advisor might lead a quarterly workshop for hospitalists, focusing on proper documentation of sepsis and acute respiratory failure, which results in measurable reductions in denial rates for these diagnoses.

Impact:
Leadership efforts improve staff confidence, enhance documentation quality, and align organizational priorities with patient care standards.

---

6. Driving Systemic Change

Physician advisors are uniquely positioned to identify inefficiencies and lead initiatives that drive systemic improvements across healthcare organizations.

Responsibilities:

- **Operational Enhancements:** Implementing projects to streamline workflows, improve patient flow, or optimize resource allocation.
- **Policy Development:** Contributing to hospital-wide policies that standardize processes, such as admission and discharge criteria.
- **Advocacy:** Promoting systemic reforms that reduce administrative burdens or enhance care delivery, both within the organization and at the industry level.

Example:
A physician advisor might spearhead a project to standardize discharge planning processes across a multi-hospital system, reducing readmissions and improving patient satisfaction.

Impact:
Systemic improvements driven by physician advisors enhance care quality, operational efficiency, and financial performance, creating a more sustainable healthcare system.

The key responsibilities of physician advisors underscore their vital role in healthcare organizations. By balancing clinical expertise with operational and regulatory insights, physician advisors ensure that care delivery aligns with payer expectations, financial goals, and regulatory standards. Their contributions drive measurable improvements in patient outcomes, hospital performance, and healthcare innovation.

Metrics for Success

The effectiveness of physician advisors is measured through key performance indicators that provide insight into their impact on clinical operations, financial performance, and patient care outcomes. These metrics allow healthcare organizations to assess the value of physician advisor programs and identify areas for improvement. Below is an expanded discussion of these metrics and their implications.

---

1. Denial Rates

Definition and Importance:
Denial rates measure the percentage of claims rejected or denied by payers, providing a direct indicator of the efficiency of utilization management and documentation practices. High denial rates can indicate gaps in compliance with payer requirements, insufficient documentation, or misalignment between clinical decisions and payer criteria.[16]

How Physician Advisors Impact Denial Rates:

- **Proactive Documentation Reviews:** Physician advisors identify potential deficiencies in documentation before claims are submitted, ensuring alignment with payer expectations.
- **Peer-to-Peer Conferences (P2Ps):** By advocating for medically necessary care during P2Ps, physician advisors can overturn inappropriate denials.
- **Trend Analysis and Education:** Tracking denial trends allows physician advisors to educate clinical teams on common issues, such as inadequate specificity in diagnosis documentation.

Key Outcomes:

- Reduced administrative burdens by minimizing the need for appeals.
- Improved cash flow through faster resolution of claims.

- Strengthened payer-provider relationships by reducing contentious denials.

---

2. Case Mix Index (CMI)

Definition and Importance:
CMI reflects the relative complexity and severity of patients treated by a healthcare organization. A higher CMI indicates more resource-intensive and clinically complex cases, which align with higher reimbursement rates under many payment models.[17]

How Physician Advisors Impact CMI:

- **Accurate Documentation:** Ensuring that all comorbidities, complications, and severity indicators are documented allows CMI to accurately represent patient acuity.
- **Collaboration with CDI Teams:** Physician advisors work with Clinical Documentation Integrity (CDI) specialists to refine documentation processes, ensuring consistency with coding and billing standards.
- **Educational Efforts:** Training clinicians on the importance of documenting specifics—such as acute organ dysfunction or the presence of chronic conditions—enhances CMI accuracy.

Key Outcomes:

- Increased reimbursement rates through proper reflection of case complexity.
- Enhanced quality scores and risk adjustment for public reporting and value-based payment models.
- Improved organizational insights into patient population acuity for strategic planning.

---

3. Length of Stay (LOS)

Definition and Importance:
LOS measures the average time patients spend in the hospital, which directly impacts operational efficiency and resource utilization. While a shorter LOS can indicate efficient care delivery, it must be balanced with patient safety and outcomes.[18]

How Physician Advisors Impact LOS:

- **Efficient Utilization Reviews:** By ensuring that patients are placed at the appropriate level of care, physician advisors prevent unnecessary delays in discharge.
- **Discharge Planning and Transitions:** Physician advisors collaborate with care teams to address barriers to discharge, such as pending diagnostic results or delays in arranging post-acute care.
- **Avoiding Readmissions:** Ensuring safe transitions to the next level of care reduces readmissions and avoids artificially inflating LOS due to complications.

Key Outcomes:

- Improved patient flow, allowing hospitals to admit and treat more patients.
- Reduced costs associated with prolonged hospital stays.
- Enhanced patient satisfaction by minimizing unnecessary hospital time.

---

4. Readmission Rates

Definition and Importance:
Readmission rates measure the percentage of patients who return to the hospital within a specific timeframe (commonly 30 days) after discharge. High readmission rates often reflect gaps in care coordination, discharge planning, or follow-up care.[18]

How Physician Advisors Impact Readmission Rates:

- **Identifying High-Risk Patients:** Physician advisors flag patients at risk of readmission and coordinate targeted interventions, such as early follow-up appointments or enhanced discharge instructions.
- **Addressing Social Determinants of Health (SDOH):** By identifying barriers such as transportation or medication affordability, physician advisors ensure that patients have the resources needed for recovery.
- **Collaborative Care Models:** Physician advisors work with interdisciplinary teams to improve transitions of care, such as connecting patients to home health services or skilled nursing facilities when appropriate.

Key Outcomes:

- Avoidance of penalties under programs like CMS's Hospital Readmissions Reduction Program.
- Improved patient outcomes and satisfaction.
- Enhanced hospital reputation in public reporting and quality rankings.

Integrating Metrics for Comprehensive Evaluation

While each metric provides valuable insights, their true power lies in combining them to create a comprehensive picture of organizational performance and physician advisor impact. For example:

- **Balancing LOS and Readmission Rates:** An extremely short LOS could lead to increased readmissions, suggesting that discharge decisions were rushed. Physician advisors balance these metrics to ensure that care transitions are both timely and safe.
- **Denial Rates and CMI:** Improving documentation to reduce denials often boosts CMI as well, reflecting a more accurate picture of patient complexity.

- **Cross-Functional Collaboration:** Metrics like denial rates and LOS require coordination between physician advisors, CDI teams, case managers, and clinical staff, emphasizing the multidisciplinary nature of the role.

Using Metrics for Continuous Improvement

Physician advisors and healthcare organizations can leverage these metrics to:

- **Identify Areas for Focus:** Metrics highlight pain points, such as common reasons for denials or patient populations with high readmission risks.
- **Guide Training and Policy Development:** Trends in metrics inform educational initiatives and the development of organizational policies.
- **Monitor Progress:** Regularly tracking these metrics provides a clear measure of the impact of interventions, helping organizations adapt strategies in real-time.

By aligning their efforts with these key metrics, physician advisors ensure their contributions translate into tangible benefits for healthcare organizations and patients alike.

Future Directions

The physician advisor role continues to evolve, with opportunities to integrate emerging technologies, expand influence beyond hospitals, and advocate for systemic reforms. By leveraging their clinical expertise and operational insights, physician advisors are well-positioned to shape the future of healthcare.

Conclusion

The physician advisor role has evolved significantly, transitioning from a narrowly defined function in utilization review to a multifaceted leadership position. Physician advisors now contribute to organizational success by bridging clinical and administrative priorities, driving operational improvements, and advocating for systemic reforms.

As healthcare continues to face challenges such as regulatory changes, workforce shortages, and payer complexities, the physician advisor role will remain pivotal. Through innovation, education, and collaboration, physician advisors will continue to enhance patient care, safeguard hospital revenue, and contribute to a more efficient and equitable healthcare system.

References

[10] Zander, Karen. "Getting more from physician advisors: building a formal job description to improve effectiveness." *Case Management Monthly*, vol. 4, no. 12, Dec. 2007, pp. 4+. *Gale OneFile: Health and Medicine*, link.gale.com/apps/doc/A171623700/HRCA?u=nysl_oweb&sid=googleScholar&xid=a49f3714. Accessed 28 Dec. 2024.

[11] Robert J. Becker, Managed care is utilization review, *American Journal of Hospital Pharmacy*, Volume 47, Issue 10, 1 October 1990, Pages 2274–2276, https://doi.org/10.1093/ajhp/47.10.2274

[12] Mitus, A. Jacqueline. (2008). *The Birth of InterQual. Professional Case Management, 13(4), 228–233.* doi:10.1097/01.pcama.0000327413.01

[13] Field, M. J., & Gray, B. H. (1989). *Should We Regulate "Utilization Management?" Health Affairs, 8(4), 103–112.* doi:10.1377/hlthaff.8.4.103

[14] Lawless, Grant D., and Dorothy B. Holt. "Developing an in-house physician advisor program." *Physician Executive*, vol. 17, no. 3, May-June 1991, pp. 29+. *Gale Academic OneFile*, link.gale.com/apps/doc/A11008711/AONE?u=nysl_oweb&sid=googleScholar&xid=c2462d1f. Accessed 29 Dec. 2024.

[15] Pahuja D, Chadha P, Sharma U, Bhatia H, Nagpal P, Jordan J, Lannom C, de Los Santos E. PHYSICIAN ADVISERS ADD UP FOR SOME HOSPITALS. Physician Leadersh J. 2017 Mar;4(2):44-48. PMID: 30571884.

[16] Hodges, Jackie. "Effective claims denial management enhances revenue: claims denial management has become a critical component of a hospital's strategic effort to offset the adverse impact of Balanced Budget Act payment reductions." *Healthcare Financial Management*, vol. 56, no. 8, Aug. 2002, pp. 40+. *Gale Academic OneFile*, link.gale.com/apps/doc/A90317285/AONE?u=nysl_oweb&sid=googleScholar&xid=579c0167. Accessed 10 Jan. 2025.

[17] Mendez, C. M., Harrington, D. W., Christenson, P., & Spellberg, B. (2014). *Impact of Hospital Variables on Case Mix Index as a Marker of Disease Severity. Population Health Management, 17(1), 28–34.* doi:10.1089/pop.2013.0002

[18] Oh, J. C., Zheng, Z. E., & Bardhan, I. R. (2017). *Sooner or Later? Health Information Technology, Length of Stay, and Readmission Risk. Production and Operations Management.* doi:10.1111/poms.12748

CHAPTER 2

Understanding Insurance Company Processes

Introduction

Insurance companies play a critical role in the financial health of hospitals, yet their processes often appear opaque and challenging to navigate. For physician advisors, understanding the workflow of insurance medical directors, the rationale behind common claim denials, and payer-specific criteria is essential to successfully overturn adverse decisions. These decisions impact patient care, organizational revenue, and hospital operations.

Denials not only delay or reduce payments but also impose significant administrative burdens on clinical and billing teams. Hospitals invest substantial resources in denial management activities, emphasizing the importance of proactive strategies. By mastering payer operations and aligning clinical arguments with insurance criteria, physician advisors can improve denial overturn rates, safeguard hospital revenue, and advocate effectively for patients.

This chapter explores the most frequent reasons for claim denials, the workflow of insurance medical directors, and strategies for tailoring clinical arguments to meet payer expectations.

Common Reasons for Claim Denials

1. Medical Necessity Denials

Medical necessity denials occur when a payer determines that the provided services do not meet established criteria for being essential for the patient's condition or safety. This type of denial is particularly prevalent in cases where clinical documentation lacks sufficient detail to demonstrate the urgency or complexity of the care provided.[19]

- **Common Scenarios:**
  - Admissions for conditions like pneumonia or sepsis where vital signs, imaging findings, or lab values do not sufficiently support the need for inpatient care.
  - Procedures performed without documenting the failure of conservative measures, such as physical therapy or medication trials.
  - Services for chronic conditions where the rationale for escalation to more intensive treatment is unclear.
- **Preventive Strategy:** To avoid medical necessity denials:

    - Include specific, objective clinical data such as abnormal vital signs (e.g., oxyger saturation <90% on room air), radiographic findings, and the patient's response (or lack thereof) to treatments.
    - Ensure that the documentation reflects any risk factors or comorbidities that elevate the severity of the patient's condition.
    - Utilize checklists or templates that align with payer-specific criteria (e.g., InterQual, MCG).

2. Insufficient Documentation

Incomplete or vague documentation is a leading cause of claim denials. Insurers require comprehensive records that clearly describe the patient's clinical condition, treatment rationale, and outcomes.[20]

- **Common Scenarios:**
    - Lack of detailed notes describing failed outpatient treatments, such as antibiotics for cellulitis or bronchodilators for asthma.
    - Missing trends in lab results or vital signs that demonstrate a patient's deterioration.
    - Progress notes that omit significant clinical findings or ongoing care needs.
- **Preventive Strategy:** To mitigate the risk of insufficient documentation:
    - Train clinicians to provide thorough, structured documentation that captures the full clinical picture, including prior treatments and their outcomes.
    - Emphasize the importance of documenting all pertinent details, such as symptom progression, response to therapies, and the reasoning behind clinical decisions.
    - Conduct regular audits to identify and address documentation gaps.

3. Coding Errors

Coding inaccuracies are a frequent cause of denials, often stemming from discrepancies between the documented clinical scenario and the assigned diagnosis or procedural codes.[21]

- **Common Scenarios:**
    - Mismatched codes for conditions such as acute myocardial infarction versus chronic ischemic heart disease, resulting in denials for high-cost interventions.
    - Ambiguous documentation leading to the use of nonspecific codes, which may not justify the level of care provided.
    - Errors in translating clinical documentation into codes that align with payer criteria for reimbursement.
- **Preventive Strategy:** Effective strategies to minimize coding errors include:
    - Strengthening collaboration among physician advisors, CDI specialists, and coding teams to ensure alignment between documentation and codes.
    - Implementing coding software that provides prompts for potential discrepancies.
    - Conducting regular training sessions for clinicians and coders to improve their understanding of coding requirements.

4. Length of Stay (LOS) Disputes

Payers frequently challenge inpatient stays that exceed expected durations, especially when progress notes fail to justify the need for ongoing hospitalization.

- **Common Scenarios:**
    - Hospital stays extended for non-medical reasons, such as awaiting placement in a skilled nursing facility, without clearly documenting the ongoing need for inpatient-level care.
    - Post-surgical admissions where pain management or mobility issues are not explicitly documented to support additional days of hospitalization.
    - Patients with chronic conditions where daily progress notes lack updates on clinical improvement or setbacks.
- **Preventive Strategy:** To address LOS disputes:
    - Document clinical reasoning for each day of hospitalization, including the patient's ongoing needs, unresolved complications, and goals for discharge.
    - Use multidisciplinary rounds to ensure that progress notes reflect all aspects of care, such as physical therapy achievements, unresolved lab abnormalities, and pain control.
    - Provide discharge planning updates in the medical record to show proactive efforts in reducing LOS.

5. Authorization Failures

Authorization denials occur when providers fail to obtain or document prior approvals for procedures or treatments that require payer review. These denials can often be automated and difficult to appeal.[22]

- **Common Scenarios:**
    - Elective surgeries or high-cost imaging studies (e.g., MRI or PET scans) performed without obtaining the necessary pre-approval.
    - Procedures requiring urgent authorization submitted without adequate supporting documentation, such as imaging results or clinical notes.
    - Changes to treatment plans, such as adding an additional procedure during surgery, not being reflected in the original authorization.
- **Preventive Strategy:** Preventing authorization failures involves:
    - Establishing a robust pre-authorization workflow that includes double-checking payer requirements and obtaining all necessary approvals before scheduling elective procedures.
    - Ensuring that authorization documentation, including payer responses, is promptly added to the patient's medical record.
    - Developing systems to track expiration dates of prior authorizations to avoid denials due to lapsed approvals.

The Workflow of Insurance Medical Directors

Insurance medical directors are central to the claim review and appeals process. Their role involves balancing payer policies, medical necessity criteria, and contractual obligations while ensuring cost-effective care. Physician advisors must understand this workflow to better advocate for patients and effectively overturn claim denials.

1. Automated Claim Screening

The initial stage in the claims review process involves automated systems that flag potential issues. These systems are powered by algorithms designed to enforce payer policies and identify claims that require further scrutiny.[23]

- **Common Screening Triggers:**
  - Missing or incomplete documentation, such as absent imaging or lab reports.
  - Admission and discharge patterns inconsistent with payer guidelines.
  - Diagnosis-procedure mismatches, where the treatment does not align with the stated diagnosis.
  - Length of stay exceeding payer-established norms for a specific condition.
- **Impact on Physician Advisors:** Automated screenings place significant emphasis on documentation quality. Physician advisors should work with clinical teams to ensure that medical records include comprehensive, precise details that align with payer expectations. For example:
  - Documenting failed outpatient treatments for conditions such as pneumonia or cellulitis.
  - Including lab trends, vital signs, and imaging findings to substantiate clinical instability.

2. Clinical Review

Claims flagged during automated screening are escalated for manual evaluation by clinical reviewers, including nurses and insurance medical directors. These professionals assess whether the provided documentation supports the medical necessity of the services rendered.

- **Role of Insurance Medical Directors:**
  - Review flagged claims against payer-specific and standardized guidelines, such as InterQual or MCG.
  - Evaluate clinical evidence to determine if it justifies the requested level of care.
  - Identify gaps or inconsistencies in the documentation that might validate a denial.
- **Challenges:**
  - Variability in payer criteria: Insurance medical directors often rely on proprietary guidelines that differ from widely used standards like InterQual.
  - Subjectivity in decision-making: Determining medical necessity can depend on interpretation, especially in ambiguous cases.
- **Key Insight for Physician Advisors:** Familiarity with payer-specific guidelines is essential. Physician advisors should maintain an up-to-date library of payer policies and ensure that documentation explicitly meets these criteria. For instance:
  - InterQual criteria for sepsis may require documentation of elevated lactic acid levels and hypotension unresponsive to fluids.
  - MCG guidelines for heart failure may emphasize failed outpatient management and the need for IV diuretics.

3. Peer-to-Peer (P2P) Conferences

P2P conferences provide a critical forum for physician advisors to advocate for their cases by engaging directly with insurance medical directors. These discussions offer an opportunity to address gaps in documentation, clarify clinical rationale, and resolve misunderstandings.

- **Structure of P2P Conferences:**
    - A payer medical director outlines the rationale for the denial, referencing specific policy guidelines.
    - The physician advisor presents the clinical justification for the claim, addressing the points raised in the denial letter.
    - Both parties collaboratively review the patient's record, focusing on key evidence that supports the claim.
- **Strategies for Success:**
    - **Thorough Preparation:** Before the conference, review the denial letter and relevant payer guidelines. Identify documentation gaps and anticipate possible objections.
    - **Data-Driven Arguments:** Focus on objective data, such as vital signs, lab results, and imaging findings, to substantiate the medical necessity of care.
    - **Address Ambiguities:** Clearly explain any gaps in documentation, providing context or supplemental evidence where needed.
    - **Professional Communication:** Maintain a collaborative tone and use payer-specific language to align arguments with the insurance medical director's perspective.

4. Escalation and External Review

Complex cases that cannot be resolved through P2P conferences may require escalation to payer specialists or external review boards. These stages represent the final opportunity to overturn a denial.[24]

- **Consultation with Payer Specialists:** For high-risk or specialized cases, insurance medical directors may consult in-house specialists (e.g., cardiologists, oncologists) to assess the claim. This process adds another layer of scrutiny and often involves a deeper analysis of the clinical evidence.
- **External Review Boards:** If internal processes fail to resolve the dispute, claims may be referred to independent external review boards. These boards operate as impartial entities, making final determinations based on the submitted documentation.
- **Challenges:**
    - Strict timelines for submitting appeals and supporting documents.
    - Limited opportunity for oral arguments, requiring a well-organized and comprehensive written appeal.
    - Variability in rulings, which may depend on differing interpretations of clinical guidelines.
- **Strategies for Success:**
    - Provide a detailed, evidence-based appeal that addresses all potential objections.

- Reference authoritative specialty guidelines (e.g., American Heart Association or American Thoracic Society) to supplement payer criteria.
    - Include a clear, concise narrative summarizing the clinical rationale and supporting data.
- **Impact on Physician Advisors:** Escalation represents the last chance to secure approval for the claim. Physician advisors must be thorough in their preparation, leveraging all available resources to present compelling, evidence-supported arguments.

Summary

By understanding the workflow of insurance medical directors—from automated screenings to external reviews—physician advisors can proactively align clinical documentation with payer expectations. This knowledge enables them to navigate denials more effectively, advocate for patient care, and protect hospital revenue. Through strategic documentation, thorough preparation, and clear communication, physician advisors can successfully engage with insurance medical directors and achieve favorable outcomes.

Key Metrics for Evaluating Success

Hospitals track various metrics to assess trends in denials and measure the effectiveness of intervention strategies. These metrics not only provide insight into performance but also identify opportunities for improvement. Below are the key indicators and their significance:

1. **Denial Rate**
   This metric represents the percentage of claims initially denied out of the total claims submitted within a specific timeframe. A high denial rate can indicate systemic issues such as documentation gaps, coding errors, or payer policy misunderstandings. Monitoring the denial rate over time helps identify patterns, such as specific payers, procedures, or departments contributing to the problem.
   Calculation Example:

   To calculate the Denial Rate (%):

   Denial Rate (%) = (Number of Denied Claims / Total Claims Submitted) × 100

   Example:
   - Number of Denied Claims: 50
   - Total Claims Submitted: 1,000

   Denial Rate (%) = (50 / 1,000) × 100 = 5%

Goal: A low and consistently decreasing denial rate indicates effective prevention strategies and improved compliance with payer requirements.

2. **Appeal Success Rate**
   This measures the percentage of denied claims successfully overturned through the appeals process. It is a direct reflection of the effectiveness of peer-to-peer discussions, appeal letters, and supporting documentation. A high success rate demonstrates the capability to challenge inappropriate denials effectively and recover lost revenue.
   Calculation Example:

   To calculate the Appeal Success Rate (%):

   Appeal Success Rate (%) = (Successful Appeals / Total Denials Appealed) × 100

   Example:
   - Successful Appeals: 30
   - Total Denials Appealed: 50

   Appeal Success Rate (%) = (30 / 50) × 100 = 60%

   Goal: Strive for an appeal success rate above 70%, which indicates strong advocacy and robust denial management practices.

3. **Time to Resolution**
   The average time required to resolve denials, either through payment or a final determination, is a critical metric. Long resolution times can lead to cash flow disruptions and higher administrative burdens. Tracking this metric helps identify bottlenecks in processes, such as delays in submitting appeals or responding to payer requests for additional information.
   Calculation Example:

   To calculate the Average Time to Resolution (days):

   Average Time to Resolution (days) = Sum of Time to Resolve Each Denial / Total Resolved Denials

   Example:
   - Sum of Time to Resolve Each Denial: 300 days
   - Total Resolved Denials: 50

   Average Time to Resolution (days) = 300 / 50 = 6 days

Goal: Reducing the time to resolution while maintaining high appeal success rates improves revenue cycle efficiency and ensures quicker access to funds.

4. **Root Cause Categorization of Denials**
Categorizing denials by their root causes—such as insufficient documentation, authorization issues, or coding errors—provides actionable insights. This metric allows targeted education and process improvements to address specific weaknesses.
Example Categories:
    - Medical Necessity
    - Coding Errors
    - Lack of Authorization
    - Timely Filing Issues
    Goal: Prioritize interventions in high-frequency denial categories and track reductions over time.

5. **Net Denial Rate**
Unlike the initial denial rate, the net denial rate considers only claims that remain denied after all appeal efforts. This metric reflects the ultimate impact of denials on hospital revenue.
Calculation Example:

    To calculate the Net Denial Rate (%):

    Net Denial Rate (%) = (Claims Remaining Denied After Appeals / Total Claims Submitted) × 100

    Example:
    - Claims Remaining Denied After Appeals: 20
    - Total Claims Submitted: 1,000

    Net Denial Rate (%) = (20 / 1,000) × 100 = 2%

    Goal: Achieve a net denial rate below 5%, indicating strong revenue recovery practices.

6. **Revenue Recovered Through Appeals**
This is the total dollar amount recovered through successful appeals. It provides a direct measure of financial impact and underscores the importance of denial management programs.
Calculation Example:

    To calculate Revenue Recovered ($):

Revenue Recovered ($) = Total Amount of Overturned Denials

Example:
- Total Amount of Overturned Denials: $150,000

Revenue Recovered ($) = $150,000

Goal: Maximize recovered revenue while minimizing resources expended during the appeal process.

7. **Payer-Specific Denial Trends**
   Analyzing denial rates and appeal success rates by payer identifies payer-specific issues, such as stricter policies or unique documentation requirements. This allows tailored strategies, such as preemptive documentation audits for high-denial payers.
   Goal: Reduce payer-specific denials through proactive engagement and contract negotiation.
8. **Cost of Denial Management**
   Denial management involves resource allocation, including staff time and technology investments. Calculating the cost per denial managed or per denial overturned provides insights into the financial efficiency of the program.
   Goal: Optimize resource utilization to ensure the denial management program provides a strong return on investment (ROI).
9. **Preventable Denial Rate**
   A subset of the denial rate, this metric focuses on denials that could have been avoided with better processes, documentation, or compliance.
   Goal: Implement preventive measures to eliminate the most common causes of preventable denials.
10. **Provider Education Impact**
    Tracking improvements in denial rates and appeal success rates after implementing provider education programs demonstrates their effectiveness.
    Goal: Foster a culture of continuous improvement through data-driven education and feedback.

By systematically tracking and analyzing these metrics, hospitals can not only improve their denial management processes but also foster a culture of accountability and continuous improvement. Integrating these metrics into regular reporting provides leadership with a clear picture of progress and areas needing further attention.

Enhancing Payer Collaboration

Effective collaboration with payers is essential for minimizing claim denials and improving the efficiency of the appeals process. While the payer-provider relationship can often be adversarial, proactive engagement and mutual understanding can transform it into a partnership aimed at achieving the best outcomes for patients. Below are detailed strategies physician advisors can use to enhance payer collaboration:

1. **Engaging Payer Representatives Early to Clarify Expectations**
   Early engagement sets the tone for a productive relationship. Physician advisors should initiate contact with payer representatives to understand the nuances of their policies and criteria.
   - **Preemptive Discussions:** Conduct pre-authorization calls or joint case reviews for high-risk or complex patients to ensure alignment on expectations before care is delivered.
   - **Policy Education:** Request updates on changes to coverage policies, prior authorization requirements, or documentation standards to stay informed and prevent denials.
   - **Escalation Protocols:** Establish clear channels for escalation in case of disputes, ensuring a smooth resolution process.
     Example: Hosting quarterly meetings with payer representatives to discuss policy updates and review trends in denied claims.
2. **Building Trust Through Transparent Communication**
   Trust is the cornerstone of any collaborative relationship. Physician advisors can build credibility with payers by demonstrating a commitment to fairness, compliance, and patient-centered care.
   - **Consistency:** Provide complete and accurate documentation to avoid discrepancies or misinterpretations.
   - **Honesty:** Acknowledge errors promptly and take corrective actions to prevent recurrence.
   - **Regular Updates:** Share outcomes of interventions, such as reduced denial rates or improved compliance with specific criteria, as evidence of progress.
     Example: Sending regular reports to payers on trends in claim approvals and highlighting efforts to address previously identified issues, such as insufficient documentation.
3. **Proactively Addressing Common Denial Patterns with Targeted Strategies**
   Analyzing denial data to identify trends is critical for proactive collaboration. Physician advisors should work with payers to address recurring issues and implement process improvements.
   - **Root Cause Analysis:** Partner with payers to investigate the underlying causes of common denials, such as unclear medical necessity criteria or misinterpretation of documentation.
   - **Standardization:** Develop mutually agreed-upon templates for frequently requested documentation to reduce back-and-forth communication.
   - **Provider Education:** Collaborate on educational initiatives for clinical staff, focusing on documentation practices that align with payer requirements.
     Example: If "medical necessity" is a frequent denial reason, physician advisors can

work with payers to create checklists ensuring that documentation addresses specific criteria upfront.

4. **Establishing Collaborative Feedback Loops**
   Continuous improvement requires regular feedback from payers about areas of non-compliance or best practices. Establishing a feedback mechanism ensures that the hospital can refine its processes in real-time.
   - **Monthly Reviews:** Set up recurring meetings or calls to review denial trends, appeal outcomes, and payer concerns.
   - **Mutual Accountability:** Encourage payers to provide detailed explanations for denials to ensure that hospitals can address them appropriately.
   - **Shared Metrics:** Agree on key performance indicators (KPIs), such as time to resolution or appeal success rates, to track the effectiveness of the collaboration.
     Example: Jointly analyzing data on overturned appeals to identify denial categories where payer policies may need clarification or revision.
5. **Fostering a Culture of Partnership, Not Adversity**
   The relationship between providers and payers is often marked by mistrust and conflict. Physician advisors can change this dynamic by fostering a culture of partnership focused on shared goals, such as improving patient outcomes and reducing unnecessary administrative burdens.
   - **Highlighting Common Goals:** Emphasize that both parties aim to ensure appropriate care while managing costs effectively.
   - **Open Communication:** Encourage dialogue to resolve disputes rather than relying solely on formal appeal processes.
   - **Celebrating Successes:** Recognize and celebrate milestones, such as a decrease in unnecessary denials or faster appeal resolutions, to reinforce a positive working relationship.
     Example: Acknowledging a payer's flexibility in reconsidering a disputed case during a peer-to-peer call can build goodwill and set a cooperative tone for future interactions.
6. **Leveraging Technology for Seamless Communication**
   Implementing technology solutions can facilitate better collaboration with payers by streamlining communication and reducing manual errors.
   - **Shared Portals:** Utilize shared electronic platforms for documentation submission, authorization requests, and status tracking.
   - **Automation:** Use artificial intelligence or machine learning tools to identify claims at high risk of denial and proactively address potential issues.
   - **Analytics Dashboards:** Share denial analytics with payers to foster transparency and collaboration in tackling systemic issues.
     Example: Partnering with payers to develop a secure electronic submission system that flags missing documentation in real-time.
7. **Engaging in Policy Advocacy**
   At times, collaboration requires challenging payer policies that may be overly restrictive or misaligned with clinical guidelines. Physician advisors should advocate for policy changes that benefit both patients and providers.

- **Clinical Evidence:** Present robust clinical evidence to support requests for policy revisions.
- **Peer Advocacy:** Join industry groups or committees to advocate for more transparent and fair payer practices.
- **Collaborative Pilots:** Partner with payers on pilot programs to test alternative payment models or coverage policies.
  Example: Advocating for a broader interpretation of "medical necessity" based on updated clinical guidelines.

By implementing these strategies, physician advisors can transform the payer-provider relationship into one characterized by mutual respect, collaboration, and shared commitment to improving patient care and financial sustainability.

How Technology is Transforming Denial Management

Advances in technology have fundamentally reshaped denial management, providing tools to improve efficiency, reduce errors, and optimize resources. These innovations enable hospitals and healthcare systems to proactively address denial challenges, streamline operations, and recover revenue more effectively. Below are some of the most impactful technologies and their applications in denial management:

1. **Artificial Intelligence (AI)**
   AI is revolutionizing denial management by bringing advanced analytical capabilities to the revenue cycle.
   - **Pattern Recognition:** AI algorithms analyze vast amounts of historical claims data to identify patterns in denials. For example, they can flag documentation gaps or procedural mismatches that commonly result in denials.[25]
   - **Real-Time Assistance:** AI-driven tools can provide real-time feedback to clinicians during documentation, ensuring that payer-specific criteria are met. This reduces errors and lowers the likelihood of denials before claims are even submitted.[26]
   - **Automation of Peer-to-Peer Discussions:** AI tools can generate data-driven insights and prepare comprehensive summaries to support physician advisors in peer-to-peer appeals.
     Example: Natural language processing (NLP) is used to extract critical information from clinical notes, ensuring alignment with payer policies.
2. **Automation**
   Automation streamlines many aspects of denial management, reducing the administrative burden and speeding up workflows.
   - **Automated Workflows:** Claims flagged for denial can automatically trigger workflows for appeal preparation, including document retrieval, template population, and submission.
   - **Document Submission:** Robotic process automation (RPA) facilitates the rapid and accurate submission of required documentation to payers, reducing delays.[27]

- **Status Tracking:** Automated systems provide real-time updates on claim status, ensuring timely follow-up and reducing the risk of missed deadlines.
  Example: Automation can pre-populate appeal letters with patient and denial-specific information, cutting down on manual preparation time.

3. **Predictive Analytics**
   Predictive analytics uses historical and real-time data to identify trends and anticipate challenges in denial management.[28]
   - **High-Risk Claim Identification:** Hospitals can flag claims at high risk of denial, allowing teams to intervene proactively by ensuring that documentation is complete and accurate before submission.[29]
   - **Resource Allocation:** By predicting denial trends, healthcare organizations can allocate resources—such as appeal teams or physician advisors—more strategically.[30]
   - **Outcome Forecasting:** Predictive models can estimate the likelihood of appeal success based on factors such as denial reason, payer, and case specifics.
     Example: A hospital using predictive analytics might identify that surgical claims from a particular payer have a 40% denial rate due to incomplete pre-authorization, prompting targeted process improvements.
4. **Blockchain Technology**
   Blockchain, while still in its early stages in healthcare, offers transformative potential for denial management by enhancing data integrity and transparency.
   - **Immutable Records:** Blockchain ensures that claims data is secure and tamper-proof, reducing disputes over documentation or submission dates.
   - **Streamlined Communication:** Smart contracts embedded in blockchain systems can automate approvals and flag discrepancies in real-time, minimizing delays.
   - **Interoperability:** Blockchain enables seamless and secure sharing of data between providers and payers, fostering trust and reducing administrative burdens.[31]
     Example: A blockchain system could create a single, shared ledger for claims, providing all stakeholders with a real-time view of the claim's progress and eliminating duplicate documentation requests.
5. **Data Visualization and Dashboards**
   Advanced data visualization tools provide actionable insights into denial trends and performance metrics.[32]
   - **Interactive Dashboards:** Customizable dashboards allow denial management teams to track KPIs, such as denial rates, appeal success rates, and revenue recovered, in real time.
   - **Trend Analysis:** Visualizations of denial data by payer, department, or reason help pinpoint areas for improvement.
     Example: A dashboard showing monthly denial rates by department could reveal that the emergency department has higher denial rates due to coding inconsistencies, prompting targeted interventions.
6. **Machine Learning (ML)**
   ML enhances denial management by enabling systems to "learn" and improve over time.

- **Denial Prediction:** ML models analyze historical data to continuously refine their ability to predict which claims are likely to be denied and why.[33]
- **Dynamic Decision-Making:** ML algorithms adapt to changing payer policies, ensuring compliance without manual updates.
  Example: An ML model might identify that claims involving specific procedures are increasingly being denied due to recent changes in payer guidelines, prompting preemptive adjustments.

7. **Telehealth and Remote Access Integration**
   With the rise of telehealth, technology is evolving to support denial management for virtual care.
   - **Documentation Tools:** Integrated telehealth platforms include documentation templates designed to meet payer requirements for virtual services.
   - **Real-Time Audits:** Systems audit telehealth encounters to ensure compliance with billing and coding guidelines.[34]
     Example: A telehealth platform integrated with denial management tools can flag missing telehealth-specific modifiers in billing codes.
8. **Cloud-Based Solutions**
   Cloud technology offers scalability, accessibility, and cost-effectiveness for denial management.[35]
   - **Centralized Data Access:** Cloud-based systems allow revenue cycle teams to access denial data and collaborate from anywhere, ensuring continuity in remote or hybrid work environments.
   - **Data Security:** Robust encryption and compliance with regulations like HIPAA make cloud platforms a secure option for managing sensitive claims data.
     Example: A cloud-based denial management system can integrate seamlessly with electronic health records (EHRs) to pull necessary documentation for appeals.
9. **Collaboration Platforms**
   Technology now enables better collaboration among interdisciplinary teams working on denial management.[36]
   - **Integrated Communication:** Tools like Slack or Microsoft Teams facilitate communication between physician advisors, coders, and revenue cycle teams in real-time.
   - **Case Sharing:** Secure portals allow instant sharing of denial cases and related documentation among team members.
     Example: A physician advisor could use a collaboration platform to instantly consult with a coder about a denial related to clinical documentation.
10. **Emerging Innovations**
    Future technologies, such as quantum computing and advanced AI, promise even greater capabilities for denial management.

- **Quantum Computing:** Could revolutionize data analysis by processing enormous datasets in seconds, uncovering complex patterns in denials.
- **AI-Augmented Appeals:** Advanced AI systems may soon draft entirely customized appeal letters based on payer-specific guidelines, with minimal human oversight.

By leveraging these technological advancements, healthcare organizations can significantly reduce denial rates, expedite appeals, and recover lost revenue, all while minimizing the administrative burden on staff. These innovations are not just tools—they are strategic enablers of financial and operational excellence in the ever-evolving healthcare landscape.

Conclusion

Understanding and navigating the complex processes of insurance companies is a critical skill for physician advisors. This chapter has outlined the key challenges posed by insurance workflows, including common reasons for claim denials, the structured review process employed by payers, and effective strategies to align clinical arguments with payer criteria. By focusing on these foundational elements, physician advisors are better equipped to advocate for patients, recover denied revenue, and safeguard the financial health of their organizations.

The importance of a proactive approach cannot be overstated. Mastering payer-specific criteria and addressing documentation gaps before claims are submitted are essential steps in minimizing denials. Through preventive strategies, such as tailoring documentation to meet established guidelines and leveraging checklists or templates, physician advisors can preemptively address many of the issues that result in claim rejections. Additionally, empowering clinical teams with ongoing education ensures that best practices become routine.

Collaboration is another cornerstone of effective denial management. By fostering trust and open communication with payers, physician advisors can shift the payer-provider relationship from adversarial to collaborative. Engaging payers early, addressing recurring denial patterns, and establishing feedback loops build partnerships focused on shared goals: cost-effective care and optimal patient outcomes. These relationships, coupled with strategic policy advocacy, enable providers to influence and adapt to payer requirements more effectively.

The transformative role of technology further enhances these efforts. Tools such as artificial intelligence, predictive analytics, and automation have redefined denial management, enabling real-time decision-making, streamlined workflows, and data-driven strategies. Emerging innovations like blockchain and machine learning continue to promise new opportunities for increasing transparency, improving efficiency, and reducing administrative burdens.

Metrics provide the feedback loop necessary for continuous improvement. Tracking key indicators such as denial rates, appeal success rates, and time to resolution allows hospitals to measure the impact of their efforts and identify areas requiring further attention. These insights help refine processes, allocate resources efficiently, and ensure the sustainability of denial management programs.

Ultimately, **the role of the physician advisor is as much about advocacy as it is about strategy.** By combining clinical expertise with a deep understanding of payer operations, physician advisors can bridge the gap between healthcare providers and insurers. This advocacy protects patients from inappropriate denials, enhances the patient care experience, and ensures that hospitals are compensated fairly for the services they provide.

In an era where denials can threaten financial stability and patient outcomes, mastering the intricacies of insurance processes is not just an operational necessity—it is a strategic imperative. Physician advisors who embrace this challenge will not only improve their own effectiveness but will also contribute to a culture of accountability, innovation, and continuous improvement within their organizations. As the healthcare landscape evolves, their role will remain pivotal in shaping a system that balances fiscal responsibility with the delivery of high-quality patient care.

References

[19] Olaniyan, Olakunle, et al. "Managing medical necessity and notification denials: are denials for medical necessity and notification issues eroding your bottom line? Implementing a denials mitigation process can help." *Healthcare Financial Management*, vol. 63, no. 8, Aug. 2009, pp. 62+. *Gale Academic OneFile*, link.gale.com/apps/doc/A210170120/AONE?u=nysl_oweb&sid=googleScholar&xid=e3746965. Accessed 10 Jan. 2025.

[20] Rode, Dan. "Medicare audits and the importance of documentation." *Healthcare Financial Management*, vol. 51, no. 10, Oct. 1997, pp. 82+. *Gale Academic OneFile*, link.gale.com/apps/doc/A20168613/AONE?u=nysl_oweb&sid=googleScholar&xid=89961abe. Accessed 10 Jan. 2025.

[21] Kilanko, V. (2023). Leveraging Artificial Intelligence for Enhanced Revenue Cycle Management in the United States. *International Journal of Scientific Advances*, *4*(4), 505-14.

[22] Gaines, M. E., Auleta, A. D., & Berwick, D. M. (2020). *Changing the Game of Prior Authorization. JAMA.* doi:10.1001/jama.2020.0070

[23] Doss, D., Pawar, R., & Maddireddy, R. (2022). Intelligent Automation in Health Insurance. *Bimaquest*, *22*(1). Retrieved from http://www.bimaquest.niapune.org.in/index.php/bimaquest/article/view/120

[24] PMLA , Volume 127 , Issue 4: Special Topic Work , October 2012 , pp. 983 - 986

DOI: https://doi.org/10.1632/pmla.2012.127.4.983

[25] Pal, S. *et al.* (2022). Driving Impact in Claims Denial Management Using Artificial Intelligence. In: Singh, M., Tyagi, V., Gupta, P.K., Flusser, J., Ören, T. (eds) Advances in Computing and Data Sciences. ICACDS 2022. Communications in Computer and Information Science, vol 1613. Springer, Cham. https://doi.org/10.1007/978-3-031-12638-3_10

[26] Lee C, Britto S, Diwan K. Evaluating the Impact of Artificial Intelligence (AI) on Clinical Documentation Efficiency and Accuracy Across Clinical Settings: A Scoping Review. Cureus. 2024 Nov 19;16(11):e73994. doi: 10 7759/cureus.73994. PMID: 39703286; PMCID: PMC11658896.

[27] Nimkar P, Kanyal D, Sabale SR. Increasing Trends of Artificial Intelligence With Robotic Process Automation in Health Care: A Narrative Review. Cureus. 2024 Sep 18;16(9):e69680. doi: 10.7759/cureus.69680. PMID: 39429258; PMCID: PMC11489308.

[28] Quan, Zhiyu; Valdez, Emiliano A. . (2018). *Predictive analytics of insurance claims using multivariate decision trees. Dependence Modeling, 6(1), 377–407.* doi:10.1515/demo-2018-0022

[29] Robinson, E. . (2024). Predictive Modeling of Health Insurance Claims in the United States. *American Journal of Statistics and Actuarial Sciences*, *5*(1), 35–46. https://doi.org/10.47672/ajsas.1994

[30] Arun Chandramouli, "Leveraging Predictive Analytics to Minimize Claim Denials in Healthcare Revenue CycleManagement", *J. Tech. Innovations*, vol. 2, no. 4, Jan. 2025, doi: 10.93153/3aexs190.

[31] Hölbl, Marko; Kompara, Marko; Kamišalić, Aida; Nemec Zlatolas, Lili . (2018). *A Systematic Review of the Use of Blockchain in Healthcare. Symmetry, 10(10), 470–.* doi:10.3390/sym10100470

[32] Sanderson, Brian, et al. "3 pillars of a successful denials management program: with a strong denials management process in place, hospitals can reduce their denial rate and increase cash recoveries." *Healthcare Financial Management*, vol. 62, no. 9, Sept. 2008, pp. 74+. *Gale Academic OneFile*, link.gale.com/apps/doc/A188898722/AONE?u=nysl_oweb&sid=googleScholar&xid=bcef7653. Accessed 12 Jan. 2025.

[33] Russell, S., Montes Suros, F., & Kumar, A. (2024). Exploiting Machine Learning Bias: Predicting Medical Denials. *Proceedings of the AAAI Symposium Series*, *3*(1), 58-63. https://doi.org/10.1609/aaaiss.v3i1.31181

[34] Bajowala, S.S., Milosch, J. & Bansal, C. Telemedicine Pays: Billing and Coding Update. *Curr Allergy Asthma Rep* **20**, 60 (2020). https://doi.org/10.1007/s11882-020-00956-y

[35] Zouo, S. J. C., & Olamijuwon, J. (2024). Financial data analytics in healthcare: A review of approaches to improve efficiency and reduce costs.

[36] Paz, Anthony J., "Online collaboration platforms: Communication implications for workplace virtual teams" (2020). *Senior Honors Theses and Projects*. 678. https://commons.emich.edu/honors/67

CHAPTER 3

Understanding Inpatient Level of Care Across Different Payer Types

Introduction

One of the most challenging responsibilities for a physician advisor—and indeed for anyone involved in hospital revenue cycle or utilization management—is determining whether a patient truly meets "inpatient level of care." This challenge is compounded by the fact that there is no universal, single definition of what inpatient status entails. Medicare, Medicaid, and Commercial insurers each apply distinct rules and frameworks, which can vary not only from one payer to another, but also from state to state or region to region.[37]

Given these complexities, an admission that easily qualifies for inpatient status under Medicare guidelines might be questioned or denied by a particular Commercial plan that operates under more restrictive proprietary guidelines. Meanwhile, Medicaid programs in different states may have unique prior-authorization requirements, special coverage determinations, or other policies that differ sharply from federal (Medicare) principles.

Physician advisors must therefore possess a broad, versatile understanding of how different payers define "inpatient," the guidelines that shape those definitions, and how to align clinical documentation with those criteria. This chapter offers a comprehensive examination of these variations, focusing on the crucial interplay between insurance regulations and the day-to-day clinical realities of patient care. We will also look at how standard reference tools—especially MCG (formerly Milliman Care Guidelines) and InterQual—fit into the picture, and how best to reference them when advocating for payment or appealing a denial.

The Concept of "Inpatient Level of Care"

Before diving into specific payers, it is useful to define the term "inpatient level of care." Conceptually, this generally means that a patient's clinical condition or the complexity of interventions performed is such that they require round-the-clock monitoring and hospital-based resources that cannot be provided in a less intensive setting. This might include the need for constant bedside nursing intervention, repeated administration of IV medications, advanced respiratory support, or any number of treatments or evaluations that demand on-site medical staff and hospital equipment.

However, beyond this broad conceptualization, payers formalize the notion of "inpatient" through an assortment of policies, procedures, and guidelines. These policies interpret and codify medical necessity in specific, often rigid ways. The trick for physician advisors is to ensure that patient documentation not only reflects the genuine clinical facts, but also meets the specific thresholds that a given payer or guideline has set.

An illustrative anecdote might involve two patients hospitalized for acute exacerbations of COPD. From a purely clinical standpoint, both patients present with elevated respiratory rates, difficulty maintaining adequate oxygenation on room air, and require nebulizer treatments. However, subtle differences—one patient's repeated need for IV steroids rather than oral ones, for instance—might be enough to tilt the scales from observation to inpatient status under certain guidelines. For the physician advisor, recognizing these critical nuances and documenting them is the key to securing an appropriate level of care designation.

Medicare and the Two-Midnight Rule

Historical Context

Of all the payer systems in the United States, Medicare is often considered the archetype from which many other insurers build their own policies. Historically, Medicare has undergone multiple evolutions in determining how inpatient admissions are justified and reimbursed. The inception of the Two-Midnight Rule in 2013 represented a significant shift, as it provided hospitals with a more structured guide to deciding whether a patient's stay qualified for inpatient billing.[38]

The Two-Midnight Rule Explained

Under the Two-Midnight Rule, an admission is generally deemed appropriate for inpatient status if the physician reasonably expects the patient to require hospital care that spans at least two midnights. The rationale is that conditions that require extended, around-the-clock monitoring typically indicate higher acuity—i.e., more than just an overnight observation.

The official rule states that if, at the time of the admission order, the physician expects the patient will need at least two midnights of hospital care, and if that expectation is backed by proper clinical documentation, then the hospital can bill Medicare for an inpatient stay.[39] It is vital that the record clearly shows the severity of the patient's illness and any complicating factors that support that decision.

1. **Key Exceptions**
   While the two-midnight metric serves as the general yardstick, CMS provides exceptions for cases involving new-onset mechanical ventilation, certain inpatient-only procedures that cannot be safely performed on an outpatient basis, and unpredictable escalations in patient complexity. For instance, a patient might come in with suspected sepsis, receive advanced IV therapies, and stabilize more rapidly than anticipated—leaving the hospital in fewer than two midnights. If the documentation justifies that the patient's presentation was severe enough to warrant inpatient admission at the outset, Medicare often allows that stay to remain billed as inpatient.[40]
2. **Practical Documentation Tips**
   In real-world scenarios, an admission might begin under the assumption that the patient will require more than 48 hours of hospital-level care, but the actual clinical picture changes. Physician advisors can help front-line clinicians reflect in their documentation the reasons why an inpatient admission was appropriate on day one—even if the discharge took place earlier than expected. This could include detailed notes on how the patient failed to respond to initial outpatient treatment, or how the presence of significant comorbidities (e.g., heart failure, diabetes with poor glycemic control) elevated the risk profile.
3. **Avoiding Common Pitfalls**
   A common pitfall is to retroactively label a short stay as inpatient without ensuring that the documentation supports it. Auditors and recovery audit contractors (RACs) scrutinize the record to see if the decision for inpatient status was justified at the time of admission. If the record only mentions that the patient "needs to stay for tests" or "will be observed," it may weaken the hospital's position should a denial occur.

From a physician advisor perspective, knowledge of the Two-Midnight Rule—and its exceptions—allows proactive discussions with the clinical team at or near the point of admission. This ensures the hospital's billing stance matches the anticipated length and intensity of care required, and that charting reflects the correct level of clinical severity.

Medicaid Considerations

A Patchwork of Policies

When it comes to Medicaid, uniformity is the exception rather than the rule. Each state administers its own Medicaid program within federal guidelines, which leads to significant diversity in how inpatient status is defined or reimbursed. Some states choose to mirror Medicare's overarching framework, while others adopt entirely unique procedures.[41]

Additionally, many state Medicaid programs outsource utilization management to third-party organizations that apply their own set of criteria (often InterQual or MCG, sometimes in a modified form). In other situations, the state's Medicaid plan might rely heavily on prior authorization for certain procedures or for extended lengths of stay. For a physician advisor working in a multi-state health system—or even within a single state where different managed Medicaid plans operate—staying current is a never-ending task.

Social Determinants of Health (SDOH)

One prominent difference in many Medicaid programs is the emphasis on social determinants of health when deciding whether a patient meets inpatient level of care. For instance, if a patient's living conditions make it impossible for them to manage frequent nebulizer treatments or IV antibiotics at home, the hospital admission might be justified despite the patient's otherwise borderline clinical presentation.[42]

In practical terms, thorough documentation of SDOH can make the difference between a denied claim and an approved inpatient stay. If a patient does not have reliable transportation to get to outpatient appointments, lacks a stable home environment, or depends on daily assistance with activities of daily living (ADLs) that simply is not available outside the hospital, these elements can tilt a decision toward inpatient coverage for some Medicaid plans.

Navigating Prior Authorizations

Another factor to keep in mind is the requirement for prior authorization. Depending on the state, Medicaid may mandate pre-approval for certain admissions, particularly for scheduled procedures like elective surgeries. If the hospital neglects to secure this approval or fails to align the clinical documentation with the terms agreed upon in the authorization, denials can swiftly follow.

In this sense, Medicaid's approach can be more procedural than Medicare's. It is sometimes not just about whether the patient's condition meets inpatient criteria, but whether the appropriate boxes have been checked and the correct channels followed. Physician advisors can play a vital role in guiding their facilities to meet these administrative hurdles and in ensuring that the submission for authorization includes robust clinical rationale.

Commercial Insurance Variations

The Role of Contractual Agreements

Commercial insurance companies, sometimes referred to as private or third-party payers, function based on an intricate web of contractual agreements with hospitals and provider networks. These contracts specify, often in great detail, how admissions will be classified, which guidelines will be used (MCG, InterQual, or proprietary criteria), and what procedures or lengths of stay require pre-authorization.

Because these agreements are typically negotiated at an institutional or system level, the physician advisor must be well-versed not only in general payer policies but also in the specifics of how their particular hospital's contract with a Commercial insurer is written. What is accepted as "inpatient" at one hospital under a Blue Cross/Blue Shield contract might not be accepted under the same insurer at another hospital with a different contract.

Observation vs. Inpatient Under Commercial Plans

A recurrent challenge with Commercial insurance is that these payers often carve out extremely specific guidelines for observation stays, sometimes encouraging hospitals to keep a patient under observation status longer than might traditionally be expected. Commercial payers might argue that IV antibiotic therapy alone is not necessarily grounds for an inpatient admission, or that borderline oxygen requirements do not always justify a full admission.

Thus, to succeed in attaining or defending inpatient status, physician advisors need to highlight objective evidence of significant clinical instability or the need for interventions that cannot be safely managed in an observation context. This can include repeated IV medication administrations (especially at high frequency or with close monitoring), frequent vital sign checks, advanced respiratory support, or multi-consult coordination that realistically must happen within an inpatient framework.

Proprietary Guidelines

Some Commercial insurers use well-known resources such as MCG or InterQual, while others create their own "homegrown" criteria. These proprietary guidelines can be more restrictive, focusing on cost containment and narrower definitions of medical necessity. Appeals often involve referencing recognized medical literature or standard criteria (like MCG/InterQual) to point out that the proprietary rules deviate from established best practices. Physician advisors thus become mediators between the hospital's clinical standpoint and the insurer's often budget-centric approach.

One effective tactic is to highlight specific details—lab abnormalities, underlying comorbidities, the risk of readmission if the patient is discharged too early—and link them directly to widely accepted guidelines. When dealing with proprietary rules, it is helpful to ask for a copy of those guidelines and compare them to standard references, calling out inconsistencies in any written or peer-to-peer appeals.

MCG and InterQual: The Cornerstones of Medical Necessity Criteria

Historical and Practical Significance

MCG (Milliman Care Guidelines) and InterQual are two of the most widely adopted clinical guidelines in the United States for determining the appropriateness of admissions and levels of care. Though these guidelines are used primarily by payers, many hospitals also license them so that they can proactively evaluate the status of their patients and help shape documentation.[43]

The fundamental rationale behind both MCG and InterQual is to give a structured, evidence-based benchmark that clinicians, payers, and auditors can consult to decide whether a particular admission is clinically justified. If a patient meets the listed criteria—for example, a certain threshold of oxygen requirement, frequent abnormal labs, acute neurological deficits, or the inability to manage tasks independently—the recommendation is typically that inpatient care is warranted. If the criteria are not met, the guidelines might recommend observation or outpatient care instead.

Core Differences

While MCG and InterQual share a broad mission, they differ in layout and emphasis:

1. **MCG**
   MCG often provides detailed narratives on the clinical rationale behind each recommendation. Its guidelines typically outline thresholds for inpatient versus observation status, factoring in what they term "optimal recovery course" or "goal length of stay." MCG is sometimes perceived as more flexible, especially if a physician advisor can align the patient's documented complexities with MCG's text-based descriptions of care needs.
2. **InterQual**
   InterQual is well-known for its "Severity of Illness (SI)" and "Intensity of Services (IS)" rubrics. Each condition or procedure has a specific set of SI/IS criteria that must be met for an inpatient admission to be recommended. For example, the severity side might look at vital sign abnormalities, lab values, or imaging findings. The intensity side might examine the kind of medical or surgical interventions required. If enough SI/IS factors are present, InterQual endorses admission.

Where to Access and How to Stay Updated

Both MCG and InterQual typically require a subscription, which might be managed by a hospital's case management department, utilization review team, or medical library. Most of their content is available via web-based portals or integrated modules within an EHR system.

Staying current is crucial. Both MCG and InterQual release updates (usually annually) to reflect new evidence, changes in clinical practice, and evolving payer requirements. Physician advisors who rely on older versions risk missing newly added criteria or changes in wording that can influence an admission's justification.

The Value of Directly Citing MCG or InterQual

In many payer disputes—particularly in peer-to-peer conferences or written appeals—directly referencing a specific bullet point, numeric threshold, or condition guideline from MCG or InterQual can be very persuasive. If a payer's own denial letter cites "patient does not meet criteria for inpatient," you can respond by pinpointing exactly how your patient's chart data hits bullet point X, Y, or Z within the guidelines.

However, it's equally important to remember that these guidelines are not the final word in clinical medicine. If a patient's unique situation falls outside the typical criteria but still warrants an inpatient level of care, physician advisors can argue using specialty-specific guidelines or peer-reviewed literature to supplement or rebut MCG/InterQual's more general approach.

Practical Guidance for the Physician Advisor

Maintaining a Quick-Reference Database

Given the kaleidoscopic variety of payer arrangements, it is invaluable to keep a current repository—a spreadsheet, online drive, or intranet site—detailing which payers subscribe to MCG, which use InterQual, and which have proprietary guidelines. Include key contact information, any known quirks in the guidelines, or updated phone numbers for medical director lines. This helps you pivot quickly when you see a claim denial from a particular payer.

Customizing Prep Sheets for P2P Calls

Peer-to-peer (P2P) calls often hinge on how effectively you can reference the guidelines that the insurer respects. If you know you'll be speaking to a payer that uses InterQual, design your P2P "Prep Sheet" to highlight the patient's severity of illness and intensity of service data that map to InterQual's relevant section. If the payer uses MCG, structure your references accordingly. Being able to say, "Per MCG Inpatient & Surgical Care 26th Edition for Heart Failure, the patient meets the following three criteria," typically resonates with insurance medical directors.

Emphasizing Objective Instability

Payers generally respond best to objective clinical measures: vital signs that remain persistently abnormal (like sustained tachycardia, hypotension, or tachypnea), repeated or escalating IV medication needs (e.g., vasopressors, continuous nebulized bronchodilators), advanced wound care, or neurological changes that require immediate intervention. The more precisely the chart documents these measurements, the easier it is to demonstrate that the patient truly needs inpatient care.

Addressing Regional or Contractual Discrepancies

Even a nationwide payer like UnitedHealthcare or Aetna might adopt region-specific guidelines or ride on state-level laws that modify the standard approach. Physician advisors should regularly meet with managed-care contracting teams to glean insights on any changes in coverage or prior-authorization requirements. This is especially vital if your health system operates in multiple states where local Medicaid policies or unique contract riders can alter the coverage determination.

Staying Proactive

The best defense against denial is a proactive approach at admission. Encourage hospitalists, surgeons, and other frontline staff to consult with a utilization review nurse or physician advisor if there's any uncertainty about whether an inpatient or observation status is more appropriate. Document the reasoning thoroughly: "Patient's unstable respiratory status requires around-the-clock evaluation, repeated arterial blood gas checks, and possible transition to BiPAP if O2 saturation does not improve—these interventions exceed observation-level resources."

This kind of detail can avert a denial altogether. And if, despite best efforts, a denial happens, you already have a well-documented record to support your appeal.

Illustrative Scenarios and Case Examples

Case Example 1: Short-Stay Inpatient Under Medicare

Clinical Scenario: A 72-year-old with multiple comorbidities (COPD, diabetes, and chronic kidney disease) presents to the ED with suspected pneumonia. Her oxygen saturations dip to 88% on room air, and the ED physician anticipates that she may need IV antibiotics, repeated nebulizer treatments, and at least a 48-hour stay to stabilize.

Documentation: The admitting physician writes, "Expect inpatient stay requiring two midnights due to comorbidities that raise risk for complications; patient is not stable on oral antibiotics alone." However, by the second day, the patient dramatically improves and is discharged just shy of two midnights.

Outcome: A Medicare auditor questions the short length of stay. The physician advisor, referencing the Two-Midnight Rule exceptions, points out that at admission time, the documented expectation was a two-midnight stay given her comorbidities and low oxygen saturation. Thus, the inpatient claim stands despite the actual length of stay falling short.

Case Example 2: Medicaid Denial Over Social Factors

Clinical Scenario: A 45-year-old with poorly controlled diabetes presents with a diabetic foot infection. She lives in a rural area two hours from the nearest wound care clinic and lacks reliable transportation. The treating physician suggests that she will need daily IV antibiotics and advanced dressing changes beyond the scope of her outpatient capabilities.

Denial: State Medicaid denies inpatient admission, insisting that the patient could receive outpatient IV therapy.

Appeal Argument: Physician advisor compiles documentation of the patient's environment: no caregiver at home, minimal financial resources to travel to a clinic daily, and the high risk of complications if the wound deteriorates. Citing state Medicaid guidelines that allow consideration of social determinants for extended inpatient coverage, the advisor prevails. The claim is approved because the medical record precisely details the patient's inability to manage the condition outside the hospital.

Case Example 3: Commercial Plan with MCG

Clinical Scenario: A hospital system's major Commercial payer uses MCG for all admission determinations. A patient with acute heart failure exacerbation is initially placed in observation, but subsequently requires IV diuretics every six hours and continuous oxygen therapy.

Documentation: The clinical team details repeated episodes of respiratory distress and the need for close monitoring of electrolytes. The physician advisor references MCG's guidelines for heart failure, highlighting that the patient meets multiple criteria for inpatient care, including ongoing IV diuretic therapy and persistent dyspnea at rest.

Outcome: The payer denies the admission, claiming the patient did not appear acutely unstable. During a P2P call, the advisor counters by quoting MCG word-for-word: "Hospital admission is indicated if the patient cannot be stabilized in observation status with repeated or high-frequency IV diuresis." The payer medical director concedes, and the inpatient stay is upheld.

Conclusion: A Roadmap for Effective Advocacy

Mastering the variations in how different payers define and reimburse for inpatient care is a foundational skill for every physician advisor. The intricacies of the Medicare Two-Midnight Rule, the labyrinthine state-specific Medicaid regulations, and the shifting terrain of Commercial insurance can seem daunting at first glance, but they all revolve around the same core principle: ensuring that the level of care is genuinely appropriate to the patient's clinical needs.

By staying informed of each payer's guidelines—whether MCG, InterQual, or something proprietary—physician advisors can craft stronger documentation from the outset. Proactive discussions with clinicians upon admission, meticulous charting of clinical instability, and shrewd referencing of national guidelines in appeals or P2P conferences all greatly reduce the likelihood of a denial.

When denials do occur, the foundation laid by robust admission notes and alignment with recognized criteria can be critical to overturning the denial swiftly. Moreover, by diligently tracking local and national trends, as well as the social determinants that might uniquely affect Medicaid beneficiaries, physician advisors can keep their organizations one step ahead in the ever-changing world of utilization management.

Looking Ahead

In the next chapter, we will zoom in on the practical aspects of preparing for peer-to-peer conferences, building on the knowledge laid out here. With a solid grasp of how different payers define inpatient status and the reference guidelines at play, you will be well-equipped to anticipate payer objections, create stronger arguments, and ultimately secure more favorable outcomes for your hospital and its patients.

References

[37] Medicare Benefit Policy Manual, Chapter 1: Inpatient Hospital Services Covered Under Part A. Section 10: "Covered Inpatient Hospital Services Covered Under Part A."

[38] CMS Final Rule for Inpatient Prospective Payment System (IPPS). 78 Fed. Reg. 50495 (August 19, 2013), pages 50905–50909

[39] Centers for Medicare & Medicaid Services. Medicare Program; Hospital Inpatient Prospective Payment Systems for Acute Care Hospitals and the Long Term Care; Hospital Prospective Payment System and Fiscal Year 2014 Rates; Quality Reporting Requirements for Specific Providers; Hospital Conditions of Participation; Payment Policies Related to Patient Status; Final Rule. August 19, 2013. http://www.gpo.gov/fdsys/pkg/FR-2013-08-19/pdf/2013-18956.pdf.

[40] Dadana S, Bornovali SM. HCQM Two-Midnight Rule. [Updated 2023 Jul 30]. In: StatPearls [Internet]. Treasure Island (FL): StatPearls Publishing; 2025 Jan-. Available from: https://www.ncbi.nlm.nih.gov/books/NBK594265/

[41] Sommers, B. D., Tomasi, M. R., Swartz, K., & Epstein, A. M. (2012). Reasons For The Wide Variation In Medicaid Participat on Rates Among States Hold Lessons For Coverage Expansion In 2014. Health Affairs, 31(5), 909–919. doi:10.1377/hlthaff.2011.0977

[42] Chisolm, D.J., Brook, D.L., Applegate, M.S. et al. Social determinants of health priorities of state Medicaid programs. BMC Health Serv Res 19, 167 (2019). https://doi.org/10.1186/s12913-019-3977-5

[43] Daxa P. Clarke, MD, FAAP, Katherine Johnson, MD, FAAP, "Utilization Management", Caring for the Hospitalized Child: A Handbook of Inpatient Pediatrics, American Academy of Pediatrics Section on Hospital Medicine, Jeffrey C. Gershel, MD, FAAP, Daniel A. Rauch, MD, FAAP, SFHM

CHAPTER 4

Preparing for Peer-to-Peer Conferences

Introduction

Preparation is the cornerstone of success in peer-to-peer (P2P) conferences. These high-stakes discussions offer a pivotal opportunity for physician advisors to act as advocates for patient care, bridging the gap between clinical expertise and payer guidelines. The stakes are significant: a well-prepared P2P conference can not only secure necessary approvals for critical treatments but also strengthen relationships with insurance medical directors, fostering trust and collaboration in future cases. However, navigating these conversations requires more than medical knowledge—it demands strategic planning, effective communication, and a clear understanding of the payer's perspective.

One of the first and most crucial decisions is determining whether a case is worth pursuing in a P2P conference. Not every denial should be challenged; engaging in weak cases may not only waste valuable time and resources but could also harm the physician advisor's credibility. Repeatedly presenting poorly supported cases may lead insurance medical directors to question the reliability of future appeals, undermining the potential for success even in stronger cases.

The essence of effective preparation lies in making judicious choices and laying a solid foundation for each conference.[44] This involves:

- **Case Selection:** Identifying cases where the medical necessity for inpatient care is well-documented and aligned with payer criteria.
- **Strategic Organization:** Gathering, refining, and presenting clinical evidence that directly addresses the denial rationale.
- **Understanding the Payer's Language:** Carefully analyzing denial letters to uncover leverage points, such as payer oversights or misinterpretations.
- **Leveraging Guidelines:** Employing evidence-based literature and payer-specific or specialty guidelines to craft compelling arguments.
- **Communication Skills:** Developing a structured and concise approach that anticipates payer objections and builds trust through clarity and professionalism.

The importance of preparation extends beyond overturning denials. It reinforces the broader goals of improving operational efficiency, reducing unnecessary administrative burdens, and advocating for the best possible patient outcomes. It also positions physician advisors as key contributors to organizational success by optimizing denial management processes, improving overturn rates, and protecting revenue streams.

This chapter will delve into the essential practices that form the backbone of effective P2P preparation. From evaluating the merits of a case to creating structured, evidence-based arguments, these strategies are designed to empower physician advisors with the tools they need to navigate even the most challenging denial scenarios. By mastering these principles, physician advisors can ensure that each P2P conference becomes not just an opportunity to overturn a denial, but also a chance to showcase their expertise, advance patient care, and build stronger partnerships with payers.

Evaluating the Case for Peer-to-Peer Conferences

Before diving into preparation for a peer-to-peer (P2P) conference, physician advisors must critically assess whether the case merits escalation. Not all denials are worth the time, effort, and resources required to challenge them. Selecting battles wisely ensures that the focus remains on cases where advocacy can yield the most impactful results, whether by improving patient outcomes, influencing payer practices, or preserving financial integrity. Additionally, this discernment helps safeguard the physician advisor's professional credibility, fostering long-term trust and respect with insurance medical directors.

1. Assess the Strength of the Case

The foundation of any successful P2P conference lies in the strength of the case. Physician advisors should start by evaluating whether the clinical documentation robustly supports the necessity for inpatient care. A strong case is characterized by objective evidence and clear alignment with payer guidelines.

- **Objective Clinical Indicators:** Look for documented evidence of clinical severity, instability, or deterioration. Examples include vital sign abnormalities, laboratory results, imaging findings, or treatment escalations.
  Example: A patient with persistent hypotension (systolic blood pressure below 90 mmHg) despite fluid resuscitation presents a clear indicator of instability.
- **Alignment with Payer Criteria:** Does the case meet the specific guidelines outlined by the payer, such as InterQual or MCG criteria? Understanding the exact requirements for inpatient admission helps establish whether the denial can be reasonably challenged.
  Tip: Familiarize yourself with the most current versions of payer criteria and ensure that the clinical documentation directly corresponds to the listed requirements.

When the case lacks clear indicators or alignment with guidelines, consider whether additional documentation or clinical input might strengthen it before proceeding.

2. Consider the Potential Impact

Once the case's strength is established, physician advisors should evaluate the broader implications of pursuing the P2P conference. This step involves balancing patient-centered care with organizational and strategic considerations.

- **Patient Care:** Will overturning the denial significantly improve the patient's health outcomes? For example, approval for inpatient admission may be critical for a patient requiring intensive monitoring or complex interventions.
- **Resource Utilization:** Does the potential financial benefit justify the time and effort required to pursue the denial? While overturning high-value cases often aligns with organizational goals, the advisor must weigh this against the operational burden of P2P preparation.
- **Precedent Setting:** Could a successful outcome influence payer practices or internal policies? For instance, overturning a denial based on vague criteria might prompt the payer to revise its approach to similar cases in the future, benefiting other patients and providers.

3. Reflect on Professional Relationships

The way a physician advisor engages in P2P conferences has long-term implications for their reputation with insurance medical directors. Building and maintaining professional credibility requires a strategic approach to case selection.

- **Credibility:** Presenting poorly supported cases or challenging denials that clearly align with payer criteria can lead to skepticism. This skepticism may carry over to future P2Ps, reducing the likelihood of success even in strong cases.
- **Strategic Advocacy:** Prioritize cases with compelling clinical evidence. Consistently presenting well-prepared, strong cases enhances respect and trust, creating an environment where insurance medical directors are more likely to engage collaboratively.

4. Collaborate with the Care Team

Collaboration with the care team is essential to ensure that the case is well-prepared and the messaging is unified.

- **Additional Insights:** Attending physicians and care managers can often provide valuable information that strengthens the case. They may highlight aspects of the patient's care that were not immediately apparent or suggest additional documentation that reinforces the argument for medical necessity.
- **Unified Approach:** Consistency in messaging between the physician advisor, attending physician, and other team members minimizes the risk of conflicting statements during

the P2P conference. A cohesive narrative lends credibility to the case and reduces payer objections.

5. Document the Decision-Making Process

Finally, physician advisors should establish a process for documenting why certain cases were pursued or not. This practice serves both accountability and quality improvement purposes.

- **Accountability:** Detailed records of decision-making rationales can provide justification during audits or reviews. For example, documentation might include notes on why a case met specific payer criteria or why the potential impact warranted pursuing a P2P.
- **Continuous Improvement:** Over time, tracking decisions and outcomes can help refine criteria for selecting P2P conferences. Patterns in successful or unsuccessful cases can guide future strategies, leading to better resource allocation and higher overturn rates.

Key Considerations

By evaluating cases thoroughly before initiating a P2P conference, physician advisors can ensure their efforts are focused where they are most likely to succeed.[45] This proactive, strategic approach not only increases the likelihood of overturning denials but also fosters trust, professionalism, and efficiency within the broader denial management process.

Gathering All Relevant Medical Documentation

Thorough preparation for a peer-to-peer (P2P) conference begins with an exhaustive and focused review of the medical record. The ultimate goal is to extract and organize key data points that directly address the denial rationale, building a strong and concise argument for medical necessity. Effective documentation review is not about quantity but quality—emphasizing relevant clinical evidence that substantiates the need for inpatient care while avoiding unnecessary details that could detract from the argument.

---

1. Prioritize Critical Clinical Data

The most compelling arguments for overturning a denial are built on clear, objective clinical evidence. Physician advisors should focus on data that demonstrates the severity of the patient's condition and supports the need for inpatient-level care.

- **Vital Signs:** Persistent abnormalities in vital signs are often strong indicators of clinical instability. These can include:

    - Persistent hypotension (e.g., systolic blood pressure <90 mmHg despite fluid resuscitation).
    - Tachycardia or bradycardia reflecting hemodynamic instability.
    - Hypoxia or declining oxygen saturation requiring supplemental oxygen or non-invasive ventilation.
- **Diagnostics:** Imaging, lab results, or other diagnostic findings that confirm the severity of the patient's condition or indicate deterioration should be highlighted. Examples include:
    - CT scan findings showing bilateral pulmonary emboli.
    - Lab results such as lactic acid >4 mmol/L in a septic patient or critically low hemoglobin levels necessitating transfusion.
- **Interventions:** Focus on treatments that clearly demonstrate the complexity and intensity of care required, such as:
    - Continuous IV antibiotics for severe infections.
    - Cardiac monitoring for arrhythmias or acute coronary syndromes.
    - Ventilatory support or advanced hemodynamic monitoring.

By prioritizing this data, physician advisors can provide a clear narrative that aligns with payer expectations for inpatient care.

---

2. Focus on Failed Outpatient Management

Documenting the failure of outpatient treatments can significantly bolster the argument for hospitalization. This evidence demonstrates that lower levels of care were insufficient to manage the patient's condition, justifying the escalation to inpatient care.

- **Timeline of Interventions:** Construct a detailed timeline that shows:
    - The initiation of outpatient therapies (e.g., oral antibiotics, fluid administration).
    - Progression of the patient's condition despite these interventions.
    - Clinical decision points that led to the need for hospitalization.
- **Specific Reasons for Failure:**
    - Highlight worsening symptoms, such as escalating pain, unrelieved hypoxia, or increased hemodynamic instability.
    - Include evidence of poor response to outpatient therapies, such as persistent infection or failure to resolve dehydration after oral rehydration.

Example: "The patient's community-acquired pneumonia progressed despite a 7-day course of oral antibiotics, resulting in hypoxia requiring oxygen therapy."

---

3. Use Screenshots and Annotations

Visual aids are powerful tools for quickly conveying complex clinical information to insurance medical directors during a P2P conference. Screenshots and annotations can enhance clarity and ensure that critical data is not overlooked.

- **Annotated Lab Trends:**
  - Use color coding or highlights to draw attention to critical abnormalities, such as steadily rising creatinine levels or declining hemoglobin trends.
- **Medication Administration Records (MAR):**
  - Include data showing the escalation of therapies, such as progression from oral to IV antibiotics or the addition of pressors for hemodynamic support.
- **Vital Sign Graphs:**
  - Provide visual trends indicating clinical decline, such as sustained tachycardia or fluctuating oxygen saturation levels despite interventions.

Example: A screenshot of serial hemoglobin levels annotated to show a significant drop from 12.5 g/dL to 8.2 g/dL within 24 hours, justifying the need for transfusion and inpatient monitoring.

These visual tools not only make it easier to present evidence but also streamline discussions, allowing the physician advisor to maintain focus on the most critical data points.

---

4. Avoid Overloading with Details

While it is essential to provide thorough documentation, including irrelevant or extraneous details can dilute the strength of the argument. A concise, targeted approach ensures that the focus remains on the denial rationale.

- **Include Relevant Details:**
  - Highlight data directly tied to the payer's rationale for denial. For example, if the denial cites "lack of evidence of clinical instability," present objective indicators of instability, such as persistent tachycardia or hypotension.
- **Exclude Irrelevant Information:**
  - Avoid including unrelated details, such as a history of seasonal allergies, which do not pertain to the patient's current condition or the reason for denial.

Example: For a denial of inpatient admission for heart failure, include annotated BNP trends and oxygen requirements while excluding unrelated comorbidities like chronic back pain.

- **Streamline the Narrative:**
  - Ensure that every piece of evidence presented ties back to the central argument for medical necessity. This approach saves time during the P2P and makes the case easier to understand and defend.

Practical Tips for Gathering Documentation

- **Create a Checklist:** Develop a standardized checklist for common denial scenarios to ensure that all critical data points are gathered systematically.
- **Use Templates:** Design templates for organizing key clinical information, such as lab trends, vital signs, and diagnostic findings.
- **Collaborate:** Work with care managers and attending physicians to identify and compile relevant data. Their input can provide valuable insights into the patient's care trajectory.

By prioritizing relevant clinical data, documenting failed outpatient management, leveraging visual aids, and avoiding unnecessary details, physician advisors can effectively prepare for P2P conferences. This thorough approach ensures that the focus remains on compelling evidence, increasing the likelihood of overturning denials and securing the care patients need.

Understanding the Denial Letter and Finding Leverage Points

Denial letters are more than mere rejections; they are structured documents that provide insight into the payer's rationale, often pointing directly to the objections that need to be addressed. A carefully analyzed denial letter can serve as a blueprint for crafting a strong argument to overturn the decision. By thoroughly understanding the payer's perspective and leveraging the details provided, physician advisors can build compelling appeals that address the specific reasons for denial while highlighting errors, gaps, or inconsistencies in the payer's reasoning.

1. Identify the Specific Denial Rationale

The first step in responding to a denial is to understand why the claim was rejected. Denial letters typically outline the payer's reasoning in detail, and identifying the specific rationale is crucial for formulating an effective response. Common reasons for denial include:

- **Lack of Documented Medical Necessity:**
  This rationale often stems from insufficient documentation of the patient's clinical instability or the failure to justify inpatient-level care.
  Example: "The patient does not meet criteria for inpatient admission due to lack of evidence of persistent hemodynamic instability."
- **Failure to Meet Payer Criteria:**
  Payers frequently cite guidelines such as InterQual, MCG, or proprietary criteria to justify their decision.
  Example: "The patient's clinical presentation does not meet the outlined criteria for sepsis-related inpatient care, as lactate levels were normal."

Action: Pinpoint the exact statements or criteria the payer claims were not met. This clarity allows the physician advisor to align the rebuttal with the payer's framework.

---

2. Analyze Criteria References

Denial letters often reference specific guidelines or criteria that the payer used to evaluate the claim. Comparing the patient's documented clinical presentation against these criteria can help identify areas where the payer's assessment may be incomplete or incorrect.

- **Aligning Documentation with Criteria:**
  Review payer guidelines thoroughly and match the patient's clinical evidence to the required elements.
  Example: For a denial involving pneumonia, the advisor might cross-reference oxygen saturation levels, imaging results, and treatment escalations against InterQual criteria to demonstrate the need for inpatient care.
- **Highlighting Discrepancies:**
  If the denial claims the patient did not meet criteria, but documentation shows otherwise, call attention to these discrepancies.
  Example: If the denial states, "The patient's oxygen saturation was normal," provide evidence of subsequent readings below 90% that were omitted in the initial review.

Tip: If guidelines are proprietary or vague, seek additional support from specialty organization recommendations or evidence-based literature.

---

3. Look for Payer Errors or Oversights

Denial letters are not immune to mistakes, and identifying errors can be a powerful tool for overturning decisions. Payer oversights often stem from incomplete reviews of documentation or misinterpretations of the medical record.

- **Overlooking Trends in Vital Signs or Lab Results:**
  Payers may focus on isolated data points rather than trends that indicate clinical deterioration.
  Example: A denial citing "stable hemoglobin levels" may ignore a documented drop from 12 g/dL to 8 g/dL over 24 hours, justifying the need for transfusion and monitoring.
- **Misinterpreting Documentation:**
  Ambiguities in the medical record can lead to misinterpretations. Clarifying the intent of the documentation can correct these errors.
  Example: If the denial states that "the patient was hemodynamically stable," but the

record includes notes of multiple fluid boluses and continuous vasopressor support, this discrepancy should be highlighted.

- **Ignoring Critical Clinical Findings:**
  Key elements such as imaging results, ongoing interventions, or repeated escalations in care may not have been considered.
  Example: A denial for a heart failure admission that fails to address documented increases in BNP levels and oxygen requirements.

Action: Create a structured argument that emphasizes these overlooked or misinterpreted details, using annotated visuals or summarized timelines where applicable.

---

4. Use the Denial Letter to Create a Prep Sheet

The denial letter itself can serve as a guide for building a targeted Prep Sheet, consolidating all relevant arguments, evidence, and references into a concise, organized format.

- **Summarize the Denial Rationale:**
  Begin with a brief summary of the payer's stated objections. This provides a clear focus for addressing the denial during the P2P conference.
- **List Key Clinical Findings:**
  Identify and organize the clinical evidence that directly addresses the denial rationale. Ensure that the findings are tied to the payer's criteria or guidelines.
  Example: Include trends in vital signs, imaging results, lab findings, and interventions that substantiate the need for inpatient care.
- **Highlight Gaps in the Denial:**
  Note areas where the denial letter overlooked critical evidence, misinterpreted the medical record, or applied guidelines incorrectly.
  Example: "The denial did not account for the patient's persistent hypoxia and need for escalating oxygen therapy, as documented in nursing notes and MAR data."
- **Incorporate Relevant Guidelines:**
  Include excerpts from payer criteria or specialty guidelines that explicitly support the case for inpatient care. Use these to align the argument with the payer's expectations.

Practical Example of Denial Analysis and Leverage

Denial: "The patient does not meet criteria for sepsis-related inpatient care due to normal lactate levels and lack of organ dysfunction."
Response:

- **Clarify Documentation:** Highlight persistent tachycardia >120 bpm and hypotension requiring multiple fluid boluses, meeting criteria for hemodynamic instability.

- **Correct Oversights:** Present serial lactic acid levels showing an initial elevation above 4 mmol/L, with trends demonstrating partial but incomplete response to therapy.
- **Align with Guidelines:** Cite payer criteria indicating that hemodynamic instability, even with normal lactate levels, meets criteria for sepsis-related inpatient care.

By thoroughly analyzing denial letters, identifying objections, correcting errors, and using the information to build a robust Prep Sheet, physician advisors can transform rejection into opportunity. This focused, strategic approach not only strengthens the appeal but also demonstrates professionalism and attention to detail, enhancing the likelihood of overturning the denial.

Creating and Using a Prep Sheet

A well-organized Prep Sheet is an indispensable tool for ensuring that peer-to-peer (P2P) conferences are conducted with precision and focus. By consolidating critical information into a concise, structured format, the Prep Sheet serves as a quick reference guide, empowering physician advisors to present their case effectively and confidently. This document eliminates the need to sift through extensive medical records during the conference, allowing the advisor to concentrate on articulating arguments and addressing payer objections.

---

1. Compile Key Data

The foundation of a successful Prep Sheet lies in the careful selection and organization of relevant data. Each piece of information should directly address the denial rationale, ensuring that the discussion remains focused and evidence-based.

- **Patient Demographics:**
  Include the patient's name, age, admission date, and primary diagnosis. These details establish the context and ensure that both parties are discussing the correct case.
- **Clinical Findings Tied to the Denial Rationale:**
  Highlight objective evidence that supports the necessity of inpatient care. Examples include vital sign trends, lab results, imaging findings, and documentation of failed outpatient management.
  Example: "Patient experienced persistent hypotension (systolic BP <90 mmHg) despite fluid resuscitation and required vasopressor support."
- **Supporting Evidence from Guidelines or Specialty Recommendations:**
  Reference payer-specific criteria (e.g., InterQual or MCG) and any relevant specialty guidelines to substantiate the argument. Ensure that excerpts are concise and directly applicable to the case.

Example: "InterQual criteria for sepsis-related inpatient admission indicate that persistent tachycardia and hypotension, even with normal lactate levels, meet the threshold for instability."

---

2. Highlight Strong Arguments

Physician advisors must ensure that their most compelling points stand out in the Prep Sheet. Visual emphasis, such as bold text or color-coding, can help draw attention to critical evidence.

- **Examples of Highlighted Data:**
  - "Serial lactate levels increased from 2.5 to 6.0 mmol/L, indicating worsening sepsis."
  - "BNP levels tripled over 48 hours, consistent with acute decompensated heart failure requiring inpatient monitoring."
- **Prioritize Key Points:**
  Arrange the strongest arguments at the top of each section. This ensures that the most persuasive evidence is readily accessible, especially in situations where time is limited.

Tip: Use bold headers or bullet points to organize information logically and make the Prep Sheet easy to navigate.

---

3. Eliminate Weak or Irrelevant Language

The language used in the Prep Sheet should convey urgency and clinical significance. Avoid vague or non-specific terms that could weaken the argument.

- **Reframe Weak Statements:**
  - Replace: "Patient was mildly hypoxic."
  - With: "Patient demonstrated persistent hypoxia with oxygen saturation <88% on room air, requiring continuous supplemental oxygen."
- **Exclude Irrelevant Information:**
  Omit details unrelated to the denial rationale, such as a history of mild depression or a resolved minor ailment. Including unnecessary information can distract from the main argument.
- **Use Clear, Direct Language:**
  Avoid passive phrasing and emphasize active interventions.
  Example: "Patient required continuous IV antibiotics for MRSA bacteremia" is more compelling than "Antibiotics were administered."

---

4. Incorporate Visual Aids

Visual aids are invaluable for simplifying complex clinical scenarios and providing tangible evidence to support the argument. Properly annotated visuals can make critical points easier to understand and harder to dispute.

- **Annotated Lab Trends:**
  Include graphs or tables showing changes over time, such as worsening creatinine levels or declining oxygen saturation. Use color coding or arrows to highlight critical trends.
- **Medication Administration Records (MAR):**
  Provide screenshots or excerpts showing escalation in therapy, such as the addition of vasopressors or frequent boluses of fluids.
- **Diagnostic Imaging:**
  Annotate imaging reports to draw attention to findings that justify inpatient care.
  Example: Highlight CT scan results showing bilateral pulmonary emboli with right heart strain.
- **Vital Sign Graphs:**
  Present trends that demonstrate clinical deterioration, such as persistent tachycardia or declining blood pressure.

---

5. Practice Navigating the Prep Sheet

Familiarity with the Prep Sheet is essential to ensure that physician advisors can quickly locate and reference critical points during the P2P conference. Practice builds confidence and minimizes the risk of errors or delays.

- **Review and Rehearse:**
  Before the conference, review the Prep Sheet thoroughly and anticipate potential payer objections. Practice responding to these objections using evidence from the Prep Sheet.
- **Use Bookmarks or Tabs:**
  For virtual conferences, create bookmarks or tabs to facilitate seamless navigation through electronic Prep Sheets. This ensures that critical information is easily accessible during the discussion.
- **Conduct Mock P2Ps:**
  Simulate the conference with a colleague or care team member to refine the presentation and address any gaps in the Prep Sheet.

Tip: For in-person conferences, organize the Prep Sheet into clearly labeled sections and use highlighters or sticky notes to mark key data points.

Practical Example of a Prep Sheet Structure

Case Overview:

- Patient Name: John Doe
- Age: 65
- Admission Date: 12/01/2024
- Primary Diagnosis: Acute Decompensated Heart Failure

Denial Rationale:

- "The patient does not meet inpatient criteria due to stabilized BNP levels and no evidence of hemodynamic instability."

Supporting Clinical Data:

- BNP increased from 600 to 2,000 pg/mL over 48 hours.
- Oxygen saturation dropped to <88% on room air, requiring continuous supplemental oxygen.
- Echocardiogram showed an ejection fraction of 30%, consistent with systolic dysfunction.

Guidelines and Recommendations:

- InterQual Criteria: Inpatient care justified for patients requiring continuous cardiac monitoring and IV diuresis.
- American Heart Association: Recommends hospitalization for heart failure patients with persistent hypoxia or worsening hemodynamic status.

Visual Aids:

- Graph of serial BNP levels with annotations showing escalation.
- Screenshot of MAR documenting administration of IV diuretics and oxygen therapy.

By compiling key data, emphasizing strong arguments, eliminating irrelevant details, incorporating visual aids, and practicing navigation, physician advisors can create Prep Sheets that are both comprehensive and user-friendly. This preparation ensures a confident, efficient, and persuasive presentation during P2P conferences, significantly increasing the likelihood of overturning denials.

Conclusion

Preparation is the foundation of effective advocacy in peer-to-peer (P2P) conferences. These high-stakes interactions require a combination of clinical acumen, strategic planning, and persuasive communication to navigate the complex terrain of insurance denials. Through the practices outlined in this chapter, physician advisors can position themselves as skilled advocates for patient care while fostering trust and credibility with payers.

Key aspects of preparation include evaluating the merits of each case, gathering and organizing clinical evidence, analyzing denial letters, and creating well-structured Prep Sheets. Each of these steps is critical not only for overturning denials but also for reinforcing the physician advisor's role as a leader in denial management and patient advocacy.

Case evaluation forms the backbone of P2P preparation, ensuring that the advisor's time and resources are invested where they can make the greatest impact. By selecting cases with robust clinical documentation and clear alignment with payer criteria, physician advisors can focus their efforts on scenarios most likely to result in successful outcomes. Careful consideration of the implications of each case—whether on patient care, resource utilization, or professional relationships—ensures a balanced and strategic approach.

Thorough documentation review transforms a potentially overwhelming amount of clinical information into a clear, concise narrative. By prioritizing critical clinical data, focusing on failed outpatient management, and avoiding irrelevant details, physician advisors can present a compelling case for medical necessity. Visual aids, such as annotated lab trends and diagnostic imaging, further enhance the clarity and impact of the argument.

Understanding and leveraging denial letters enables physician advisors to craft responses tailored to the payer's objections. By identifying specific rationales, analyzing payer criteria, and exposing errors or oversights, advisors can uncover leverage points that strengthen their arguments. Denial letters also serve as a blueprint for creating targeted Prep Sheets that streamline the P2P discussion.

Prep Sheets are the culmination of the preparation process, consolidating all relevant information into a user-friendly format. These structured tools not only facilitate efficient communication during the P2P conference but also boost the advisor's confidence and credibility. By emphasizing strong arguments, incorporating visuals, and practicing navigation, advisors can ensure they are fully prepared to address payer objections with precision and professionalism.

Ultimately, preparation extends beyond the immediate goal of overturning denials. It supports broader organizational objectives, such as improving operational efficiency, protecting revenue streams, and enhancing patient outcomes. It also positions physician advisors as trusted partners to payers, demonstrating a commitment to collaboration and evidence-based decision-making.

As physician advisors master the principles outlined in this chapter, they elevate their role from transactional to transformative. Each P2P conference becomes an opportunity not only to advocate for individual patients but also to influence payer practices, improve institutional processes, and advance the standard of care. Through meticulous preparation and strategic execution, physician advisors can turn challenges into opportunities, ensuring that their efforts leave a lasting impact on both patient care and the healthcare system as a whole.

Looking ahead, Chapter 4 will build on these foundational practices by exploring the art of building persuasive arguments. It will delve into techniques for framing discussions, leveraging empathy, and anticipating counterarguments to achieve successful outcomes in even the most contentious P2P conferences.

References

[44] Pasetsky J, Garcia-Young JA, Beatley E, Yu JB. Peer-to-Peer Phone Calls and Letters Appealing Insurance Denials of Service: Practical Tips and Resources. Pract Radiat Oncol. 2024 Nov-Dec;14(6):e434-e437. doi: 10.1016/j.prro.2024.06.015. Epub 2024 Jul 30. PMID: 39089624.

[45] Fairchild KD, Duncan SD, Stroustrup A, McNamara PJ. Insurance denials: a peer-to-peer problem in neonatology. J Perinatol. 2024 Aug;44(8):1210-1211. doi: 10.1038/s41372-024-01991-7. Epub 2024 May 11. Erratum in: J Perinatol. 2024 Aug;44(8):1233. doi: 10.1038/s41372-024-02028-9. PMID: 38734803; PMCID: PMC11300295.

CHAPTER 5

Building Persuasive Arguments

Introduction

Peer-to-peer (P2P) conferences are a cornerstone of the physician advisor's role, providing a critical opportunity to advocate for patient care and challenge insurance denials. However, success in these conferences requires much more than clinical expertise. While a deep understanding of medical conditions, treatments, and guidelines is essential, it must be combined with strategic communication and persuasive advocacy to achieve favorable outcomes.[46]

At its core, a successful P2P conference relies on the ability to present compelling, evidence-based arguments. These arguments must directly address the payer's concerns while emphasizing the clinical necessity of inpatient care. Insurance medical directors, who are responsible for upholding organizational policies and payer guidelines, often approach these discussions with a focus on adherence to criteria and cost containment. Physician advisors must navigate these constraints, using language and frameworks that resonate with the payer's priorities.

Empathy plays a vital role in P2P advocacy. Recognizing the challenges and limitations faced by insurance medical directors helps physician advisors foster productive, collaborative conversations. By aligning arguments with the shared goal of ensuring appropriate care, advisors can establish a rapport that increases the likelihood of overturning denials. This empathetic approach also enhances long-term relationships with payers, making future interactions more constructive and efficient.

A structured approach is essential to maintaining clarity and focus during P2P conferences. Physician advisors must frame their discussions logically, prioritize the most compelling points, and anticipate potential objections. A well-organized argument not only streamlines the conversation but also demonstrates professionalism and preparedness, reinforcing the advisor's credibility.[47]

This chapter provides actionable strategies to enhance the effectiveness of P2P advocacy, focusing on three key areas:

1. **Framing Discussions Effectively:** Techniques to organize and present arguments in a clear, impactful way.

2. **Leveraging Clinical Evidence:** Methods for integrating objective data, payer-specific guidelines, and specialty recommendations to strengthen advocacy.
3. **Fostering a Collaborative Tone:** Approaches to building empathy, aligning with payer priorities, and creating a dialogue that emphasizes shared goals.

By mastering these practices, physician advisors can transform P2P conferences into opportunities to advocate successfully for patient care while strengthening organizational relationships with payers.[48] These skills not only increase overturn rates but also contribute to the broader objectives of improving patient outcomes and organizational performance.

Strategies for Framing Your Discussion Effectively

A well-framed argument is the foundation of a successful peer-to-peer (P2P) conference. It ensures that the key points are clear, concise, and focused on addressing the payer's concerns. By presenting information in an organized and impactful way, physician advisors can guide the discussion, engage the insurance medical director effectively, and increase the likelihood of overturning the denial. Below are detailed strategies for framing P2P discussions to achieve maximum clarity and persuasiveness.

---

1. Start with the Strongest Point

Opening with a compelling clinical fact immediately captures attention, establishes credibility, and sets a strong tone for the discussion. This approach helps direct the conversation toward the most critical elements of the case from the outset.

- **Why This Works:** Starting with a powerful point demonstrates preparedness and confidence while addressing the denial rationale directly. This strategy ensures the medical director understands the severity and urgency of the situation early in the discussion.
- **Example:**
    - "The patient required 4L oxygen to maintain SpO_2 above 90%, meeting criteria for hypoxia-related admission."
    - This succinctly highlights an objective clinical indicator, directly linking it to criteria for inpatient admission.
- **Tips:**
    - Use specific data points or clinical findings to make the strongest argument.
    - Avoid starting with peripheral details that may dilute the focus of the discussion.

---

2. Organize Arguments Logically

A structured and logical flow ensures that all critical points are covered and presented in a way that is easy to follow.[49] A disorganized argument risks losing the attention of the medical director and may weaken the overall persuasiveness of the case.

- Suggested Sequence:
    1. **Describe the Patient's Clinical Presentation and History:** Summarize the presenting symptoms and relevant medical history.
    2. **Highlight Diagnostic Findings and Treatments:** Emphasize objective data, imaging results, lab findings, and any interventions.
    3. **Connect Findings to Payer-Specific Criteria:** Link the patient's clinical data to the guidelines or criteria outlined by the payer.
- Example:
    - For a heart failure denial:
        - Begin with symptoms like worsening dyspnea and weight gain.
        - Transition to diagnostic data such as escalating BNP levels and reduced ejection fraction.
        - Conclude with alignment to inpatient admission criteria, such as the need for continuous IV diuresis and cardiac monitoring.
- Tips:
    - Create a checklist of key points to ensure nothing is overlooked.
    - Use headings or transitions to guide the listener through the argument logically.

---

3. Focus on Relevance

Physician advisors must avoid the temptation to include every detail of the patient's case. While comprehensive documentation is essential for preparation, the P2P discussion should focus exclusively on information that directly supports the argument for overturning the denial.

- **Why This Matters:** Including extraneous details can overwhelm the listener, distract from the main argument, and waste valuable time. Staying focused ensures that the discussion remains impactful and aligned with the denial rationale.
- **Relevant Information:**
    - "The patient's pneumonia failed outpatient antibiotics, progressing to hypoxia."
    - This ties the clinical course to the criteria for inpatient admission.
- **Irrelevant Information:**
    - "The patient's hypothyroidism has been well-controlled."
    - This detail does not contribute to the argument and should be excluded.

- Tips:
    - Before the discussion, identify 2-3 key points that address the denial directly.
    - Be prepared to explain why any included details are critical to the argument.

---

4. Anticipate and Address Objections

Proactively addressing potential objections demonstrates thorough preparation and strengthens the credibility of your argument. This strategy also helps prevent the discussion from being derailed by unanticipated questions or concerns.

- **Why This Works:** Anticipating objections shows the medical director that you have considered alternative perspectives and are prepared to address them.
- **Example of Rebuttal:**
    - Objection: "The denial notes improvement within 24 hours."
    - Response: "While the patient showed some improvement, the initial instability, including hypotension requiring vasopressors and oxygen dependency, justified inpatient admission."
- **Tips:**
    - Review the denial letter carefully to identify potential points of contention.
    - Prepare evidence-based responses to counter common objections.

---

5. Maintain a Collaborative Tone

Framing the conversation as a joint effort fosters a sense of partnership and encourages a productive dialogue. A collaborative tone emphasizes shared goals, such as ensuring appropriate patient care while adhering to guidelines.

- **Why This Matters:** An adversarial approach can create resistance and reduce the willingness of the medical director to reconsider the denial. Collaboration, on the other hand, builds rapport and mutual respect.
- **Example:**
    - "I appreciate your time, and I believe this case aligns with your criteria for inpatient admission based on the patient's condition."
- **Tips:**
    - Avoid language that may come across as confrontational, such as "You denied this case without sufficient evidence."
    - Use phrases that highlight shared objectives, such as "We both want to ensure this patient receives the care they need."

Practical Example of a Well-Framed Argument

Case Overview: A patient was denied inpatient admission for pneumonia, with the denial citing insufficient evidence of hypoxia or treatment escalation.

Discussion Outline:

1. **Start with the Strongest Point:** "The patient's oxygen saturation dropped to 85% on room air, requiring 4L supplemental oxygen. This meets hypoxia criteria for inpatient admission."
2. **Organize Logically:**
   - Symptoms: "The patient presented with worsening shortness of breath and fever."
   - Diagnostics: "A chest X-ray showed bilateral infiltrates consistent with pneumonia."
   - Interventions: "The patient required IV antibiotics and continuous oxygen therapy."
3. **Focus on Relevance:** Exclude minor comorbidities that do not support the argument.
4. **Address Objections:** "While the patient stabilized after 48 hours, their initial hypoxia and need for oxygen escalation necessitated inpatient care."
5. **Maintain a Collaborative Tone:** "I appreciate your review of this case and your attention to the clinical details. I believe these findings align with criteria for inpatient care."

A well-framed argument is essential for maximizing clarity, building rapport, and increasing the likelihood of overturning a denial in a P2P conference. By starting with the strongest points, organizing arguments logically, focusing on relevance, anticipating objections, and maintaining a collaborative tone, physician advisors can present their cases with confidence and professionalism. These strategies ensure that each discussion is not only persuasive but also fosters trust and mutual understanding with insurance medical directors.

Using Clinical Evidence and Guidelines to Your Advantage

Evidence-based arguments are the cornerstone of effective peer-to-peer (P2P) advocacy. By integrating objective clinical data, payer-specific criteria, and specialty guidelines into your discussions, you can establish credibility, directly address payer concerns, and build a compelling case for overturning denials. The following strategies focus on how to effectively incorporate these elements into your advocacy.

---

1. Leverage Objective Data

Objective clinical indicators provide a factual, unbiased foundation for demonstrating the necessity of inpatient care. These data points are difficult to dispute and allow physician advisors to focus discussions on clear, quantifiable evidence.

- Examples of Objective Data:
    - **Vital Signs:** Persistent abnormalities, such as heart rate >130 bpm despite IV beta blockers or systolic blood pressure <90 mmHg requiring vasopressors.
    - **Diagnostics:** Imaging findings, like a chest X-ray showing bilateral pulmonary infiltrates or a CT scan identifying significant pulmonary emboli with right heart strain.
    - **Interventions:** Advanced treatments or escalations of care, such as continuous cardiac monitoring revealing atrial fibrillation with rapid ventricular response or initiation of high-flow oxygen therapy for hypoxia.
- Why This Works:
    - Objective data are universally accepted as the foundation of clinical decision-making.
    - Highlighting these indicators reduces the likelihood of subjective interpretations or disputes over the necessity of care.
- Practical Application:
    - Start your argument with a critical data point: "The patient's heart rate remained above 130 bpm despite beta blockers, necessitating continuous cardiac monitoring and IV rate control."
    - Use graphs or trends to illustrate how these indicators evolved during the patient's stay.

---

2. Reference Payer Guidelines

Payer-specific criteria are often the basis for denials. Structuring your arguments to align with these guidelines ensures that discussions are framed in terms that resonate with the medical director's expectations.

- Why This Matters:
    - Insurance medical directors rely heavily on these guidelines to make decisions.
    - Demonstrating familiarity with the payer's criteria builds credibility and aligns your argument with their framework.
- Example:
    - **Payer Criteria:** "MCG guidelines for sepsis recommend inpatient admission for refractory hypotension."
    - **Response:** "This patient demonstrated persistent hypotension despite receiving 2L of IV fluids, meeting your criteria for inpatient admission."
- Tips:

- Quote directly from the guidelines when possible to reinforce alignment: "As noted in your policy, refractory hypotension is a clear indicator of clinical instability requiring inpatient care."
- Keep a repository of commonly used payer criteria for easy reference during P2P conferences.

---

3. Supplement with Specialty Guidelines

Specialty guidelines from reputable organizations can fill gaps when payer criteria are vague, insufficient, or too rigid. These guidelines offer additional clarity and lend authority to your argument.

- Why This Matters:
  - Specialty guidelines provide evidence-based recommendations that can counter overly narrow interpretations of payer criteria.
  - Citing respected organizations adds weight to your argument and demonstrates a broader understanding of clinical standards.
- Examples:
  - For pneumonia denials, cite **American Thoracic Society (ATS)** recommendations, such as inpatient care for patients with hypoxia, multi-lobar involvement, or CURB-65 scores ≥2.
  - For heart failure cases, refer to **American Heart Association (AHA)** guidelines emphasizing inpatient monitoring for acute decompensations, uncontrolled arrhythmias, or hemodynamic instability.
- Practical Application:
  - Example Argument: "While the payer criteria for pneumonia focus on oxygen saturation, the ATS guidelines recommend inpatient care for patients with hypoxia and multi-lobar involvement. This patient met both criteria with oxygen saturation <88% on room air and imaging showing bilateral infiltrates."
- Tips:
  - Ensure that the guidelines you cite are the most current and relevant to the patient's condition.
  - Present these guidelines as complementary rather than contradictory to payer criteria.

---

4. Address Documentation Gaps

Acknowledging and addressing missing details in the medical record demonstrates honesty and proactive problem-solving. This approach helps preempt objections and provides supplemental evidence to support your argument.

- **Why This Works:**
    - Insurance medical directors are often trained to identify gaps in documentation as a basis for denials. Acknowledging these gaps directly shows thoroughness and transparency.
    - Offering supplemental evidence mitigates the impact of missing information.
- **Example:**
    - A denial cites insufficient documentation of oxygen saturation trends: "While the record does not document oxygen saturation hourly, the patient's continuous requirement for 4L oxygen supports the need for inpatient care."
- **Strategies:**
    - Use additional evidence, such as nurse's notes or MAR data, to fill gaps: "Nursing records indicate that the patient required increasing oxygen levels to maintain saturation above 90%."
    - Highlight the clinical context to reinforce the argument: "Although exact trends were not captured, the documented interventions—continuous oxygen and IV antibiotics—demonstrate the severity of the patient's condition."
- **Tips:**
    - Proactively address potential objections in your opening statement to prevent them from derailing the discussion.
    - Emphasize how the available evidence still aligns with payer criteria despite documentation gaps.

Effectively incorporating clinical evidence and guidelines into your P2P advocacy is essential for building a strong, persuasive case. By leveraging objective data, aligning arguments with payer criteria, supplementing with specialty guidelines, and addressing documentation gaps, physician advisors can present compelling arguments that resonate with insurance medical directors.

These strategies not only improve the likelihood of overturning denials but also demonstrate a commitment to professionalism, evidence-based care, and collaboration. By mastering these techniques, physician advisors can advocate more effectively for their patients while fostering trust and mutual respect with payers.

Developing Empathy for the Insurance Medical Director's Perspective

Empathy is an often-overlooked yet critical component of successful peer-to-peer (P2P) conferences. While these discussions often focus on facts and clinical evidence, understanding the human and professional dynamics at play can make the difference between a productive dialogue and a contentious exchange. Insurance medical directors are not adversaries; they are professionals operating under significant constraints, tasked with balancing clinical appropriateness with financial stewardship and adherence to payer guidelines. Recognizing and respecting their role fosters collaboration and increases the likelihood of a favorable outcome.

1. Recognize Their Role

Insurance medical directors work within a structured framework that prioritizes organizational policies, payer guidelines, and cost-containment measures. Their decisions are not solely clinical but must align with the broader goals of their organization, such as minimizing unnecessary expenditures while ensuring patient safety.

- Why This Matters:
  - Acknowledging these constraints demonstrates respect for their position and sets a cooperative tone for the discussion.
  - Recognizing their role helps frame your argument in a way that aligns with their **responsibilities and decision-making framework.**
- Example:
  - "I understand the importance of ensuring that resources are used efficiently, and I appreciate the time you've taken to review this case."
- Tips:
  - Avoid framing the conversation as adversarial; instead, present it as a mutual effort to ensure appropriate care.
  - Respect their expertise and role in applying payer criteria, even if you disagree with their initial decision.

---

2. Use Their Language

Effective communication involves tailoring your argument to the medical director's perspective. Using terminology that aligns with payer guidelines and criteria shows familiarity with their framework and builds rapport.

- Why This Matters:
  - Aligning your language with payer-specific terminology demonstrates professionalism and understanding.
  - This approach helps bridge the gap between clinical language and the operational language used by payers.
- Examples of Aligned Terminology:
  - Replace "critical condition" with "meets criteria for escalation due to instability."
  - Replace "requires monitoring" with "qualifies for continuous telemetry based on policy."
- Practical Example:
  - Instead of saying, "The patient was very sick and needed close monitoring," reframe as, "The patient met criteria for continuous telemetry due to hemodynamic instability and escalating oxygen requirements."

- Tips:
    - Review the denial letter for specific terms or phrases used by the payer and incorporate them into your argument.
    - Avoid overloading the conversation with clinical jargon that may not directly tie to payer criteria.

---

3. Highlight Shared Goals

Both physician advisors and insurance medical directors share a common objective: to ensure patients receive appropriate care in the most effective setting. Emphasizing this shared goal establishes a collaborative tone and reduces resistance.

- Why This Matters:
    - Focusing on mutual goals helps shift the conversation from contention to collaboration.
    - Highlighting shared objectives shows that your advocacy aligns with their responsibilities and organizational mission.
- Example Statement:
    - "We both want to ensure that this patient receives the right level of care to stabilize their condition while avoiding unnecessary complications."
- Shared Goals to Emphasize:
    - **Patient Safety:** Ensuring timely and appropriate interventions to prevent deterioration.
    - **Compliance:** Aligning care decisions with established guidelines and criteria.
    - **Cost-Effectiveness:** Minimizing avoidable complications or readmissions that could increase costs.
- Tips:
    - Frame denial reversals as opportunities to support shared objectives, such as avoiding adverse outcomes or reducing long-term expenses.
    - Reinforce the idea that overturning the denial benefits both the patient and the payer.

---

4. Listen and Adapt

Active listening is a cornerstone of empathetic communication. Demonstrating that you value the medical director's perspective fosters mutual respect and creates opportunities to refine and strengthen your argument.

- Why This Matters:

- Listening attentively allows you to understand their concerns fully and address them effectively.
- A willingness to adapt your approach based on their feedback shows flexibility and professionalism.

- **Strategies for Active Listening**:
  - **Avoid Interruptions:** Allow the medical director to express their concerns fully before responding.
  - **Take Notes:** Document objections or points of clarification to ensure your response is comprehensive.
  - **Ask Clarifying Questions:** Seek additional details or context if their rationale is unclear.
- **Example Response:**
  - "I understand your concern about the short duration of instability, but the patient's persistent hypotension and failure to respond to initial treatments warrant continued inpatient care."
- **Tips:**
  - Use reflective language to validate their concerns: "It sounds like your main concern is the lack of documented oxygen saturation trends. Here's how we can address that..."
  - Maintain a calm and professional tone, even if the discussion becomes challenging.

Practical Example of Empathy in Action

Scenario: A patient's admission for pneumonia was denied due to insufficient evidence of hypoxia at presentation.

Empathy-Driven Approach:

1. **Recognize Their Role:**
   - "I appreciate your role in ensuring that care decisions align with established guidelines and resource use is appropriate."
2. **Use Their Language:**
   - "The patient qualified for escalation due to progressive oxygen desaturation, which required 4L of supplemental oxygen to maintain saturation above 90%."
3. **Highlight Shared Goals:**
   - "We both aim to ensure that patients like this receive the right care at the right time to avoid further complications."
4. **Listen and Adapt:**
   - If the medical director raises concerns about the lack of continuous oxygen saturation data, respond with, "While continuous monitoring data isn't available, the nursing notes show progressive desaturation requiring increasing oxygen levels, which aligns with criteria for escalation."

Empathy is not just a communication skill—it is a strategic tool that transforms P2P conferences from contentious debates into collaborative problem-solving sessions. By recognizing the insurance medical director's role, using language that aligns with payer criteria, highlighting shared goals, and actively listening, physician advisors can foster trust, demonstrate professionalism, and improve the likelihood of overturning denials.

This empathetic approach not only enhances the outcome of individual P2P conferences but also builds stronger, more productive relationships with payers over time, ultimately benefiting patients, providers, and insurers alike.

Conclusion

Building persuasive arguments for peer-to-peer (P2P) conferences is a multifaceted skill that requires the integration of strategic communication, evidence-based advocacy, and empathetic collaboration. These elements combine to form a structured and effective approach to addressing denials and advocating for patient care. Throughout this chapter, we have explored the critical components of successful P2P discussions, highlighting actionable strategies to improve clarity, focus, and outcomes.

Key Insights from Chapter 4

1. **Framing Discussions Effectively**
    - **Clarity and Focus:** A well-framed argument is essential to quickly capture the insurance medical director's attention and guide the discussion. Beginning with the strongest points establishes credibility and sets a decisive tone.
    - **Logical Structure:** Organizing arguments systematically—starting with clinical presentation, transitioning to diagnostic findings and treatments, and concluding with alignment to payer criteria—ensures that key points are not overlooked.
    - **Relevance:** Avoiding extraneous details keeps the conversation centered on the denial rationale, ensuring that time is spent on impactful and actionable arguments.
    - **Anticipating Objections:** Addressing potential concerns before they are raised demonstrates preparedness and strengthens the overall persuasiveness of the argument.
    - **Collaboration:** Maintaining a cooperative tone fosters a productive dialogue, positioning the physician advisor and insurance medical director as partners in achieving shared goals.

2. **Using Clinical Evidence and Guidelines**
    - **Objective Data:** Leveraging measurable, quantifiable clinical indicators provides a strong foundation for arguing medical necessity. Objective evidence is difficult to refute and anchors the argument in facts.
    - **Payer Guidelines:** Aligning arguments with specific payer criteria demonstrates a thorough understanding of the framework within which the medical director operates.

This approach enhances credibility and ensures discussions are grounded in shared reference points.

- **Specialty Guidelines:** When payer criteria fall short, specialty guidelines from reputable organizations provide additional clarity and authoritative support for complex cases.
- **Addressing Documentation Gaps:** Acknowledging and supplementing incomplete records with available evidence shows honesty, attention to detail, and a proactive approach to resolving issues.

3. **Developing Empathy for the Insurance Medical Director**

    - **Understanding Their Role:** Recognizing the constraints and responsibilities of insurance medical directors—balancing clinical appropriateness with cost containment—lays the foundation for a respectful and productive conversation.
    - **Using Their Language:** Framing arguments in terms that align with payer-specific terminology bridges the gap between clinical and operational priorities, making discussions more relatable and persuasive.
    - **Highlighting Shared Goals:** Focusing on mutual objectives, such as ensuring patient safety, adhering to guidelines, and minimizing unnecessary costs, fosters collaboration and reduces resistance.
    - **Active Listening:** Demonstrating attentiveness and adaptability during the discussion builds rapport and allows for real-time refinement of arguments to address objections effectively.

The Broader Impact of Persuasive Advocacy

Effective P2P advocacy extends beyond overturning individual denials. It serves as a critical driver for organizational success by safeguarding revenue, improving patient outcomes, and building stronger relationships with payers. Each successful conference reinforces the physician advisor's credibility, creating a foundation for future collaboration and increasing the likelihood of favorable outcomes in subsequent interactions.

By mastering the strategies outlined in this chapter, physician advisors can position themselves not only as advocates for their patients but also as trusted partners in the healthcare ecosystem. This dual role fosters a sense of mutual respect and shared purpose, benefiting all stakeholders—patients, providers, and payers alike.

Looking Ahead

The ability to craft and deliver persuasive arguments is a vital skill for physician advisors, but even the most well-prepared discussions can be derailed by avoidable missteps. In the next chapter, we will delve into *Avoiding Common Pitfalls in P2P Conferences*. This exploration will focus on maintaining clarity, preventing errors, and implementing strategies to achieve consistent success, ensuring that your advocacy efforts remain focused, effective, and impactful.

References

[46] Overton AR, Lowry AC. Conflict management: difficult conversations with difficult people. Clin Colon Rectal Surg. 2013 Dec;26(4):259-64. doi: 10.1055/s-0033-1356728. PMID: 24436688; PMCID: PMC3835442.

[47] Stephens, E., William, L., Lim, LL. et al. Complex conversations in a healthcare setting: experiences from an interprofessional workshop on clinician-patient communication skills. BMC Med Educ 21, 343 (2021). https://doi.org/10.1186/s12909-021-02785-7

[48] Enabling difficult conversations in the Australian health sector . Aust J Adv Nurs [Internet]. 2021 Aug. 26 [cited 2025 Feb. 19];38(3). Available from: https://www.ajan.com.au/index.php/AJAN/article/view/310

[49] Dorela Priftanji, John D Hill, Daniel M Ashby, Managing difficult conversations, American Journal of Health-System Pharmacy, Volume 77, Issue 21, 1 November 2020, Pages 1723–1726, https://doi.org/10.1093/ajhp/zxaa149

CHAPTER 6

Communication Techniques

Introduction

Effective communication is the foundation of successful peer-to-peer (P2P) conferences, serving as the bridge between clinical expertise and the payer's decision-making framework. While a physician advisor's deep understanding of medical conditions, treatments, and guidelines forms the backbone of their role, it is their ability to communicate this knowledge persuasively and professionally that determines the outcome of these high-stakes discussions. Success in a P2P conference often hinges on the advisor's ability to not only present a compelling case but also navigate the nuances of payer interactions with precision and empathy.

P2P conferences are more than just clinical debates—they are strategic conversations that require a balance of evidence-based arguments and interpersonal skills. Physician advisors must articulate their points clearly and concisely, ensuring that the most critical elements of the case are highlighted and linked to payer-specific criteria. They must also anticipate and address potential objections, presenting their arguments in a way that aligns with the payer's perspective while reinforcing the necessity of care. This dual focus on clinical rigor and strategic communication is what transforms a strong case into a successful appeal.

Moreover, P2P conferences often involve challenging interactions, especially when disagreements arise over medical necessity or documentation gaps. In these situations, professionalism and tact become essential tools for maintaining control of the conversation and redirecting it toward productive outcomes. Effective communication includes the ability to de-escalate tension, acknowledge the constraints faced by insurance medical directors, and frame objections as opportunities for collaboration rather than conflict.[50]

This chapter provides physician advisors with actionable strategies to enhance their communication skills in P2P settings. It explores techniques for crafting persuasive arguments, handling difficult interactions, and maintaining a collaborative tone that fosters trust and mutual respect. By mastering these approaches, physician advisors can not only increase their success in overturning denials but also build stronger relationships with payers, contributing to more effective and efficient resolution of future cases.

Ultimately, the ability to communicate effectively during P2P conferences is not just about achieving short-term victories. It is about establishing a reputation as a credible, professional, and empathetic advocate who prioritizes patient care and organizational integrity. This chapter serves as a comprehensive guide to refining these skills, equipping physician advisors to navigate even the most complex payer interactions with confidence and professionalism.

Crafting Clear and Concise Arguments

In the fast-paced environment of peer-to-peer (P2P) conferences, where time is limited, and the outcomes carry significant implications for both patient care and organizational finances, the ability to communicate effectively is paramount.[51] Physician advisors must prioritize clarity and brevity in their presentations, focusing on delivering arguments that directly address the denial rationale and align with payer criteria. A well-crafted argument not only increases the likelihood of overturning a denial but also demonstrates professionalism, preparation, and respect for the insurance medical director's time.

Below are detailed strategies to help physician advisors craft clear, concise, and impactful arguments during P2P conferences.

---

1. Structure Your Argument Logically

A logical flow of information ensures that your argument is easy to follow and effectively communicates the key points. By presenting information in a structured sequence, you can guide the insurance medical director through the case and build a compelling narrative that aligns with payer criteria.

- **Step 1: Clinical Presentation** Begin by summarizing the patient's condition and history to establish context.
    - Example: "The patient presented to the emergency department with complaints of worsening shortness of breath, fever, and productive cough over the past 72 hours."
- **Step 2: Key Findings** Highlight the objective data, including diagnostic results, clinical indicators, and treatments that substantiate the need for inpatient care.
    - Example: "The chest X-ray revealed bilateral infiltrates, and the patient required 4L oxygen to maintain saturation above 90%. Initial lab work showed a white blood cell count of 16,000 and a CRP level indicative of inflammation."
- **Step 3: Alignment with Criteria** Conclude by explicitly connecting these findings to the payer's guidelines, demonstrating how the patient's clinical presentation meets the criteria for inpatient care.
    - Example: "The combination of hypoxia, elevated inflammatory markers, and failed outpatient antibiotics clearly meets the criteria for pneumonia-related inpatient admission as outlined in your policy."

Tip: Use a checklist to ensure that your argument touches on all critical aspects—clinical presentation, key findings, and criteria alignment.

---

2. Avoid Overloading with Details

While thorough documentation is vital during the preparation phase, including every detail in your argument can overwhelm the medical director and dilute the impact of your key points. Focus on the most relevant information and exclude extraneous details that do not directly support your case.

- **Focus on Critical Facts** Highlight the elements that directly address the denial rationale and support the necessity of inpatient care.
  - Example: "The patient required continuous IV antibiotics and supplemental oxygen due to persistent hypoxia and worsening infection."
- **Exclude Unrelated Details** Avoid mentioning aspects of the case that do not impact the argument, such as stable comorbidities or resolved symptoms.
  - Irrelevant Example: "The patient's dyslipidemia has been well-controlled for years."

Tip: Before the conference, identify the top 2-3 points that are most critical to the denial rationale and ensure these are the focal points of your argument.

---

3. Eliminate Weak Language

The language used in your argument can significantly impact how your points are perceived. Weak or ambiguous phrasing may undermine the strength of your case. Instead, use precise and assertive language that emphasizes the severity of the patient's condition and the necessity of inpatient care.

- **Examples of Weak vs. Strong Language:**
  - Weak: "The patient had a small infiltrate."
  - Strong: "The patient had a left lower lobe infiltrate consistent with pneumonia, confirmed on imaging."
  - Weak: "The patient seemed unwell upon admission."
  - Strong: "The patient presented with clinical signs of sepsis, including tachycardia, hypotension, and elevated lactate levels."

Tip: Review your argument for vague or non-specific terms and replace them with definitive statements supported by objective data.

4. Reinforce Key Points

Restating the critical findings and conclusions at the end of the discussion leaves a lasting impression and ensures that the medical director remembers the most important aspects of your argument. This summary serves as your final opportunity to reinforce the necessity of inpatient care and tie the case back to payer criteria.

- **How to Reinforce Key Points:**
    - Briefly recap the clinical presentation, key findings, and criteria alignment in 1-2 sentences.
    - End with a strong statement that directly addresses the denial rationale.
    - Example: "To summarize, the patient's persistent hypoxia, worsening infiltrates, and failed outpatient management meet the criteria for inpatient admission under your guidelines for pneumonia management. I believe this aligns with both the clinical evidence and the policy requirements for appropriate care."

Tip: Practice your summary beforehand to ensure it is concise, impactful, and confidently delivered.

Practical Example: Crafting a Clear and Concise Argument

Case: A patient denied inpatient admission for pneumonia due to insufficient evidence of clinical instability.

- **Start with Clinical Presentation:**
    - "The patient, a 68-year-old male with a history of COPD, presented to the emergency department with acute dyspnea, fever, and productive cough. Symptoms had progressed over the past 72 hours despite a 7-day course of oral antibiotics."
- **Highlight Key Findings:**
    - "Chest X-ray revealed bilateral infiltrates, consistent with pneumonia. The patient required 4L oxygen to maintain saturation above 90%, and lab results indicated leukocytosis with a WBC count of 16,500."
- **Align with Criteria:**
    - "These findings, combined with the failure of outpatient antibiotics and evidence of hypoxia, clearly meet the criteria for inpatient admission under your pneumonia guidelines."
- **Conclude with a Summary:**
    - "In summary, the patient's persistent hypoxia and need for oxygen escalation, combined with imaging and lab findings, support the necessity of inpatient care to prevent further deterioration."

Crafting clear and concise arguments is a vital skill for physician advisors in P2P conferences. By structuring arguments logically, focusing on relevance, using strong language, and reinforcing key points, advisors can present their cases effectively and persuasively. These techniques not only improve the clarity and impact of the discussion but also increase the likelihood of achieving favorable outcomes for patients and organizations alike.

Managing Difficult Conversations

Peer-to-peer (P2P) conferences often involve high-stakes discussions where emotions can run high, especially when there are disagreements over medical necessity or gaps in documentation. Navigating these contentious moments with professionalism and empathy is essential for fostering productive dialogue and achieving favorable outcomes. Physician advisors must remain adaptable, maintain composure, and employ strategies to redirect and de-escalate challenging interactions. Below are expanded strategies for managing difficult conversations effectively.[52]

1. Maintain Composure

Remaining calm and professional in the face of adversity is critical to ensuring that discussions remain constructive. When the insurance medical director's tone becomes adversarial or dismissive, losing composure can derail the conversation and reduce the likelihood of reaching a resolution.

- Why This Matters:
  - Staying calm demonstrates professionalism and helps keep the discussion focused on clinical facts.
  - Emotional responses can escalate tension and make it more difficult to achieve a favorable outcome.
- How to Maintain Composure:
  - **Pause Before Responding:** Take a moment to collect your thoughts before replying to inflammatory comments. This brief pause can help you formulate a measured response.
  - **Focus on Clinical Facts:** Redirect the conversation to objective data, avoiding personalizing disagreements.
  - **Avoid Defensive Reactions:** Instead of countering hostility with defensiveness, acknowledge the medical director's perspective and guide the discussion back to the case.
- Example Statement:
  - "I understand your perspective, and I appreciate the opportunity to discuss this case further. Let's review the clinical evidence together."
- Tips:

- Practice deep breathing or other stress management techniques if you anticipate challenging interactions.
- Prepare a few neutral phrases to use as a buffer if the discussion becomes heated.

---

2. Acknowledge Their Constraints

Insurance medical directors operate within a framework of payer guidelines, cost containment priorities, and organizational policies. Recognizing these constraints not only demonstrates empathy but also helps to defuse tension by showing respect for their role.

- Why This Matters:
  - Acknowledging the medical director's responsibilities fosters mutual respect and sets a collaborative tone.
  - It reframes the conversation as a joint effort to resolve the case rather than an adversarial debate.
- How to Acknowledge Constraints:
  - Use language that validates their role and the challenges they face.
  - Avoid dismissing their perspective, even if you disagree with their decision.
- Example Statement:
  - "I understand your role involves balancing clinical appropriateness with cost containment, and I value your insights."
- Tips:
  - Avoid implying that their decision was made in bad faith.
  - Frame your argument as a way to support their objectives, such as ensuring appropriate care while aligning with guidelines.

---

3. Redirect Negative Responses

When conversations veer off course or become overly contentious, redirecting the focus back to clinical facts can help refocus the discussion on resolving the denial. Pivoting away from emotionally charged or unproductive exchanges ensures that the conversation remains aligned with its purpose.

- Why This Matters:
  - Redirecting negative responses prevents the conversation from escalating and keeps it centered on the case.
  - It reinforces your role as an advocate for the patient, guided by evidence and professionalism.
- How to Redirect Negative Responses:
  - Acknowledge the comment without dwelling on it, then pivot back to the clinical evidence.
  - Use specific data points to steer the discussion in a productive direction.

- **Example Statement:**
    - "I hear your concern about the documentation gaps. Let's focus on how the patient's persistent hypoxia aligns with your criteria for inpatient admission."
- **Tips:**
    - Avoid engaging in back-and-forth debates about subjective opinions.
    - Keep your tone neutral and your language centered on clinical facts.

---

4. De-Escalate Tension

Using neutral, non-confrontational language can prevent conflicts from escalating and help reset the tone of the conversation. Even if the discussion becomes contentious, a calm and measured response can encourage collaboration.

- **Why This Matters:**
    - De-escalation techniques help maintain a professional tone and preserve the opportunity for a resolution.
    - Neutral language fosters a sense of partnership rather than opposition.
- **How to De-Escalate Tension:**
    - Replace confrontational phrases with collaborative alternatives.
    - Avoid blaming or accusatory language, even if the denial seems unjustified.
- **Examples of Neutral Language:**
    - Avoid: "This denial is completely unjustified."
    - Use: "I'd like to clarify how the clinical findings support inpatient care based on your guidelines."
- **Tips:**
    - Use phrases like "Let's explore this further" or "Can we agree on..." to encourage dialogue.
    - If emotions run high, pause briefly to allow both parties to regroup.

---

5. Be Prepared to Pause

Sometimes, despite your best efforts, a P2P conference may become unproductive due to unresolved disagreements or missing documentation. Suggesting a pause for additional review can reset the tone and allow both parties to approach the case with fresh perspectives and better preparation.

- **Why This Matters:**
    - Pausing the discussion shows that you are committed to resolving the issue thoroughly and are open to addressing gaps in information.
    - It demonstrates professionalism and avoids forcing a resolution prematurely.
- **How to Pause Effectively:**

    - Suggest revisiting the case after obtaining additional documentation or reviewing the payer criteria in more detail.
    - Frame the pause as an opportunity to strengthen your case and ensure alignment with guidelines.
- **Example Statement:**
    - "Would it be helpful if I provided further documentation to clarify the clinical picture? I'd be happy to follow up after gathering additional details."
- **Case Example:**
    - In a heated call regarding a denial for heart failure admission, the physician advisor proposed pausing the discussion to gather documentation of serial BNP trends. The second discussion, with additional evidence in hand, led to a more constructive dialogue and a successful resolution.
- **Tips:**
    - Follow up promptly after the pause to maintain momentum and demonstrate your commitment to resolving the case.
    - Use the additional time to refine your argument and address any gaps identified during the initial discussion.

Practical Example: Managing a Difficult Conversation

Scenario: A denial for pneumonia admission becomes contentious when the medical director dismisses the need for inpatient care, citing a rapid improvement in oxygen saturation.

1. Maintain Composure:
    - "I appreciate your perspective, and I'd like to discuss this case further to clarify the clinical findings."
2. Acknowledge Constraints:
    - "I understand your role involves ensuring that care aligns with guidelines while managing resources effectively."
3. Redirect Negative Responses:
    - "Let's revisit the patient's initial presentation, which included oxygen saturation levels dropping to 85% on room air, requiring supplemental oxygen."
4. De-Escalate Tension:
    - "I'd like to focus on how these findings align with criteria for inpatient care based on your policy."
5. Be Prepared to Pause:
    - "If additional documentation of oxygen saturation trends would be helpful, I'd be happy to provide that for further review."

Managing difficult conversations in P2P conferences requires a combination of professionalism, empathy, and strategic thinking. By maintaining composure, acknowledging the constraints of medical directors, redirecting negative responses, de-escalating tension, and knowing when to pause, physician advisors can transform contentious discussions into productive dialogues. These techniques not only improve the likelihood of overturning denials but also foster stronger relationships with payers, paving the way for more collaborative and successful future interactions.

Handling Objections and Pivoting

In peer-to-peer (P2P) conferences, objections from insurance medical directors are a natural part of the discussion. These objections often stem from perceived gaps in documentation, misinterpretation of clinical data, or strict adherence to payer guidelines. Effectively addressing these objections requires preparation, adaptability, and strategic thinking. By anticipating potential challenges and using them as opportunities to reinforce the argument, physician advisors can navigate objections with confidence and professionalism.

---

1. Anticipate Objections

Preparation is key to handling objections effectively. By reviewing the denial letter, payer guidelines, and the patient's medical record in advance, physician advisors can anticipate the most likely objections and craft evidence-based responses. This proactive approach not only strengthens the argument but also demonstrates professionalism and thoroughness.

- Why This Matters:
    - Anticipating objections allows the physician advisor to address concerns swiftly and confidently, keeping the discussion focused and productive.
    - It prevents the conversation from becoming derailed by unanticipated challenges.
- How to Anticipate Objections:
    - Identify common objections, such as:
        - "Observation status would have been sufficient."
        - "The patient stabilized too quickly for inpatient care."
    - Prepare concise, evidence-based responses for each objection.
- Example Objection and Response:
    - **Objection:** "Observation status would have been sufficient."
    - **Response:** "The patient's continuous oxygen requirement and IV antibiotics clearly necessitated inpatient monitoring to prevent further clinical decline."
- Tips:
    - Focus on objective data that directly refutes the objection.
    - Avoid becoming defensive; instead, use objections as opportunities to highlight key clinical findings.

---

2. Address Documentation Gaps

Documentation gaps are a common reason for denials. Acknowledging these gaps while providing supplemental evidence demonstrates honesty and a proactive approach to resolving concerns. This strategy helps preempt further objections and shifts the focus back to the strengths of the case.

- Why This Matters:
  - Transparency about documentation gaps builds trust with the medical director.
  - Providing supplemental evidence reinforces the argument and mitigates the impact of incomplete records.
- How to Address Documentation Gaps:
  - Acknowledge the gap without undermining the overall argument.
  - Provide alternative evidence to fill the gap, such as nurse's notes, medication administration records (MAR), or trends in clinical data.
- Example Statement:
  - "While the record does not include hourly oxygen readings, the patient's need for 4L oxygen to maintain SpO_2 >90% is well-documented in nursing notes and MAR data."
- Tips:
  - Tie supplemental evidence directly to payer criteria to maintain focus on the denial rationale.
  - Reiterate how the available evidence supports the need for inpatient care despite documentation gaps.

---

3. Pivot to Stronger Points

When an objection cannot be fully addressed or resistance persists, shifting the focus to the strongest aspects of the case can redirect the discussion. This approach ensures that the conversation remains constructive and emphasizes the most compelling evidence.

- Why This Matters:
  - Pivoting allows the physician advisor to steer the discussion away from weaker aspects of the case and highlight the most persuasive clinical findings.
  - It keeps the conversation aligned with the ultimate goal: overturning the denial.
- How to Pivot Effectively:
  - Transition smoothly from the objection to a stronger argument.
  - Use objective data to redirect attention to critical clinical findings.
- Example Statement:
  - "I'd like to emphasize the patient's persistent hypotension and need for IV vasopressors, which meet your criteria for inpatient care."
- Tips:
  - Avoid dismissing the objection outright; instead, acknowledge it briefly before pivoting.
  - Highlight data that directly aligns with the payer's criteria for inpatient admission.

---

4. Ask Clarifying Questions

Asking clarifying questions can uncover the underlying rationale for an objection and provide an opportunity to address specific concerns. This approach fosters a collaborative tone and ensures that all aspects of the denial rationale are fully understood.

- **Why This Matters:**
    - Clarifying questions demonstrate active listening and a willingness to engage in dialogue.
    - They provide insights into the medical director's thought process, allowing for more targeted responses.
- **How to Use Clarifying Questions:**
    - Use open-ended questions to encourage detailed responses.
    - Frame questions in a way that invites collaboration rather than confrontation.
- **Example Question:**
    - "Could you elaborate on which aspect of the criteria you feel was not met? I'd like to address your specific concerns."
- **Tips:**
    - Avoid interrupting or challenging the medical director's response; instead, use their feedback to refine your argument.
    - Use follow-up questions to drill down into specific points if needed.

---

5. Frame Objections as Opportunities

Rather than viewing objections as barriers, reframing them as opportunities to strengthen the argument can foster a collaborative tone and demonstrate a problem-solving mindset. This approach positions the physician advisor as a partner in achieving the shared goal of ensuring appropriate patient care.

- **Why This Matters:**
    - Reframing objections emphasizes collaboration and reduces the likelihood of adversarial interactions.
    - It shifts the focus from disagreement to problem-solving, increasing the chances of a favorable outcome.
- **How to Reframe Objections:**
    - Acknowledge the objection without defensiveness.
    - Highlight how addressing the objection strengthens the overall argument.
- **Example Statement:**
    - "I appreciate your point about observation care, but the patient's worsening symptoms and failed outpatient management demonstrate the need for inpatient monitoring to prevent further complications."
- **Tips:**
    - Use language that emphasizes shared goals, such as "ensuring appropriate care" or "aligning with guidelines."

- Avoid framing objections as mistakes; instead, focus on how they provide an opportunity to clarify or reinforce the case.

Practical Example: Handling Objections and Pivoting

Scenario: A denial for heart failure admission cites "rapid stabilization" as the primary reason for observation status instead of inpatient care.

1. **Anticipate Objections:**
    - Prepare to address stabilization timelines with clinical data: "The patient demonstrated persistent dyspnea and elevated BNP levels despite initial IV diuresis."
2. **Address Documentation Gaps:**
    - "While serial oxygen saturations are not documented hourly, the nursing notes show progressive improvement with supplemental oxygen, which aligns with inpatient care needs."
3. **Pivot to Stronger Points:**
    - "I'd also like to emphasize the need for continuous telemetry due to the patient's atrial fibrillation with rapid ventricular response, which meets criteria for monitoring in an inpatient setting."
4. **Ask Clarifying Questions:**
    - "Can you clarify whether the stabilization timeline or the lack of documentation is the primary concern? I'd like to address your specific points."
5. **Frame Objections as Opportunities:**
    - "Thank you for pointing that out. This gives me the opportunity to highlight how the initial instability and continuous monitoring requirements align with your criteria for inpatient care."

Handling objections and pivoting effectively is a critical skill for physician advisors during P2P conferences. By anticipating objections, addressing documentation gaps, pivoting to stronger points, asking clarifying questions, and reframing objections as opportunities, physician advisors can maintain control of the discussion and strengthen their case. These strategies not only improve the likelihood of overturning denials but also foster a collaborative dynamic that enhances relationships with insurance medical directors and payers over time.

Maintaining Empathy and Collaboration

Empathy and collaboration are fundamental to the success of peer-to-peer (P2P) conferences. Physician advisors who foster trust and build rapport with insurance medical directors increase the likelihood of achieving favorable outcomes. By understanding the priorities and constraints of medical directors, aligning discussions with shared goals, and maintaining a respectful and professional tone, physician advisors can create a cooperative environment that benefits both parties and, most importantly, the patient.[52]

---

1. Recognize Their Role

Insurance medical directors operate within a structured framework that prioritizes resource allocation, regulatory compliance, and clinical appropriateness. They are tasked with making decisions that not only align with payer guidelines but also ensure that healthcare resources are used efficiently. Acknowledging these responsibilities demonstrates respect for their role and fosters a collaborative dynamic.

- Why This Matters:
  - Recognizing their role reduces potential tension and adversarial tones in discussions.
  - It establishes mutual respect, setting the stage for a more productive exchange.
- How to Acknowledge Their Role:
  - Express understanding of their responsibilities and constraints.
  - Avoid dismissing or challenging their expertise; instead, frame your argument as supportive of their objectives.
- Example Statement:
  - "I understand that your role involves balancing resource allocation with clinical appropriateness, and I appreciate the thoughtful review you've given to this case."
- Tips:
  - Use neutral and respectful language, even when there is disagreement.
  - Show appreciation for the medical director's expertise and the complexity of their role.

---

2. Emphasize Shared Goals

Both physician advisors and insurance medical directors share a common objective: ensuring that patients receive appropriate care in the right setting. Highlighting these shared goals helps reinforce the idea that the discussion is a partnership rather than a conflict.

- Why This Matters:
  - Emphasizing common objectives helps reduce resistance and fosters alignment between both parties.
  - t shifts the focus from contentious issues to collaborative problem-solving.
- How to Highlight Shared Goals:
  - Identify specific outcomes that benefit both the patient and the payer, such as avoiding readmissions, ensuring compliance with guidelines, or promoting timely and effective care.
- Example Statement:
  - "We both want to ensure this patient receives the care they need in the most appropriate setting to support recovery and avoid readmission."

- Shared Goals to Highlight:
    - **Patient Safety:** Ensuring timely interventions and monitoring to prevent complications.
    - **Cost-Effectiveness:** Avoiding unnecessary care while ensuring necessary treatments are provided.
    - **Compliance:** Aligning care decisions with evidence-based guidelines and payer criteria.
- Tips:
    - Frame arguments in terms of mutual benefits, such as reducing long-term costs by preventing readmissions.
    - Reinforce the idea that the payer's criteria and the clinical needs of the patient can coexist harmoniously.

---

3. Collaborate on Solutions

Transforming discussions into collaborative efforts creates a sense of partnership and encourages productive dialogue. By positioning your arguments as contributions to a shared goal, you can defuse tension and build rapport.

- Why This Matters:
    - A collaborative approach fosters goodwill and makes the medical director more likely to engage constructively.
    - It reframes the discussion from a debate into a mutual effort to achieve the best outcome.
- How to Collaborate Effectively:
    - Present your perspective while inviting feedback and suggestions.
    - Frame solutions as joint efforts to align with payer expectations and patient needs.
- Example Statement:
    - "Based on the clinical findings, I believe inpatient care was necessary, but I'd like to discuss how we can align future cases more closely with your expectations."
- Practical Strategies:
    - Discuss ways to improve documentation for future cases to avoid similar denials.
    - Offer to follow up with additional data or clarification if needed.
- Tips:
    - Avoid a combative tone; instead, use language that invites collaboration, such as "let's discuss" or "how can we work together to resolve this?"
    - Emphasize a willingness to adapt and learn from the experience to enhance future interactions.

---

4. Express Gratitude

Concluding the discussion with a note of appreciation leaves a positive impression and strengthens relationships for future interactions. Gratitude demonstrates professionalism and respect for the medical director's time and effort, regardless of the outcome.

- **Why This Matters:**
  - Expressing gratitude fosters goodwill and creates a foundation for more constructive interactions in the future.
  - It reinforces a collaborative tone and leaves the door open for ongoing dialogue.
- **How to Express Gratitude:**
  - Thank the medical director for their time and input, emphasizing the value of their feedback.
  - Maintain a professional and respectful tone, even if the outcome is not favorable.
- **Example Statement:**
  - "Thank you for taking the time to review and discuss this case. Your feedback is invaluable, and I appreciate the collaborative dialogue."
- **Tips:**
  - Be sincere in your expression of gratitude; avoid sounding perfunctory or insincere.
  - Use the conclusion as an opportunity to reiterate your willingness to engage constructively in future discussions.

Practical Example: Empathy and Collaboration in Action

Scenario: A patient's admission for pneumonia was denied due to insufficient evidence of hypoxia at presentation.

1. **Recognize Their Role:**
   - "I understand the importance of your role in ensuring care aligns with established guidelines and resources are allocated appropriately. I appreciate your careful review of this case."
2. **Emphasize Shared Goals:**
   - "We both aim to ensure this patient receives timely and appropriate care to prevent further complications and avoid unnecessary readmissions."
3. **Collaborate on Solutions:**
   - "Based on the clinical findings, inpatient care was necessary, but I'd like to hear your perspective on how we might improve documentation for similar cases in the future."
4. **Express Gratitude:**
   - "Thank you for discussing this case with me. Your insights are valuable, and I look forward to collaborating on ensuring the best outcomes for our patients."

Maintaining empathy and collaboration during P2P conferences is essential for fostering productive discussions and building trust with insurance medical directors. By recognizing their role, emphasizing shared goals, collaborating on solutions, and expressing gratitude, physician advisors can create a cooperative dynamic that improves the likelihood of favorable outcomes.

These strategies not only enhance the success of individual P2P conferences but also contribute to stronger, more constructive relationships with payers over time, benefiting patients, providers, and insurers alike.

Empathy and collaboration are not just communication techniques but essential strategies for achieving success in peer-to-peer (P2P) conferences. In an environment often marked by competing priorities—clinical needs versus resource allocation—physician advisors must find common ground with insurance medical directors to advocate effectively for patient care. This chapter has underscored the importance of fostering trust, building rapport, and approaching discussions with a mindset focused on shared goals and mutual respect.

Key Lessons from the Chapter

1. **Recognizing Their Role**
    - Insurance medical directors operate under significant constraints, including adherence to payer guidelines, regulatory requirements, and organizational cost-containment policies. Acknowledging these responsibilities sets a respectful tone and establishes a foundation for productive dialogue.
    - Physician advisors can strengthen their arguments by framing them in ways that align with these constraints, showing respect for the medical director's expertise while reinforcing the clinical necessity of their case.

    Takeaway: Respect for the medical director's role fosters an environment of mutual respect, reducing potential adversarial tones and opening the door for collaboration.

2. **Emphasizing Shared Goals**
    - Both physician advisors and insurance medical directors share a vested interest in ensuring that patients receive the right care in the right setting. Highlighting these shared goals—such as patient safety, compliance with guidelines, and cost-effective care—helps shift the conversation from contention to collaboration.
    - By aligning clinical advocacy with payer priorities, advisors can create a partnership dynamic that emphasizes mutual benefits, such as preventing readmissions or avoiding complications.

    Takeaway: Focusing on shared objectives creates alignment and makes it easier to find common ground, even in contentious discussions.

3. **Collaborating on Solutions**
    - Transforming a P2P conference into a cooperative effort builds rapport and fosters a sense of partnership. By inviting feedback and presenting solutions as joint efforts, physician advisors can turn even challenging discussions into opportunities for constructive dialogue.

- Offering to improve processes, documentation, or communication for future cases demonstrates a willingness to learn and adapt, which can strengthen relationships with payers over time.

Takeaway: Collaboration reinforces trust and transforms the discussion from a debate into a shared effort to achieve the best outcome for the patient.

4. **Expressing Gratitude**
   - Ending the discussion with genuine appreciation for the medical director's time and insights helps maintain professionalism and leaves a positive impression for future interactions. Even in cases where the outcome is not favorable, expressing gratitude demonstrates respect and keeps the relationship open for future opportunities.
   - Gratitude also humanizes the conversation, reminding both parties of their shared commitment to patient care.

   Takeaway: Expressing gratitude fosters goodwill and builds the foundation for more constructive interactions in the future.

Putting It All Together: Empathy and Collaboration in Practice

A successful P2P conference relies on the integration of empathy, respect, and strategic advocacy. Physician advisors who recognize the constraints faced by medical directors, focus on shared goals, invite collaboration, and express appreciation can turn even contentious discussions into opportunities for productive outcomes.

- **Scenario in Practice:** A patient's admission for pneumonia is denied due to insufficient documentation of hypoxia.
  - **Recognize Their Role:** "I appreciate your role in ensuring care aligns with established guidelines and resources are allocated appropriately. I value your thoughtful review of this case."
  - **Emphasize Shared Goals:** "We both aim to ensure this patient receives timely and appropriate care to prevent further complications and avoid unnecessary readmissions."
  - **Collaborate on Solutions:** "Based on the clinical findings, inpatient care was necessary, but I'd like to hear your perspective on how we might improve documentation for similar cases in the future."
  - **Express Gratitude:** "Thank you for discussing this case with me. Your insights are valuable, and I look forward to collaborating on ensuring the best outcomes for our patients."

This empathetic and collaborative approach not only enhances the likelihood of overturning denials but also builds long-term trust and respect between physician advisors and insurance medical directors.

Why Empathy and Collaboration Matter

Empathy and collaboration extend beyond individual P2P conferences. They are cornerstones of broader payer-provider relationships, influencing how insurers view the healthcare organization and its representatives. When physician advisors consistently approach these interactions with professionalism, respect, and a shared commitment to patient care, they contribute to:

- **Improved Outcomes:** Increased overturn rates for denials that directly impact patient care.
- **Stronger Relationships:** Trust and rapport that enhance future interactions with payers.
- **Operational Efficiency:** Smoother resolution processes that save time and resources for both parties.

Looking Ahead

The skills outlined in this chapter—recognizing roles, emphasizing shared goals, collaborating on solutions, and expressing gratitude—serve as the foundation for effective P2P advocacy. However, physician advisors must also be aware of potential pitfalls that can undermine these efforts. The next chapter will explore common mistakes to avoid in P2P conferences, such as overloading discussions with irrelevant details, misinterpreting payer criteria, and neglecting the operational implications of denials. By understanding these pitfalls, physician advisors can refine their approach and further enhance their effectiveness in P2P advocacy.

In summary, empathy and collaboration are not just tactics—they are integral to the physician advisor's role in creating meaningful, productive dialogues with payers that prioritize patient care while respecting organizational constraints. These strategies ensure that P2P conferences are not only about overturning denials but also about building bridges between clinical teams and payers for the greater good of healthcare delivery.

References

[50] Prober CG, Grousbeck HI, Meehan WF 3rd. Managing Difficult Conversations: An Essential Communication Skill for All Professionals and Leaders. Acad Med. 2022 Jul 1;97(7):973-976. doi: 10.1097/ACM.0000000000004692. Epub 2022 Jun 23. PMID: 35767404.

[51] Stephens, E., William, L., Lim, LL. et al. Complex conversations in a healthcare setting: experiences from an interprofessional workshop on clinician-patient communication skills. BMC Med Educ 21, 343 (2021). https://doi.org/10.1186/s12909-021-02785-7

[52] West, M.A. (2020). Compassionate and Collective Leadership for Cultures of High-Quality Care. In: Montgomery, A., van der Doef, M., Panagopoulou, E., Leiter, M.P. (eds) Connecting Healthcare Worker Well-Being, Patient Safety and Organisational Change. Aligning Perspectives on Health, Safety and Well-Being. Springer, Cham. https://doi.org/10.1007/978-3-030-60998-6_13

CHAPTER 7

Avoiding Common Pitfalls

Introduction

Peer-to-peer (P2P) conferences are essential components of the physician advisor's role, serving as pivotal opportunities to advocate for patients and challenge insurance denials. These interactions often hold the key to ensuring that patients receive the appropriate level of care while safeguarding the financial stability of healthcare organizations. However, achieving success in P2P conferences requires more than clinical expertise—it demands strategic communication, precise application of payer policies, and an awareness of the broader implications of denials.

Even the most experienced physician advisors can fall into common traps that undermine the effectiveness of their arguments. These pitfalls may include presenting excessive or irrelevant information, misinterpreting or overlooking payer guidelines, or allowing external influences to erode clinical confidence. Additionally, physician advisors who neglect the financial and operational consequences of denials risk missing critical opportunities to strengthen their advocacy.

Understanding and addressing these challenges is crucial. When physician advisors recognize and proactively avoid these pitfalls, they can present clearer, more persuasive arguments that resonate with insurance medical directors. This not only increases the likelihood of overturning denials but also builds credibility and fosters more productive relationships with payers.

This chapter serves as a practical guide to avoiding the most common pitfalls encountered during P2P conferences. It explores actionable strategies to help physician advisors refine their approach, including:

- Streamlining arguments to focus on the most relevant clinical data.
- Mastering the nuances of payer-specific criteria to align discussions with their expectations.
- Resisting external pressures that may compromise clinical judgment or decision-making.
- Addressing the financial and operational dimensions of denials to emphasize the broader impact.

By adopting these strategies, physician advisors can elevate their performance in P2P conferences, improving both individual outcomes and the overall effectiveness of their organization's denial management processes.

Overloading with Irrelevant Details

One of the most common and detrimental mistakes in peer-to-peer (P2P) conferences is presenting excessive or tangential information. While thorough documentation is essential for preparation, overloading the discussion with unnecessary details can distract from the main argument, confuse the insurance medical director, and ultimately weaken the case. The goal is to communicate a focused, compelling narrative that directly addresses the denial rationale and emphasizes the clinical necessity of care. By streamlining the presentation and honing in on the most relevant data, physician advisors can ensure their arguments are clear, impactful, and persuasive.[53]

1. Focus on Key Data Points

The cornerstone of a strong argument is presenting only the most critical clinical findings that directly support the need for inpatient care or overturning the denial. Highlighting these points ensures the conversation remains focused on the key aspects of the patient's case and the payer's criteria.

- Examples of Key Data Points:
    - **Persistent Hypotension:** "The patient remained hypotensive with a mean arterial pressure <65 despite two liters of IV fluid resuscitation."
    - **Critical Diagnostics:** "CT scan confirmed bilateral pulmonary emboli with evidence of right heart strain."
    - **Interventions:** "The patient required IV vasopressors and continuous oxygen therapy to stabilize hemodynamics."
- Why It Matters:
    - Presenting focused data builds a logical and evidence-based argument that is easier for the insurance medical director to follow.
    - It avoids diverting attention to less pertinent details, ensuring the conversation remains impactful.
- Tips for Success:
    - Identify the top three to five data points that directly address the payer's denial rationale.
    - Ensure these points align closely with the criteria outlined by the payer or relevant clinical guidelines.

2. Use Concise Language

Clear and succinct communication is essential in P2P conferences, where time is limited. Brevity ensures that critical points are delivered efficiently without losing their impact.

- Example of Revised Language:
  - **Before:** "The patient initially presented with mild dyspnea, which gradually worsened over several hours, eventually requiring oxygen therapy to maintain adequate saturation levels."
  - **After:** "The patient required 4L oxygen to maintain SpO_2 >90% due to worsening dyspnea."
- Why It Matters:
  - Concise language eliminates ambiguity, making it easier for the medical director to grasp the clinical significance of the case.
  - It reduces the risk of overwhelming the listener with excessive or irrelevant details.
- Tips for Success:
  - Use short, direct sentences that prioritize critical clinical facts.
  - Avoid overly technical jargon unless it is necessary to address the denial rationale.

---

3. Eliminate Redundant Information

Including irrelevant or redundant information can dilute the strength of your argument and make the discussion less persuasive. Physician advisors should focus exclusively on details that align with the denial rationale and payer criteria.

- Include:
  - "Serial hemoglobin levels dropped from 12.5 to 7.8 g/dL, necessitating a transfusion."
  - "The patient's oxygen saturation declined to 85% on room air, requiring escalation to high-flow oxygen."
- Exclude:
  - "The patient also has a history of osteoarthritis, which is unrelated to the current admission."
  - "The patient's prior medical history includes appendectomy as a child."
- Why It Matters:
  - Redundant or irrelevant details can distract from the main argument, reducing the overall persuasiveness of the case.
  - Keeping the focus on pertinent information ensures the medical director can quickly grasp the critical aspects of the patient's condition.
- Tips for Success:
  - Review the patient's medical record before the conference and extract only the information relevant to the denial rationale.
  - Practice presenting the case to identify and eliminate unnecessary details.

---

4. Leverage Prep Sheets

Prep Sheets are invaluable tools for organizing and streamlining arguments. They provide a structured format that ensures all critical points are addressed concisely while avoiding extraneous information.

- How to Organize Prep Sheets:
    - Section 1: Clinical Presentation
        - Include a brief summary of the patient's symptoms and presenting condition.
    - Section 2: Diagnostic Findings
        - Highlight key results from imaging, labs, or other tests that substantiate the need for inpatient care.
    - Section 3: Alignment with Payer Criteria
        - Directly connect clinical findings to the payer's specific guidelines or criteria.
- Why It Matters:
    - Prep Sheets help physician advisors stay focused and organized during the P2P conference, ensuring they present a clear and concise argument.
    - They reduce the risk of forgetting critical points or veering into irrelevant details.
- Tips for Success:
    - Use bullet points or numbered lists to enhance readability and ensure critical data is easy to reference.
    - Prepare separate sections for anticipated objections or supplemental evidence to address potential challenges during the discussion.

Avoiding the pitfall of overloading P2P conferences with irrelevant details is crucial for maintaining clarity and persuasiveness. By focusing on key data points, using concise language, eliminating redundant information, and leveraging Prep Sheets, physician advisors can present streamlined, impactful arguments that directly address payer concerns. These strategies not only enhance the efficiency of the discussion but also improve the likelihood of achieving favorable outcomes for both patients and healthcare organizations.[54]

Misinterpreting Medical Policies

Misinterpreting or misapplying payer criteria is a critical pitfall that can erode the credibility of a physician advisor during peer-to-peer (P2P) conferences. These errors not only weaken arguments but may also result in missed opportunities to overturn denials. Ensuring a thorough understanding of payer policies and applying them accurately is essential for crafting strong, effective appeals. By combining a detailed knowledge of guidelines with precision in communication, physician advisors can build compelling cases that align with payer expectations.

---

1. Review Payer Guidelines Thoroughly

Understanding the specific criteria outlined in payer policies is foundational to successful P2P advocacy. Each payer may have unique guidelines, and even slight variations in wording can significantly influence how a case should be presented. Physician advisors must ensure that their arguments directly address these criteria.

- Example of Accurate Application:
  - *Policy:* "Inpatient admission for sepsis requires persistent hypotension despite 2L of IV fluid resuscitation."
  - *Application in Argument:* "This patient remained hypotensive with a mean arterial pressure <65 despite receiving two liters of fluid, meeting the criteria for inpatient admission."
- Why It Matters:
  - Adhering closely to payer-specific criteria demonstrates professionalism and credibility.
  - Addressing the denial rationale directly increases the likelihood of a favorable outcome.
- Tips for Success:
  - Review payer guidelines before the conference and cross-reference them with the patient's clinical data.
  - Maintain an up-to-date repository of frequently used payer criteria for quick access during preparation.

---

2. Avoid Overgeneralizations

Precision in applying medical policies is critical. Broad or vague statements can make arguments appear superficial or poorly prepared, reducing their persuasiveness. Instead, align specific clinical findings with the detailed language of the payer criteria.

- Examples:
  - **Incorrect:** "The patient meets criteria for inpatient care due to pneumonia."
  - **Correct:** "The patient's oxygen requirement of 4L and failed outpatient antibiotics meet criteria for inpatient admission under standard guidelines."
- Why It Matters:
  - Overgeneralizations risk undermining credibility by failing to demonstrate a clear understanding of the criteria.
  - Precision reassures the medical director that the argument is grounded in a comprehensive knowledge of policies.
- Tips for Success:
  - Use exact phrasing from the guidelines whenever possible to demonstrate alignment.
  - Avoid making assumptions about how the payer interprets criteria—base arguments on documented evidence.

---

3. Seek Clarification When Necessary

Not all payer guidelines are clear or comprehensive, and ambiguity can complicate P2P discussions. When faced with unclear or contradictory policies, asking clarifying questions can provide valuable insights and ensure mutual understanding between the physician advisor and the insurance medical director.

- **Example Questions:**
  - "Does your policy account for worsening trends in lactic acid levels for sepsis cases?"
  - "Could you clarify whether oxygen requirements are considered in isolation or in conjunction with imaging findings for pneumonia-related admissions?"
- **Why It Matters:**
  - Seeking clarification demonstrates thoroughness and a willingness to collaborate, which can build rapport with the medical director.
  - Clarified guidelines can provide a stronger foundation for the current appeal and inform future cases.
- **Tips for Success:**
  - Prepare a list of potential clarification questions before the conference, based on the denial rationale and payer criteria.
  - Phrase questions in a neutral, collaborative tone to avoid appearing confrontational.

---

4. Supplement with Specialty Guidelines

When payer criteria are vague, referencing specialty guidelines from reputable medical organizations can add weight and context to arguments. Specialty guidelines provide evidence-based recommendations that may address gaps or ambiguities in payer policies.

- **Example of Specialty Guidelines in Action:**
  - *Argument:* "While the payer criteria for pneumonia are vague, the American Thoracic Society (ATS) recommends inpatient care for patients with severe pneumonia, oxygen requirements ≥4L, and multi-lobar involvement. This patient met all three criteria."
- **Why It Matters:**
  - Specialty guidelines lend credibility and authority to the argument, particularly in cases where payer criteria are insufficient or open to interpretation.
  - Citing well-regarded medical organizations reinforces the perception of the physician advisor as an informed advocate.
- **Tips for Success:**
  - Regularly review specialty guidelines from organizations like the American Heart Association (AHA), American Thoracic Society (ATS), or the Infectious Diseases Society of America (IDSA).
  - Ensure that the specialty guidelines used are the most current and relevant to the patient's condition.

Misinterpreting medical policies is a preventable pitfall that can significantly undermine the effectiveness of P2P advocacy. By thoroughly reviewing payer guidelines, avoiding overgeneralizations, seeking clarification when necessary, and supplementing arguments with specialty guidelines, physician advisors can build precise, credible cases that align with payer expectations. These strategies not only enhance the likelihood of overturning denials but also establish the physician advisor as a knowledgeable and trustworthy advocate for patient care.

Allowing External Influence to Alter Clinical Judgment

In peer-to-peer (P2P) conferences, insurance medical directors may attempt to influence physician advisors' clinical judgment by framing discussions or interpreting policies in ways that weaken the advisor's negotiating position. While collaboration is vital, it is equally important for physician advisors to maintain professional integrity and confidence in their clinical expertise. Allowing external influence to dictate clinical judgment can lead to adverse outcomes for both patients and healthcare organizations, including revenue loss, compromised care standards, and operational inefficiencies.

---

1. Recognize Conditioning Tactics

Insurance medical directors may use subtle but deliberate strategies to influence physician advisors, conditioning them to accept payer-centric interpretations of policies. Recognizing these tactics is the first step in countering undue influence.

- Common Conditioning Tactics:
    - **Misleading Statements:** Providing distorted interpretations of payer policies, such as inaccurately stating that inpatient admission requires a minimum of 48 hours of care under the Medicare Two-Midnight Rule.
    - **Confidence Undermining:** Suggesting that the advisor's interpretation of clinical necessity is less valid or weaker than it actually is.
    - **Focus Shifting:** Emphasizing minor documentation gaps to detract from the core clinical rationale for admission.
- Why This Matters:
    - Unchecked conditioning can erode the advisor's confidence and lead to compromised decisions that may not align with clinical best practices or patient needs.
    - Recognizing these tactics helps advisors remain steadfast in their clinical assessments.
- Tips for Success:
    - Pay attention to patterns of communication that undermine clinical judgment.
    - Address misleading statements directly and diplomatically by referencing authoritative sources.

---

2. Maintain Confidence in Clinical Judgment

Confidence in one's clinical expertise is a cornerstone of effective advocacy. Physician advisors must trust their assessments and decisions, particularly when faced with external pressures to downgrade care or alter clinical judgment.

- How to Maintain Confidence:
  - **Stay Informed:** Regularly review institutional policies, payer guidelines, and relevant medical literature to ensure your arguments are grounded in current standards.
  - **Peer Support:** Discuss challenging cases with colleagues to validate decisions and reinforce confidence in clinical reasoning.
  - **Focus on Advocacy:** Remember that your role as a physician advisor is to advocate for the patient's best interests based on sound clinical judgment.
- Tip:
  - Prepare thoroughly for P2P conferences by reviewing the denial rationale and gathering robust evidence to support your position. Confidence comes from being well-prepared.
- Example Statement:
  - "Based on the patient's oxygen requirement and failure of outpatient treatment, this admission aligns with established guidelines for inpatient care and is medically necessary."

---

3. Rely on Established Guidelines

Using established, authoritative guidelines strengthens arguments and counters attempts to misrepresent policies. These guidelines provide an objective framework for clinical decision-making and advocacy.

- Why Established Guidelines Matter:
  - They offer evidence-based support for your decisions, making it harder for external parties to discredit your arguments.
  - Referencing guidelines directly aligns your argument with recognized standards, enhancing credibility.
- Example:
  - *Medicare Two-Midnight Rule:* "According to Medicare policies, the expectation of a two-midnight stay is sufficient for inpatient admission, regardless of actual time spent."
- Tips for Effective Use of Guidelines:
  - Reference specific criteria or language from payer guidelines to demonstrate alignment.
  - Supplement payer policies with specialty guidelines when additional context is needed, such as those from the American Thoracic Society or American College of Cardiology.

---

4. Understand the Financial Implications of Changing Strategy

Allowing external influence to alter clinical judgment can have significant financial and operational repercussions for healthcare organizations. Physician advisors must consider these implications when evaluating denials or responding to payer feedback.

- Financial and Operational Risks:
    - **Revenue Loss:** Downgrading inpatient admissions to observation or outpatient care unnecessarily can result in underpayment and lost revenue.
    - **Increased Administrative Burden:** Adjustments in case management strategies can lead to more appeals, delays in reimbursement, and inefficiencies in the revenue cycle.
- Example:
    - A hospital experienced a noticeable decline in reimbursement rates after physician advisors began downgrading inpatient admissions based on payer feedback. These changes created operational bottlenecks and strained administrative resources.
- Strategies to Mitigate Financial Risks:
    - **Evaluate the Impact:** Analyze the financial outcomes of downgrading care or altering strategies based on payer influence.
    - **Advocate for Appropriate Care:** Highlight the broader implications of denied claims, such as delays in treatment or increased readmission rates.
    - **Educate the Team:** Provide training on the financial consequences of altered evaluation strategies and reinforce the importance of adhering to clinical guidelines.

Staying vigilant against undue influence from insurance medical directors is crucial for physician advisors to uphold clinical standards and advocate effectively for patient care. By recognizing conditioning tactics, maintaining confidence in clinical judgment, relying on established guidelines, and understanding the financial and operational implications of changing strategy, physician advisors can resist external pressures and make decisions that prioritize patient well-being and organizational stability. These practices not only safeguard the integrity of clinical decision-making but also contribute to the financial health and operational efficiency of healthcare organizations.

Failing to Address Financial and Operational Implications of Denials

Denials are not isolated clinical decisions; they carry extensive financial and operational consequences for healthcare organizations. These implications affect hospital revenue cycles, administrative workloads, and patient care quality. Physician advisors who address these broader considerations in their arguments not only strengthen their case for overturning denials but also highlight the cascading impacts on the organization's ability to deliver high-quality care.

Incorporating financial and operational factors into P2P advocacy provides a more comprehensive argument, demonstrating the interconnectedness of clinical necessity and institutional sustainability.

---

1. Highlight Administrative Costs

Denials create a substantial operational burden, increasing the workload for administrative and clinical teams tasked with managing appeals. Physician advisors can emphasize these inefficiencies to underscore the importance of timely resolutions.

- Why This Matters:
  - Prolonged denials result in increased documentation requests, follow-up communications, and additional appeals, diverting resources from patient care.
  - Administrative delays extend the time to reimbursement, straining the hospital's cash flow.
- Example Statement:
  - "Delays in resolving denials strain hospital resources, requiring additional documentation requests, appeals, and follow-up, which increases administrative workload and diverts staff from more critical tasks."
- Tips:
  - Highlight specific operational challenges caused by denials, such as increased workload for case management teams or delays in revenue cycle processes.
  - Use data to illustrate the scope of the problem, such as average time spent managing denials or the percentage of claims requiring additional documentation.

---

2. Frame Denials as Patient Safety Risks

Beyond operational burdens, denials can have direct consequences for patient safety. By emphasizing how denied or delayed care compromises patient outcomes, physician advisors can make a compelling case for approval.

- Why This Matters:
  - Timely approvals ensure continuity of care, preventing treatment delays that could exacerbate medical conditions or lead to readmissions.
  - Payers share a vested interest in avoiding preventable complications and ensuring positive patient outcomes.
- Example Statement:
  - "Approving this claim ensures continuity of care, reducing the risk of readmission and preventing delays in treatment that could worsen the patient's condition."
- Case Example:
  - A patient with heart failure denied inpatient admission experienced worsening volume overload due to delayed diuresis, ultimately requiring a longer hospital stay and increasing the risk of complications.
- Tips:
  - Provide examples of how denial-related delays have negatively impacted patient safety in similar cases.

    - Connect the resolution of the denial to improved patient outcomes and reduced long-term costs for both the hospital and the payer.

---

3. Quantify Financial Impact

Using metrics to illustrate the financial implications of denials adds weight to the argument and highlights the broader repercussions for the organization. Physician advisors should present these impacts in a clear and concise manner during P2P conferences.

- **Why This Matters:**
    - Quantifying the financial consequences of denials demonstrates their impact on the hospital's bottom line, making the argument more persuasive to payer representatives who are also focused on cost management.
- **Examples of Financial Metrics:**
    - "A single denied inpatient stay costs the hospital an average of $10,000 in unreimbursed care and administrative expenses."
    - "Similar denials have delayed reimbursements by an average of 30 days, negatively affecting cash flow and operational efficiency."
- **Tips:**
    - Use hospital-specific data to provide concrete examples of the financial impact of denials.
    - Present comparisons to highlight how timely approvals reduce costs and streamline operations, benefiting both parties.

---

4. Link Financial Outcomes to Clinical Justification

Connecting financial and operational implications to clinical justification reinforces the importance of approving medically necessary claims. Demonstrating how denials hinder the hospital's ability to provide care strengthens the overall argument.

- **Why This Matters:**
    - Financial sustainability directly impacts the hospital's capacity to invest in staffing, equipment, and resources needed to manage complex cases effectively.
    - Linking financial outcomes to clinical care helps align payer and provider priorities, creating a more collaborative dialogue.
- **Example Statement:**
    - "Approving claims like this one ensures that our hospital can allocate resources effectively for managing complex cases, ultimately benefiting patient outcomes."
- **Practical Application:**
    - Relate the financial impact of a denial to the hospital's ability to maintain quality improvement initiatives or meet regulatory standards.
    - Emphasize that timely reimbursement supports the hospital's commitment to delivering evidence-based, patient-centered care.

Addressing the financial and operational implications of denials in P2P conferences elevates the argument beyond clinical justification, providing a more comprehensive case for approval. By highlighting administrative costs, framing denials as patient safety risks, quantifying financial impact, and linking these factors to clinical care, physician advisors can strengthen their advocacy efforts. This approach not only increases the likelihood of overturning denials but also reinforces the broader importance of collaboration between payers and providers to achieve shared goals of sustainability and quality care delivery.

Strategies for Avoiding Pitfalls

Avoiding common pitfalls in peer-to-peer (P2P) conferences requires a proactive and methodical approach. Physician advisors who invest time in preparation, practice, and ongoing improvement can significantly enhance their effectiveness in overturning denials. This section outlines actionable strategies to help physician advisors navigate potential challenges and achieve successful outcomes.

1. Prepare Thoroughly

Comprehensive preparation lays the foundation for a successful P2P conference. Organizing key data and familiarizing yourself with payer criteria ensure that your argument is well-supported and directly addresses the denial rationale.

- Why This Matters:
    - Thorough preparation minimizes the risk of omitting critical details and strengthens the clarity of your argument.
    - It demonstrates professionalism and respect for the medical director's time.
- Steps for Effective Preparation:
    - **Review Documentation:** Analyze the patient's medical record to identify relevant data points, including clinical presentation, diagnostics, and treatments.
    - **Understand Payer Criteria:** Study payer-specific guidelines, such as InterQual or MCG, to ensure your argument aligns with their expectations.
    - **Create a Prep Sheet:** Organize information into a concise format with sections for clinical findings, diagnostic results, and alignment with criteria.
- Example:
    - For a sepsis denial, your Prep Sheet might include:
        1. Persistent hypotension with a mean arterial pressure <65 mmHg.
        2. Elevated lactic acid levels >4 mmol/L.
        3. IV fluid resuscitation exceeding 2 liters without resolution.
- Tips:
    - Use templates to standardize Prep Sheets for different conditions.

- Prioritize information directly related to the denial rationale to keep discussions focused.

---

2. Practice Arguments in Advance

Rehearsing discussions ahead of time can identify potential gaps in your argument and ensure a polished, confident presentation during the P2P conference.

- Why This Matters:
  - Practice reduces the likelihood of overloading the discussion with unnecessary details or overlooking critical points.
  - It builds confidence, particularly when anticipating payer objections.
- How to Practice:
  - **Role-Play Scenarios:** Simulate P2P conferences with colleagues acting as medical directors to practice responses to common objections.
  - **Time Your Presentation:** Ensure that your key points can be delivered succinctly within the time constraints of the conference.
- Benefits of Practice:
  - Highlights areas where documentation or evidence may be lacking.
  - Helps refine phrasing and tone for better clarity and professionalism.
- Tip: Use constructive feedback from practice sessions to adjust and improve your approach.

---

3. Strengthen Clinical Confidence

Maintaining confidence in your clinical judgment is critical for countering undue influence and advocating effectively for patient care. Confidence comes from staying informed and engaged in professional development.

- Why This Matters:
  - Confidence reinforces credibility and ensures that your arguments are persuasive and authoritative.
  - It helps you remain composed when facing challenging objections or adversarial tones.
- Strategies to Build Confidence:
  - **Continuous Education:** Attend conferences, webinars, or training sessions to stay current on medical advancements and payer guidelines.
  - **Peer Collaboration:** Discuss complex cases with colleagues to gain insights and alternative perspectives.
- Example:
  - Joining a physician advisor peer group allows you to share strategies and learn from others' experiences, enhancing both your knowledge and confidence.

- **Tip:** Regularly review past P2P conferences to identify successes and areas for improvement, reinforcing your expertise.

---

4. Solicit Feedback from Peers

Peer reviews of past P2P conferences provide valuable insights into the effectiveness of your arguments and presentation style. Constructive feedback can identify blind spots and suggest strategies for improvement.

- **Why This Matters:**
  - Feedback helps refine your approach and avoid repeating mistakes in future conferences.
  - It provides a different perspective on how your arguments are received and interpreted.
- **How to Solicit Feedback:**
  - **Review Sessions:** Schedule regular team meetings to analyze complex cases and discuss the outcomes of P2P conferences.
  - **Direct Feedback:** Ask colleagues to listen to your presentation and provide specific suggestions for improvement.
- **Case Example:**
  - A peer review highlighted that a physician advisor frequently included redundant details, prompting them to streamline their arguments for greater clarity and impact.
- **Tip:** Use feedback to create a checklist of best practices for structuring and delivering arguments effectively.

---

5. Stay Updated on Guidelines

Keeping up-to-date with payer criteria and specialty guidelines ensures that your arguments remain aligned with current standards. This is essential for maintaining credibility and crafting persuasive cases.

- **Why This Matters:**
  - Guidelines frequently change, and staying informed helps you avoid using outdated or inaccurate information.
  - It reinforces your ability to counter payer objections with evidence-based arguments.
- **Resources for Staying Informed:**
  - **Professional Organizations:** Follow updates from societies such as the American Thoracic Society or the American Heart Association.
  - **Payer Communications:** Regularly review newsletters, webinars, and updates from payers.
  - **Internal Training:** Participate in hospital or network training sessions on payer-specific policies.
- **Example:**

- After reviewing updated ATS guidelines, an advisor incorporated new criteria for CURB-65 scores into their arguments for pneumonia cases.
- Tip: Maintain a repository of the most current guidelines for quick reference during preparation and P2P conferences.

---

6. Adapt Based on Feedback

Learning from past experiences and payer feedback is key to refining your strategies and avoiding repeated mistakes. Adaptability demonstrates professionalism and a commitment to continuous improvement.

- Why This Matters:
  - Flexibility allows you to address recurring objections or challenges more effectively in future conferences.
  - It shows payers that you are responsive to their concerns and willing to adjust your approach.
- Steps to Adapt:
  - **Identify Patterns:** Analyze trends in payer objections or denial rationales.
  - **Develop Targeted Solutions:** Create supplemental documentation or refine your arguments to address common challenges.
  - **Implement Changes:** Test new strategies in upcoming P2P conferences and evaluate their effectiveness.
- Example:
  - After receiving feedback about unclear alignment with payer criteria, an advisor began including direct quotes from guidelines in their Prep Sheets to improve clarity.
- Tip: Document lessons learned from each conference to create a personal toolkit of strategies for future use.

By employing these strategies, physician advisors can avoid common pitfalls and build stronger, more persuasive arguments during P2P conferences. Thorough preparation, continuous learning, and adaptability not only increase the likelihood of overturning denials but also enhance overall advocacy skills. These practices contribute to better patient outcomes, improved payer relationships, and greater financial and operational stability for healthcare organizations.

Conclusion

Peer-to-peer (P2P) conferences are a cornerstone of the physician advisor's role, serving as a critical interface between healthcare providers and payers. These interactions are not only about overturning denials but also about upholding clinical standards, ensuring patient care, and safeguarding the financial health of healthcare organizations. However, even seasoned physician advisors can encounter pitfalls that weaken their advocacy, delay resolutions, and strain relationships with payers.

The pitfalls covered in this chapter highlight the complexity of P2P advocacy. From overloading arguments with irrelevant details to misinterpreting payer policies, succumbing to external influence, or neglecting financial and operational impacts, each challenge underscores the importance of preparation, precision, and adaptability. Addressing these issues requires a methodical approach that blends clinical expertise with strategic communication and organizational awareness.

Key Strategies to Avoid Pitfalls

1. **Streamline Arguments with Clarity and Relevance**
    - Avoid overloading discussions with extraneous details by focusing on the most critical data points.
    - Use concise language to present clinical findings and align them with payer criteria effectively.
    - Eliminate redundancy to keep the argument sharp and impactful.
2. **Understand and Apply Payer Guidelines Accurately**
    - Familiarize yourself with payer-specific policies and reference them directly in arguments.
    - Avoid broad generalizations and ensure arguments are precise and evidence-based.
    - Supplement vague or ambiguous payer criteria with reputable specialty guidelines when necessary.
3. **Maintain Clinical Confidence and Resist External Influence**
    - Trust your expertise and rely on established guidelines to counteract undue influence from insurance medical directors.
    - Recognize tactics that may undermine confidence or shift focus away from clinical necessity.
    - Advocate firmly for patient care while balancing professionalism and collaboration.
4. **Address Financial and Operational Implications**
    - Highlight the administrative burden, financial impact, and patient safety risks associated with denials.
    - Quantify the costs of delayed reimbursements and denied claims to emphasize the broader implications.
    - Link financial outcomes to clinical justification to strengthen the overall case and align payer and provider priorities.

5. **Continuously Adapt and Improve**

    - Prepare thoroughly with organized data and payer-specific Prep Sheets to support focused presentations.
    - Practice arguments in advance to refine delivery and anticipate objections.
    - Solicit feedback from peers and adjust strategies based on past experiences and payer input.
    - Stay updated on evolving payer guidelines and specialty recommendations to ensure arguments remain aligned with current standards.

The Broader Impact of Avoiding Pitfalls

When physician advisors proactively address these challenges, they not only enhance their success in overturning denials but also reinforce their credibility as knowledgeable and effective advocates. These efforts contribute to:

- **Improved Patient Outcomes:** Ensuring timely and appropriate care for patients by securing necessary approvals for admissions or treatments.
- **Stronger Relationships with Payers:** Building rapport through clear, collaborative, and evidence-based advocacy.
- **Financial Stability for Healthcare Organizations:** Minimizing revenue losses, reducing administrative burdens, and supporting the sustainability of healthcare operations.

Final Thoughts

Avoiding common pitfalls in P2P conferences is not merely a matter of skill—it is a commitment to continuous learning, professional integrity, and organizational success. By adopting the strategies outlined in this chapter, physician advisors can navigate even the most challenging discussions with confidence and effectiveness.

This proactive approach ultimately benefits all stakeholders—patients receive the care they need, payers experience more collaborative interactions, and healthcare organizations sustain their ability to deliver high-quality care. As physician advisors master these techniques, they position themselves as pivotal players in the evolving landscape of healthcare advocacy, ensuring that clinical excellence and operational efficiency remain at the forefront of denial management.

References

[53] O'Daniel M, Rosenstein AH. Professional Communication and Team Collaboration. In: Hughes RG, editor. Patient Safety and Quality: An Evidence-Based Handbook for Nurses. Rockville (MD):

Agency for Healthcare Research and Quality (US); 2008 Apr. Chapter 33. Available from: https://www.ncbi.nlm.nih.gov/books/NBK2637/?utm_source=chatgpt.com

[54] Mata, Á.N.S., de Azevedo, K.P.M., Braga, L.P. et al. Training in communication skills for self-efficacy of health professionals: a systematic review. Hum Resour Health 19, 30 (2021). https://doi.org/10.1186/s12960-021-00574-3

CHAPTER 8

Case Studies

Introduction

Case studies provide a unique lens into the intricacies of peer-to-peer (P2P) conferences, offering real-world examples that illustrate the challenges and opportunities inherent in the denial management process. These practical scenarios allow physician advisors to explore successful strategies, learn from common pitfalls, and apply these lessons to their own P2P advocacy. By analyzing detailed case examples, physician advisors can better understand how to navigate complex payer interactions, craft compelling arguments, and secure favorable outcomes for patients and healthcare organizations.

Each P2P conference is as unique as the patient case it represents, yet recurring themes and challenges often emerge. These include effectively addressing denial rationales, aligning arguments with payer-specific criteria, and overcoming objections through clear, evidence-based advocacy. Case studies provide an invaluable opportunity to dissect these dynamics, showcasing both successful approaches and instances where outcomes fell short due to avoidable missteps.

This chapter highlights a series of illustrative case studies designed to:

- **Demonstrate Effective Techniques:** Learn how to structure arguments, address payer objections, and align with clinical and payer guidelines.
- **Identify Key Challenges:** Understand the common pitfalls that can derail a P2P conference, such as inadequate documentation or misinterpreted guidelines.
- **Provide Actionable Insights:** Extract lessons that physician advisors can apply to refine their strategies and enhance the effectiveness of their advocacy.

Through these examples, physician advisors will gain deeper insights into the nuances of P2P conferences, enhancing their ability to navigate payer interactions with confidence and precision. By understanding what works—and what doesn't—physician advisors can elevate their practice, ensuring that patients receive appropriate care while maintaining the financial stability of their organizations.

Case Study 1: Pneumonia with Hypoxia

Overview

A 72-year-old patient arrived at the emergency department with acute dyspnea, fever, and hypoxia. A chest X-ray revealed bilateral infiltrates consistent with pneumonia. The patient required oxygen therapy at 4L/min to maintain SpO_2 >90%, indicating significant respiratory compromise. Despite these findings, the payer denied inpatient admission, asserting that the severity of the patient's condition did not meet their criteria for medical necessity.

This case presented a common scenario in P2P conferences: addressing payer skepticism regarding the need for inpatient care in cases of pneumonia. The challenge lay in demonstrating the clinical necessity of close monitoring and active management to mitigate risks associated with severe pneumonia and hypoxia.

Approach

The physician advisor prepared for the P2P conference by constructing a targeted and evidence-based argument. Key strategies included:

- **Highlighting Critical Clinical Data:** The physician advisor emphasized the patient's objective findings, including:
    - Persistent hypoxia requiring supplemental oxygen at 4L/min to maintain SpO_2 >90%.
    - Imaging evidence of bilateral infiltrates on the chest X-ray, indicating extensive pulmonary involvement.
    - Symptoms of fever and dyspnea, reflective of systemic infection and respiratory compromise.

  These data points were crucial in illustrating the severity of the patient's condition and the need for intensive monitoring and treatment.

- **Connecting the Clinical Presentation to Guidelines:** The advisor aligned the patient's clinical presentation with widely recognized guidelines for pneumonia management, which support inpatient admission for:
    - Oxygen dependency exceeding 2L/min.
    - Imaging findings consistent with multi-lobar involvement or extensive infiltrates.
    - Risk factors for complications, including advanced age and hypoxia at presentation.

  By referencing these criteria, the advisor underscored the medical necessity of inpatient care and addressed the payer's initial rationale for denial.

- **Framing the Risks of Complications:** The advisor emphasized the potential complications of untreated or under-monitored severe pneumonia, including:
    - Worsening hypoxia leading to respiratory failure.

- Sepsis and systemic inflammatory response syndrome (SIRS).
- The heightened risk of adverse outcomes due to the patient's advanced age.

This approach highlighted the critical role of inpatient care in mitigating these risks and ensuring timely intervention.

Outcome

The denial was successfully overturned after the physician advisor presented a clear and compelling argument. By focusing on the patient's significant hypoxia and the clinical need for close monitoring and IV antibiotics, the advisor effectively demonstrated that observation care would not suffice in managing the patient's condition. The payer approved the inpatient admission, recognizing the necessity of the care provided.

Lessons Learned

This case underscores key takeaways for physician advisors engaged in P2P conferences:

- **Objective Data is Key:** The use of quantifiable clinical findings, such as oxygen saturation levels and imaging results, provides an irrefutable foundation for advocating inpatient care. Objective data should always form the core of the argument.
- **Guideline Alignment Strengthens Arguments:** Referring to widely accepted clinical guidelines, such as those from the American Thoracic Society or payer-specific criteria, enhances credibility and aligns the argument with the payer's expectations. This approach bridges the gap between clinical necessity and payer policy.
- **Risk Framing is Effective:** Highlighting the potential consequences of insufficient monitoring or delayed intervention can persuade payers by emphasizing the shared goal of minimizing complications and optimizing patient outcomes.

By adopting these strategies, physician advisors can effectively navigate similar cases, ensuring patients receive the appropriate level of care while addressing payer concerns with precision and professionalism.

Case Study 2: Atrial Fibrillation with Rapid Ventricular Response

Overview

A 65-year-old patient with a history of hypertension and diabetes presented to the emergency department (ED) with palpitations, fatigue, and lightheadedness. On initial assessment, the patient was found to be in atrial fibrillation (AF) with a rapid ventricular response (RVR) at a heart rate of 140 beats per minute (bpm).

Attempts to achieve rate control with oral beta-blockers in the ED were unsuccessful, prompting the administration of IV diltiazem. Despite these interventions, the payer denied inpatient admission, asserting that the patient's condition could have been managed in an observation setting.

The challenge for the physician advisor in this case was to demonstrate that the patient's condition, characterized by sustained hemodynamic instability and the need for IV medications, warranted inpatient care due to the potential for severe complications.

Approach

The physician advisor utilized a structured, evidence-based approach to advocate for inpatient admission during the P2P conference:

- **Highlighting the Failure of Outpatient Management:** The advisor emphasized the clinical progression of the patient's condition, pointing out that:
    - Oral medications, including beta-blockers, were insufficient to control the patient's heart rate, necessitating IV therapy.
    - Persistent tachycardia at 140 bpm posed a risk of hemodynamic instability, especially in a patient with comorbidities such as hypertension and diabetes.

  By framing the case as a failure of standard outpatient interventions, the advisor reinforced the need for escalation to inpatient-level care.

- **Focusing on Clinical Risks:** The advisor articulated the significant risks associated with inadequately managed AF with RVR, including:
    - **Thromboembolic Events:** AF increases the risk of stroke, particularly when rate control is not achieved promptly.
    - **Heart Failure Exacerbation:** Prolonged tachycardia can lead to decreased cardiac output and worsening heart failure.
    - **Electrical Instability:** Persistent RVR increases the risk of ventricular arrhythmias, which can be life-threatening.

  These potential complications underscored the necessity for continuous monitoring and rapid intervention, both of which exceed the capabilities of observation care.

- **Referencing Guidelines**: The advisor aligned the patient's clinical presentation with established payer criteria and professional guidelines, which support inpatient admission for:

- AF with RVR unresponsive to outpatient-level therapies.
- Hemodynamic instability requiring IV rate control agents.
- Risk stratification for thromboembolism using tools like the CHA_2DS_2-VASc score, which indicated a high risk of stroke in this patient.

Outcome

The payer overturned the denial after the physician advisor presented a clear and compelling argument. The approval of inpatient status was based on the demonstrated instability of the patient's condition, the failure of outpatient management, and the clinical risks necessitating close monitoring and IV therapy. The case highlighted the importance of inpatient care in achieving prompt stabilization and mitigating potential complications.

Lessons Learned

This case offers valuable insights into effective strategies for handling similar denials:

- **Failed Outpatient Management Validates Medical Necessity: Demonstrating** that standard outpatient interventions were inadequate is a strong indicator that a higher level of care is warranted. In this case, the inability to control heart rate with oral medications justified the need for inpatient admission.
- **Articulating Potential Risks Strengthens Arguments:** Highlighting the clinical risks associated with AF with RVR—such as stroke, heart failure exacerbation, and arrhythmias—adds weight to the argument for inpatient care. These risks resonate with payers who share a vested interest in avoiding adverse outcomes.
- **Guideline Alignment Builds Credibility:** Referencing clinical guidelines, such as those from the American Heart Association (AHA), reinforces the argument by demonstrating alignment with evidence-based practices and standard criteria for medical necessity.

Takeaway

This case demonstrates the importance of presenting a focused, risk-oriented argument that addresses the payer's denial rationale while emphasizing the patient's clinical instability. Physician advisors who combine objective data with evidence-based guidelines can effectively advocate for inpatient care, ensuring that patients receive the appropriate level of treatment for their condition.

Case Study 3: Postoperative Bleeding

Overview

A 58-year-old patient underwent an outpatient laparoscopic cholecystectomy for symptomatic cholelithiasis. Two days post-discharge, the patient was readmitted to the emergency department (ED) with severe abdominal pain, dizziness, and a significant drop in hemoglobin levels. Initial lab work revealed a hemoglobin decrease from 13 g/dL preoperatively to 8.5 g/dL on admission. Vital signs indicated tachycardia with a heart rate of 110 bpm and borderline hypotension (blood pressure 90/60 mmHg).

Despite evidence of clinical instability and active management for suspected postoperative bleeding, the payer denied inpatient status, arguing that observation care would have been sufficient. This case required the physician advisor to demonstrate that the patient's condition necessitated inpatient monitoring and intervention to ensure safety and avoid complications.

Approach

The physician advisor adopted a methodical and data-driven strategy to advocate for inpatient admission during the P2P conference:

1. **Highlighting Clinical Instability:**
    - The advisor emphasized the patient's **ongoing hemodynamic instability,** including:
        - Serial hemoglobin drops from 13 g/dL preoperatively to 8.5 g/dL upon readmission, with further decreases noted during hospitalization.
        - Tachycardia and borderline hypotension, indicative of inadequate perfusion and potential volume depletion from blood loss.
    - These findings underscored the severity of the patient's condition and the risk of rapid decompensation, which required continuous monitoring and timely interventions.
2. **Addressing Payer Concerns Directly:**
    - The payer's denial cited insufficient documentation of persistent instability. The advisor countered this by providing **additional clinical context**, such as:
        - Trends in vital signs showing sustained tachycardia and fluctuating blood pressure, necessitating IV fluid resuscitation and close observation.
        - Evidence of ongoing blood loss, including progressive decreases in hemoglobin despite stabilization attempts.
    - The advisor also clarified the **clinical risks** associated with delayed or insufficient monitoring, such as hypovolemic shock, need for transfusion, and potential reoperation.
3. **Framing the Argument Around the Need for Inpatient-Level Care:**

- The advisor highlighted that observation care is typically limited to short-term monitoring and minor interventions, which were insufficient for this patient's needs. In contrast, inpatient admission allowed for:
    - Frequent reassessment of hemodynamic stability and hemoglobin levels.
    - Immediate access to advanced interventions, including transfusion and potential surgical re-exploration, should the patient deteriorate.

4. **Reinforcing the Argument with Guidelines:**
    - While payer-specific criteria were addressed, the advisor also referenced general clinical guidelines for postoperative management of significant bleeding, which recommend inpatient care for patients with:
        - Hemodynamic instability.
        - Suspected ongoing hemorrhage requiring close monitoring and possible escalation of care.

Outcome

The denial was overturned after the physician advisor presented a clear and compelling case demonstrating that the patient's condition met the criteria for inpatient admission. The payer acknowledged that the level of monitoring and intervention required exceeded the capabilities of observation care, validating the necessity of inpatient status.

Lessons Learned

This case highlights several critical takeaways for managing similar denials:

1. **Thorough Documentation of Instability Strengthens Arguments:**
    - Serial trends in lab values (e.g., hemoglobin) and vital signs provide objective evidence of clinical instability, making it harder for payers to dispute the need for inpatient care.

2. **Addressing Payer Concerns Directly Resolves Gaps:**
    - Proactively clarifying documentation gaps or ambiguities in the medical record can preempt payer objections and strengthen the argument.

3. **Framing the Argument Around Escalation Needs Adds Weight:**
    - Highlighting the potential for rapid deterioration and the need for inpatient-level resources effectively underscores the limitations of observation care.

Takeaway

This case demonstrates the importance of combining clinical data, patient-specific risks, and a clear understanding of payer expectations to advocate effectively for inpatient care. Physician advisors who provide comprehensive, evidence-based arguments and address payer concerns directly can improve the likelihood of overturning denials, ensuring patients receive the appropriate level of care.

Case Study 4: Heart Failure Exacerbation with Kidney Injury

Overview

A 70-year-old patient with a history of chronic heart failure (HF) presented to the emergency department (ED) with worsening dyspnea, fatigue, and orthopnea. Physical examination revealed bilateral rales, jugular venous distension (JVD), and pitting edema. Laboratory findings showed a markedly elevated brain natriuretic peptide (BNP) level of 2,500 pg/mL, indicative of acute decompensated heart failure (ADHF). A chest X-ray confirmed pulmonary edema, and echocardiography revealed a reduced ejection fraction (EF) consistent with the patient's known systolic heart failure.

The patient was initiated on IV furosemide for aggressive diuresis but subsequently developed acute kidney injury (AKI) as evidenced by a rise in creatinine from 1.2 mg/dL at baseline to 2.5 mg/dL within 24 hours. Despite the need for intensive monitoring and adjustments to treatment, the payer denied inpatient admission, citing insufficient documentation of medical necessity and suggesting that observation care would have sufficed.

Approach

The physician advisor took a structured and evidence-based approach to advocate for inpatient care during the peer-to-peer (P2P) conference:

1. **Emphasizing the Clinical Risks of Dual Conditions:**
    - The advisor highlighted the complexity of managing **acute decompensated heart failure** with concurrent **acute kidney injury**, which requires balancing aggressive diuresis to relieve pulmonary congestion while avoiding further renal compromise.
    - Key clinical risks outlined included:

        - **Worsening AKI:** Volume depletion from diuresis posed a significant risk of further renal injury, potentially leading to the need for renal replacement therapy (RRT).
        - **Cardiogenic Shock:** Without adequate diuresis, the patient was at risk of progressing to cardiogenic shock due to persistent fluid overload.
        - **Electrolyte Imbalances:** Aggressive diuresis increases the risk of severe hypokalemia or hyponatremia, requiring close monitoring and immediate intervention.
2. **Highlighting the Need for Close Monitoring:**
    - The advisor emphasized the importance of continuous hemodynamic and renal function monitoring, which exceeded the scope of observation care. Specific monitoring needs included:
        - **Frequent lab tests:** Serial monitoring of electrolytes, creatinine, and BNP levels to guide diuretic therapy.
        - **Vital signs:** Continuous blood pressure and heart rate monitoring to detect early signs of hemodynamic instability.
        - **Input/output tracking:** Accurate measurement of fluid balance to optimize diuresis while protecting renal function.
3. **Leveraging the Interplay Between Comorbid Conditions:**
    - The advisor connected the clinical presentation to the payer's criteria, illustrating that the combination of heart failure exacerbation and AKI constituted a **complex, high-risk scenario** requiring inpatient-level care.
    - The advisor emphasized that observation care, typically intended for less complex and short-duration conditions, was insufficient to address the interplay of heart failure and AKI in this patient.
4. **Supporting the Argument with Guidelines:**
    - While addressing the denial rationale, the advisor referred to established guidelines for heart failure management:
        - **Heart Failure Guidelines:** Recommendations for managing ADHF highlight the need for inpatient care in cases with significant comorbidities or complications such as AKI.
        - **Nephrology Input:** The advisor pointed out that nephrology guidelines also advocate for close inpatient monitoring when AKI occurs in the context of aggressive diuresis.

Outcome

The denial was overturned after the physician advisor presented a comprehensive argument demonstrating the medical necessity of inpatient admission. The payer acknowledged that the complexity of managing acute heart failure with concurrent AKI required the level of care provided in an inpatient setting.

Lessons Learned

This case offers important insights for physician advisors navigating similar scenarios:

1. **Highlighting the Interplay Between Comorbid Conditions Adds Depth:**
   - Articulating the clinical challenges posed by overlapping conditions, such as heart failure and AKI, underscores the complexity of care and strengthens the argument for inpatient admission.
2. **Addressing Clinical Risks Strengthens Advocacy:**
   - Proactively outlining potential complications, such as worsening renal function or cardiogenic shock, demonstrates the necessity of inpatient-level monitoring and interventions.
3. **Guidelines Provide Credibility:**
   - Referring to specialty guidelines reinforces the argument and establishes the physician advisor as a knowledgeable and evidence-based advocate.

Takeaway

This case underscores the importance of a tailored, multidimensional approach when advocating for inpatient care in patients with complex presentations. Physician advisors who effectively communicate the risks, comorbidities, and limitations of observation care can build compelling cases that result in better outcomes for patients and healthcare organizations.

Case Study 5: Postoperative Respiratory Failure

Overview

A 74-year-old patient underwent an elective knee replacement surgery without immediate complications during the procedure. However, within hours of being transferred to the recovery unit, the patient developed acute respiratory failure characterized by tachypnea, oxygen desaturation to 82% on room air, and labored breathing. The clinical team initiated BiPAP (bilevel positive airway pressure) to stabilize the patient, addressing worsening oxygenation and work of breathing.

The payer denied inpatient admission, arguing that the respiratory distress could have been managed within the scope of observation care, which they claimed was appropriate for postoperative complications requiring less than 24 hours of intervention.

Approach

The physician advisor structured their argument around the complexity and acuity of the patient's condition, illustrating the necessity of inpatient-level care to ensure safety and recovery:

1. **Emphasizing the Acute Onset of Respiratory Failure:**
    - The advisor highlighted the **sudden deterioration** in the patient's respiratory status, which represented a significant deviation from the expected postoperative course.
    - Critical clinical data presented included:
        - **Oxygen desaturation:** SpO_2 <85% on supplemental oxygen, requiring BiPAP support to maintain oxygenation.
        - **Respiratory distress:** Increased work of breathing, accessory muscle use, and tachypnea (>30 breaths per minute).
    - The advisor emphasized that such acute decompensation indicated an **underlying pathology** (e.g., fluid overload, atelectasis, or pulmonary embolism) that warranted inpatient monitoring.
2. **Articulating the Need for BiPAP and Advanced Monitoring:**
    - The advisor pointed out that the initiation of BiPAP represented a **higher level of respiratory support** than what is typically managed in observation care.
    - BiPAP involves:
        - Continuous monitoring of oxygenation and ventilation parameters.
        - Adjustments to settings based on patient response.
        - Potential escalation to invasive mechanical ventilation if respiratory failure progresses.
    - The advisor underscored that this patient required close monitoring in a **step-down unit** with staff trained to manage advanced respiratory therapies, exceeding the capabilities of observation care.
3. **Differentiating Inpatient Needs from Observation Care:**
    - The advisor explained that **observation care** is generally intended for short-term evaluations or treatments with predictable clinical trajectories, such as mild postoperative complications or transient hypoxia.
    - By contrast, this patient's condition involved:
        - Significant **clinical instability** that required ongoing reassessment and immediate intervention if BiPAP failed.
        - A heightened risk of complications, including aspiration, cardiac stress, or need for intubation, necessitating **multidisciplinary inpatient care.**

    - The advisor demonstrated that the complexity of managing respiratory failure safely in this case went beyond the scope of observation care.
4. **Preemptively Addressing Payer Concerns:**
    - Anticipating the payer's argument that observation care could provide adequate support, the advisor clarified:
        - BiPAP is not typically considered a standard intervention in observation care due to the need for continuous respiratory therapy oversight.
        - The acute risk of rapid deterioration in respiratory status necessitated **inpatient-level resources** to prevent further decompensation.

Outcome

The payer overturned the denial and approved inpatient admission after the physician advisor effectively demonstrated that the patient's **acute respiratory failure** and need for BiPAP exceeded the capabilities of observation care. The payer acknowledged that the patient's condition warranted inpatient monitoring and management to ensure safe recovery.

Lessons Learned

This case highlights critical strategies for advocating effectively in scenarios involving postoperative complications:

1. **Acute Respiratory Failure is a Strong Clinical Indicator of Inpatient Necessity:**
    - Conditions requiring non-invasive ventilation or other advanced respiratory therapies typically fall under inpatient care due to the need for specialized monitoring and interventions.
2. **Clearly Distinguishing Inpatient Needs from Observation Capabilities Strengthens Arguments:**
    - Physician advisors should clearly outline why the patient's clinical presentation exceeds the scope of observation care, using specific examples of required resources and monitoring.

Takeaway

This case illustrates the importance of tailoring arguments to the specific clinical needs and acuity of the patient. By emphasizing the complexity of managing acute respiratory failure and clearly differentiating inpatient from observation-level care, physician advisors can effectively advocate for medically necessary admissions, ensuring optimal outcomes for patients and alignment with institutional standards of care.

Case Study 6: Pediatric Diabetic Ketoacidosis (DKA)

Overview

A 12-year-old patient presented to the emergency department with severe diabetic ketoacidosis (DKA), a life-threatening metabolic condition characterized by **marked acidosis** (pH 7.1), **elevated blood glucose** (540 mg/dL), and **positive serum ketones.** The patient exhibited symptoms including lethargy, tachypnea, and abdominal pain, indicative of significant metabolic derangement. Immediate management included **intravenous insulin therapy**, aggressive **fluid resuscitation,** and **continuous cardiac monitoring** to prevent complications such as arrhythmias and electrolyte imbalances.

Despite the severity of the presentation, the payer denied inpatient admission, asserting that observation care would have sufficed for the management of this case. This denial raised concerns about the payer's underestimation of the complexity and risks associated with pediatric DKA.

Approach

The physician advisor structured their argument around the **clinical severity of the condition**, the **unique risks associated with pediatric DKA**, and the **inadequacy of observation-level care** for managing these risks effectively:

1. **Highlighting the Severity of Metabolic Derangements:**
    - The advisor emphasized the critical lab findings, including:
        - **Severe acidosis:** A pH of 7.1, indicating significant disruption of acid-base balance.
        - **Elevated glucose levels:** 540 mg/dL, demonstrating poor glycemic control requiring immediate intervention.
        - **Ketosis:** Positive serum ketones consistent with ongoing fat metabolism due to insulin deficiency.
    - The advisor underscored that these findings represented a **severe form of DKA,** necessitating **intensive monitoring** and treatment that could not be adequately provided in an observation setting.

2. **Focusing on the Risks of Complications:**
    - Pediatric DKA poses unique risks, particularly **cerebral edema**, a potentially fatal complication that can develop during fluid and electrolyte replacement therapy.
        - The advisor highlighted that **continuous neurological assessment** and **careful fluid management** are essential to mitigate this risk.
    - Other risks addressed included:
        - **Hypokalemia or hyperkalemia:** Resulting from shifts in potassium during insulin therapy.
        - **Cardiac arrhythmias:** Linked to electrolyte imbalances, requiring continuous cardiac monitoring.
        - **Hypoglycemia:** A potential consequence of insulin therapy if blood glucose is not carefully monitored.
3. **Demonstrating the Need for Continuous Monitoring:**
    - The advisor explained that effective management of severe DKA requires:
        - **Frequent lab monitoring:** To track changes in glucose, pH, bicarbonate, and electrolytes.
        - **Intravenous insulin infusion:** Titrated based on blood glucose levels, necessitating inpatient-level supervision.
        - **Cardiac monitoring:** To detect arrhythmias secondary to electrolyte disturbances.
    - They emphasized that observation-level care is typically designed for conditions with predictable and short-term management needs, which is not applicable to a **dynamic condition like DKA.**
4. **Referencing Pediatric Safety Standards:**
    - The advisor cited guidelines and standards of care emphasizing that **pediatric patients with severe DKA** should be treated in an inpatient setting due to their higher susceptibility to complications and the need for intensive monitoring.
    - They noted that many leading institutions and organizations recommend inpatient admission for all pediatric DKA cases involving severe acidosis (pH <7.2), altered mental status, or significant dehydration.

Outcome

The denial was overturned after the physician advisor successfully demonstrated that the patient's **severity of illness, risk profile**, and need for **intensive, continuous monitoring** exceeded the capabilities of observation care. The payer recognized the necessity of inpatient-level care to ensure the patient's safety and recovery.

Lessons Learned

1. **Pediatric Cases Require Additional Emphasis on Safety Risks:**
    - Pediatric patients, particularly those with conditions like DKA, are more vulnerable to complications than adults, necessitating heightened attention to safety and monitoring requirements.
    - Physician advisors should emphasize the **age-specific risks** and ensure that arguments are tailored to the unique vulnerabilities of pediatric populations.
2. **Presenting the Potential Complications of Inadequate Care Strengthens Arguments:**
    - Clearly articulating the potential consequences of insufficient monitoring—such as cerebral edema, arrhythmias, or hypoglycemia—reinforces the need for inpatient admission.
    - Linking these risks to specific management requirements (e.g., frequent lab monitoring, titration of IV insulin) demonstrates that observation care would be insufficient.

Takeaway

This case underscores the importance of tailoring arguments to the unique clinical and safety needs of pediatric patients. By emphasizing the dynamic nature of DKA management and the heightened risks of complications, physician advisors can effectively advocate for the **intensive resources and monitoring** that inpatient care provides, ensuring optimal outcomes for young patients facing life-threatening conditions.

Conclusion of Case Studies

The case studies presented in this chapter illustrate the intricate dynamics of peer-to-peer (P2P) conferences and underscore the critical importance of preparation, strategic framing, and evidence-based advocacy. These scenarios offer practical insights into the common challenges physician advisors face and the innovative strategies they can employ to overturn denials effectively. From addressing complex medical presentations to navigating payer objections, these examples highlight how focused, patient-centered arguments can bridge the gap between clinical necessity and payer policies.

Through these diverse cases, key themes emerge: the importance of leveraging clinical data, aligning arguments with payer and specialty guidelines, and directly addressing payer concerns. Each example demonstrates how physician advisors can refine their techniques to better advocate for patients and ensure appropriate resource utilization within healthcare organizations.

Key Takeaways

1. **Preparation is Foundational**
    - Success begins with **meticulous preparation**—gathering, organizing, and synthesizing clinical data to construct a compelling, evidence-based argument.
    - Tools such as **Prep Sheets** and detailed summaries help streamline discussions and ensure all critical points are addressed succinctly.
2. **Alignment with Guidelines Enhances Credibility**
    - Aligning arguments with **payer-specific guidelines** and supplementing with **specialty standards** reinforces the validity of medical necessity.
    - Demonstrating familiarity with and adherence to these criteria builds trust and professionalism during P2P conferences.
3. **Documentation of Instability is Key**
    - Highlighting **objective evidence of instability**—such as lab trends, vital sign abnormalities, or failed outpatient management—provides a strong foundation for inpatient care advocacy.
    - Concrete data, such as oxygen saturation levels or hemoglobin changes, solidifies the case and reduces ambiguity.
4. **Proactively Addressing Payer Concerns Improves Outcomes**
    - Anticipating and addressing payer objections fosters collaboration and bridges gaps in understanding.
    - Acknowledging documentation gaps or clarifying perceived ambiguities demonstrates a professional and solution-focused approach.
5. **Risk Framing Adds Persuasive Weight**
    - Emphasizing the **potential consequences of inadequate care**, such as adverse clinical outcomes or complications, underscores the necessity of inpatient-level resources.

Acknowledging Valid Denials and the Importance of Picking Your Battles

While the case studies above demonstrate how strategic advocacy can successfully overturn denials, it is equally important to recognize that not all denials are unwarranted. There are instances where the insurance company's medical director may accurately identify a patient whose clinical presentation does not meet inpatient criteria. In those cases, upholding the denial is an appropriate and evidence-based decision. Understanding when to concede—and when to advocate more aggressively—is an essential skill for physician advisors.

Preserving Credibility and Professional Relationships

A physician advisor who contests every denial with the same degree of force risks damaging credibility with payer representatives. Insurance medical directors often review hundreds of cases a month; they develop an expectation of thoughtful, data-driven arguments from the hospital's physician advisors. By respectfully acknowledging the validity of a well-reasoned denial, physician advisors:

- **Demonstrate Objectivity:** Showing a willingness to accept legitimate denials conveys professional fairness and impartiality. This strengthens trust with payers, who learn to view your appeals as balanced rather than automatically adversarial.
- **Build Goodwill:** Medical directors appreciate honesty. When you concede that a denial is correct in borderline or less defensible cases, it can foster a collaborative environment for future discussions.
- **Enhance Future Negotiations:** If you consistently battle for inpatient status on tenuous grounds, your arguments may carry less weight when truly critical cases arise. Conversely, a reputation for discernment can help ensure that when you push back on a denial, payers listen more carefully.

Focusing Efforts Where They Matter Most

Just as clinical decision-making requires prioritization, denial management is most effective when you target your efforts on the cases where overturning a denial is both justifiable and critically important. A strategic approach involves:

- **Assessing Clinical Documentation Thoroughly:** Before appealing, ensure the clinical evidence strongly supports inpatient admission. If the record clearly aligns with inpatient criteria, you can argue confidently; if not, consider whether the denial has merit.
- **Evaluating Payer Guidelines:** Cross-reference the payer's policies and any relevant national guidelines. If the patient's clinical scenario objectively fails to meet criteria, continuing a fruitless appeal can undermine future credibility.
- **Aligning with Organizational Goals:** Hospitals must balance patient advocacy with resource allocation. Long, drawn-out appeals for marginal cases can strain administrative teams and may not significantly impact patient outcomes or hospital revenue.
- **Leveraging Lessons from Past Interactions:** Reflect on previous encounters with each payer. If certain patterns emerge—like consistently losing appeals for borderline cases—it may be more prudent to channel resources into improving documentation practices or focusing on cases with stronger clinical merit.

Long-Term Gains Over Short-Term Wins

In the complex landscape of payer relationships, consistently picking battles wisely yields greater benefits in the long run than fighting every denial indiscriminately. Physician advisors who judiciously decide when to challenge a denial—and when to acknowledge that an insurer's reasoning is sound—often achieve higher overturn rates on the truly contentious cases and develop a more collaborative rapport with payers. This ultimately benefits patients by streamlining approvals for necessary services, reducing administrative friction, and helping ensure that the appeals process targets the cases in greatest need of advocacy.

Metrics for Evaluation

To refine strategies and measure success in P2P advocacy, organizations should track key performance indicators:

- **Overturn Rates:**
  Measure the percentage of denials overturned during P2P conferences and subsequent appeals. This metric offers direct insight into the effectiveness of physician advisor strategies.
- **Time to Resolution:**
  Monitor the duration from initial denial to resolution to identify process inefficiencies and optimize advocacy efforts.
- **Feedback from Payers:**
  Collect qualitative feedback from insurance medical directors regarding the clarity, relevance, and professionalism of arguments. Use this feedback to enhance training and refine future approaches.

Looking Ahead

These case studies serve as a guide for physician advisors seeking to navigate the complexities of denial management more effectively. By focusing on **preparation, precision, and collaboration,** physician advisors can increase their success rates in overturning denials, reduce the operational burdens of unresolved cases, and strengthen their relationships with payers.

The lessons learned from these cases are not static but dynamic—each experience contributes to the collective expertise of physician advisors, enhancing their ability to address future challenges with confidence and skill. As these practices evolve, physician advisors will play an increasingly pivotal role in ensuring patients receive the care they need while supporting the financial and operational sustainability of healthcare organizations.

Through continuous improvement and the integration of actionable insights, physician advisors can elevate their advocacy, foster trust with payers, and contribute to a more efficient, patient-centered healthcare system.

CHAPTER 9

Measuring Success and Improving Skills

Introduction

Peer-to-peer (P2P) conferences are not just a tool for overturning insurance denials; they are a cornerstone of the physician advisor's role in ensuring patients receive appropriate care while safeguarding the financial health of healthcare organizations. These conferences provide a platform for physician advisors to advocate for clinical decisions, address payer concerns, and collaborate to achieve optimal outcomes. However, their value extends far beyond individual cases. They serve as a lens through which organizations can evaluate performance, identify trends, and implement strategies that drive continuous improvement.

For physician advisors, success in P2P advocacy demands more than clinical expertise. It requires strategic communication, thorough preparation, and an ability to adapt to evolving payer policies. While overturn rates are a fundamental metric for evaluating performance, they are only part of the equation.[55] Meaningful improvement stems from a holistic approach that includes analyzing patterns of success, fostering collaboration, and mentoring peers. These efforts not only elevate individual effectiveness but also contribute to the collective strength of the physician advisor team and the broader healthcare organization.

Central to this approach is the ability to learn from each P2P interaction—whether successful or not—and use those insights to refine skills and processes. By tracking metrics such as overturn rates, denial trends, and time to resolution, physician advisors can pinpoint areas of improvement and implement targeted solutions.[56] Equally important is staying updated on payer criteria, clinical guidelines, and industry best practices to remain an informed and credible advocate in discussions with insurers.

This chapter delves into practical strategies for measuring success, fostering professional growth, and leveraging data to improve outcomes. Key topics include:

- **Evaluating Performance:** How to assess overturn rates and other metrics to gain actionable insights into P2P effectiveness.
- **Continuous Improvement:** Strategies for learning from feedback, mentoring peers, and fostering a culture of collaboration within physician advisor teams.
- **Staying Informed:** The importance of staying current with payer policies, clinical guidelines, and technological advancements to remain effective in denial management.

- **Holistic Success:** How to measure the broader impact of P2P advocacy on organizational goals, including patient outcomes, financial sustainability, and team performance.

By embracing these methods, physician advisors can move beyond individual successes to drive systemic improvements in denial management and peer-to-peer advocacy. These practices empower advisors to navigate the complexities of payer interactions with confidence and precision, ensuring optimal outcomes for patients and healthcare organizations alike.

---

1. Tracking Overturn Rates and Patterns of Success

Measuring success in peer-to-peer (P2P) conferences is a cornerstone of assessing the effectiveness and impact of physician advisors. Overturn rates—defined as the percentage of denials successfully reversed during these conferences—serve as a primary metric for evaluation. However, these rates should not be viewed in isolation. For overturn rates to provide actionable insights, they must be contextualized within the nuances of case complexity, payer criteria, and the physician advisor's advocacy approach. A more nuanced analysis enables organizations to identify areas of strength and opportunities for improvement, ensuring that physician advisors are maximizing their impact.

Monitor Overturn Rates with Context

Overturn rates are most valuable when analyzed alongside the specific characteristics of cases and the strategies employed. Contextualizing these rates helps uncover patterns that might otherwise go unnoticed, offering a deeper understanding of success or challenges.

- **High Overturn Rates:**
    - A high overturn rate may reflect meticulous preparation, compelling advocacy, and a focus on cases with strong documentation. However, it could also suggest a conservative approach to case selection, with advisors avoiding borderline or complex cases that might yield lower success rates.
    - **Example Insight:** A physician advisor with an 85% overturn rate might primarily handle cases with clear documentation and alignment with payer criteria, ensuring a high likelihood of approval.
- **Low Overturn Rates:**
    - A low overturn rate might indicate an advisor's willingness to engage in more challenging cases, advocating aggressively for borderline admissions or cases with documentation gaps. Alternatively, it could signal a need for better preparation or improved alignment with payer expectations.

- **Example Insight:** A physician advisor with a 50% overturn rate may be taking on a wider variety of cases, including those with less obvious clinical indications, in an effort to secure approvals for cases that require additional justification.

Key Insight: Striking a Balance

A balanced overturn rate is the hallmark of a strategic and effective physician advisor. This balance reflects a thoughtful approach to case selection, where advisors focus on cases with a reasonable likelihood of success while also advocating for borderline cases that require a more nuanced argument.

- **Ideal Overturn Rate:**
    - Neither too high (suggesting excessive selectivity) nor too low (indicating potential inefficiencies or overreach), an ideal overturn rate reflects a combination of strong preparation, robust documentation, and effective communication.
    - **Guidance:**
        - **Too High (>85%):** Consider expanding case selection to include more complex or borderline cases.
        - **Too Low (<50%):** Narrow focus to cases with clear documentation and strong alignment with payer criteria.

Leveraging Overturn Rates for Continuous Improvement

Regularly reviewing overturn rates and patterns of success provides valuable insights into trends and areas for development. Organizations and physician advisors can use this data to refine case selection strategies, improve documentation, and enhance P2P advocacy techniques.

- **Case Type Analysis:**
    - Break down overturn rates by case type (e.g., sepsis, heart failure, post-operative complications) to identify specific denial categories where success is high or where additional training might be beneficial.
    - **Example Insight:** If overturn rates for heart failure cases are consistently lower than for other conditions, this may indicate the need for better documentation of fluid management or the interplay of comorbidities.
- **Payer-Specific Trends:**
    - Analyze overturn rates by payer to uncover patterns in denial rationale or criteria application. Understanding these trends can help tailor arguments to the unique requirements of different insurers.
    - **Example Practice:** Develop payer-specific templates to streamline preparation for cases that frequently arise with certain insurers.

- **Training Opportunities:**
    - Use overturn rate data to identify training needs within the physician advisor team. For example, if overturn rates for certain conditions are consistently low, targeted education on documentation and advocacy techniques can help improve outcomes.

Metrics Beyond Overturn Rates

While overturn rates are a valuable metric, additional data points can provide a more comprehensive view of success and areas for improvement:

- **Denial Categories:** Track success rates by denial type to identify areas of strength or challenge.
- **Time to Resolution:** Monitor the average time required to resolve denials, aiming to reduce delays and streamline processes.
- **Feedback from Payers:** Collect qualitative feedback to understand how arguments are received and identify recurring documentation gaps or misalignments with payer criteria.

Takeaway

Tracking overturn rates and analyzing patterns of success offers physician advisors and their organizations a powerful tool for continuous improvement. By contextualizing overturn rates within case complexity and payer dynamics, physician advisors can identify trends, refine their strategies, and enhance their overall effectiveness. A balanced approach—advocating for a mix of straightforward and complex cases—ensures that physician advisors maximize their impact while maintaining credibility and building stronger relationships with payers.

---

2. Avoid Over-Reliance on High Overturn Rates

An exceptionally high overturn rate may seem like a mark of success, but it can also signal an overly cautious or selective approach to peer-to-peer (P2P) conferences. While focusing on clear-cut cases ensures a higher likelihood of overturning denials, it may leave complex or borderline cases unchallenged. These unaddressed cases could represent missed opportunities to recover additional revenue or secure approvals for medically necessary care that would benefit patients and the organization.

The Drawbacks of Exclusively Targeting Clear-Cut Cases

Missed Revenue Opportunities:

- Cases that require more nuanced arguments or additional documentation may still have a reasonable chance of success when appropriately advocated for. Avoiding these cases may result in the loss of potential reimbursements that could positively impact the hospital's financial health.
- **Example:** A borderline sepsis denial that hinges on lactic acid trends or fluid resuscitation documentation might be overturned with sufficient preparation, yet could be overlooked if only the most straightforward cases are selected.

Stagnant Skill Development:

- Engaging only in cases with a high probability of success limits opportunities for physician advisors to hone their skills in addressing challenging denials. Handling complex cases builds expertise in crafting persuasive arguments, navigating payer objections, and mastering nuanced criteria.

Missed Patient Advocacy Opportunities:

- Borderline cases often involve patients whose medical needs may not align perfectly with payer criteria but still warrant inpatient care based on clinical judgment. Leaving these cases unchallenged could compromise the quality of care patients receive.

Why Balancing Case Selection Matters

High overturn rates often reflect selective engagement with cases that are well-documented and closely aligned with payer criteria. While this approach bolsters success metrics, it does not address the broader goal of comprehensive denial management. Expanding the scope to include more complex cases ensures that:

- The hospital optimizes its financial recovery.
- Physician advisors grow their skills and confidence in handling a wider range of scenarios.
- Patients benefit from a more robust advocacy process that champions appropriate care in all situations.

Guidance for Broadening Case Selection

Set a Target Range for Overturn Rates:

- Consistently exceeding an overturn rate of 85% may indicate an overly narrow focus on straightforward cases. Aim for a balanced range (e.g., 70–85%) that reflects both strong preparation and a willingness to engage with more complex scenarios.

Identify Borderline Cases for Inclusion:

- Collaborate with case managers and utilization review teams to identify cases that:
  - Have potential for successful overturn with additional documentation.
  - Require nuanced arguments to align clinical data with payer criteria.
- **Example:** A case of acute kidney injury (AKI) with borderline lab values may initially seem unlikely to meet criteria but could succeed with clarification of fluid management and monitoring needs.

Prioritize Strong Documentation and Preparation:

- Broadening the scope should not compromise the quality of preparation. Ensure that even complex cases are supported by thorough documentation and a clear, compelling argument that aligns with payer policies.

Track Outcomes for Expanded Case Selection:

- Monitor overturn rates and financial recovery from complex cases to evaluate the effectiveness of this approach. Use this data to refine strategies and adjust the balance of case selection as needed.

Example Practice for Balanced Engagement

- **Before Expansion:** A physician advisor focuses exclusively on denials with clear documentation of inpatient criteria, achieving a 90% overturn rate.
- **After Expansion:** The advisor begins including complex denials, such as cases requiring additional evidence of medical necessity. Although the overturn rate decreases to 80%, the hospital recovers additional revenue and addresses more patient needs.

Takeaway

While high overturn rates demonstrate effective preparation and advocacy, they can mask missed opportunities for broader impact. Physician advisors should strive for a balance, engaging in a mix of straightforward and complex cases. By broadening case selection while maintaining rigorous preparation, advisors can enhance their skills, recover more revenue, and advocate for patients more comprehensively. This balanced approach supports both individual and organizational success in denial management.

---

3. Identify Risks of Low Overturn Rates

Low overturn rates are an important metric to monitor, as they can highlight potential inefficiencies or weaknesses in the denial management process. When overturn rates fall below acceptable levels—typically under 50%—it may indicate issues such as inadequate preparation, lack of alignment with payer criteria, or overly broad case selection. Persistent low success rates not only hinder financial recovery efforts but can also damage the credibility of physician advisors with insurance medical directors, making future cases even harder to win.

Key Risks Associated with Low Overturn Rates

Inadequate Preparation:

- Low success rates may reflect insufficient review of the clinical record, resulting in incomplete or unfocused arguments during peer-to-peer (P2P) conferences.
- **Example:** A physician advisor fails to highlight critical clinical data, such as lactic acid trends or persistent hypoxia, leaving the denial rationale unchallenged.

Indiscriminate Case Selection:

- Pursuing every denial, regardless of the strength of documentation or alignment with payer criteria, can lead to wasted resources and diminished credibility.
- Insurance medical directors may begin to view the advisor as indiscriminate or unprepared, potentially dismissing even well-supported cases in the future.

Failure to Align with Payer Guidelines:

- Arguments that fail to address specific payer criteria or inconsistencies in documentation weaken the advisor's case. Without a tailored approach, appeals are less likely to succeed.
- **Example:** Using broad language like "meets criteria for inpatient care" without explicitly tying clinical data to the payer's guidelines undermines the argument.

Reputational Damage:

- Persistent low overturn rates may signal to payers that the organization's P2P processes are unreliable or inconsistent. This could result in increased scrutiny or a diminished willingness to overturn denials, even for strong cases.

Operational Inefficiencies:

- Engaging in P2P conferences for poorly documented cases diverts valuable time and resources away from cases with higher chances of success, creating inefficiencies in denial management workflows.

Recommendations to Improve Overturn Rates

Analyze Patterns in Unsuccessful Cases:

- Conduct a detailed review of unsuccessful appeals to identify common issues. Focus on patterns such as:
    - Missing or incomplete documentation.
    - Failure to align with payer-specific criteria.
    - Weak argument framing or lack of evidence-based support.
- **Example Practice:** "A review of upheld denials for sepsis revealed that lactate levels were not documented consistently, prompting a process improvement initiative."

Strengthen Documentation Processes:

- Collaborate with case managers, clinical documentation integrity (CDI) teams, and other stakeholders to ensure that critical clinical data is captured accurately and comprehensively.
- **Key Focus Areas:**
    - Trends in vital signs and lab results (e.g., serial hemoglobin levels for bleeding cases).
    - Documentation of failed outpatient treatments.
    - Details of interventions, such as IV medications or oxygen therapy.

Enhance Argument Framing:

- Reframe arguments to focus on the most compelling clinical data and directly address the payer's rationale for denial.
- Use concise, evidence-based language that aligns with both payer policies and widely accepted clinical guidelines.
- **Example:** Instead of saying, "The patient had pneumonia," state, "The patient's oxygen requirement of 4L/min and bilateral infiltrates on chest X-ray meet inpatient criteria due to the risk of respiratory failure."

Focus on Selective Engagement:

- Avoid pursuing cases with little chance of success, such as those with documentation gaps or those that do not meet payer criteria. Focus on cases where the denial rationale can be effectively countered.
- Use data analytics to identify the types of cases most likely to succeed in P2P conferences.

Provide Targeted Training:

- Offer regular training sessions to physician advisors on payer criteria, clinical documentation standards, and advanced advocacy techniques.
- Incorporate feedback from unsuccessful cases into training materials to address specific weaknesses.
- **Example Training Focus:** "Improving Documentation for Heart Failure Cases" or "Mastering MCG and InterQual Criteria."

Leverage Peer Reviews:

- Conduct collaborative case reviews with fellow physician advisors to gain insights into unsuccessful cases and identify opportunities for improvement.
- **Example Practice:** Use team-based reviews to compare how different advisors would frame the same denial, promoting shared learning and best practices.

Building Credibility with Payers

- By improving overturn rates through better preparation and targeted case selection, physician advisors can rebuild trust with insurance medical directors.
- Demonstrating a consistent focus on well-supported cases and presenting compelling, guideline-driven arguments signals professionalism and reliability, enhancing the payer's perception of the organization's denial management processes.

Takeaway

Low overturn rates are a signal to reevaluate denial management strategies. By analyzing patterns in unsuccessful cases, improving documentation, refining arguments, and focusing on cases with a strong basis for appeal, physician advisors can enhance their success rates. A balanced approach that combines strategic case selection with rigorous preparation not only improves financial outcomes but also strengthens the advisor's credibility and fosters better relationships with payers.

4. Leverage Overturn Rates for Continuous Improvement

Overturn rates are more than just a measure of success—they are a valuable tool for identifying trends, pinpointing weaknesses, and driving targeted improvements in peer-to-peer (P2P) advocacy. By regularly reviewing these metrics, physician advisors and their teams can uncover patterns that indicate areas for training, process refinement, or systemic change. This approach turns overturn rates into actionable insights, fostering a culture of continuous improvement and enhancing overall performance.

How to Use Overturn Rates for Improvement

Identify Trends by Denial Category:

- Break down overturn rates by specific types of denials, such as sepsis, pneumonia, length of stay, or post-surgical complications. Analyze whether certain categories consistently have lower success rates.
- **Example Insight:**
  - A hospital team discovers that overturn rates for sepsis-related denials are significantly lower than other categories. Further investigation reveals that documentation of lactic acid levels and fluid resuscitation outcomes is inconsistent, contributing to weaker arguments.

Focus on Systemic Gaps:

- Patterns in upheld denials may indicate systemic issues, such as:
  - Documentation inconsistencies.
  - Lack of familiarity with updated payer criteria.
  - Misalignment between clinical practice and payer expectations.
- **Example Solution:**
  - A pattern of upheld pneumonia denials prompts the development of a standardized documentation template that ensures all required clinical data—such as oxygen requirements, imaging findings, and response to treatment—is captured and aligned with payer guidelines.

Enhance Training Programs:

- Use overturn rate data to identify knowledge gaps within the physician advisor team. Focus training sessions on the specific denial categories or payer policies with the lowest success rates.
- **Example Practice:**

    - After identifying low overturn rates for length-of-stay denials, the team organizes a training session on documenting discharge barriers and aligning arguments with payer-specific criteria for extended care.

Refine Argumentation Strategies:

- Analyze patterns in payer feedback to understand why certain cases are upheld. Use this feedback to refine argument structures and improve case preparation.
- **Example Insight:**
    - Consistent feedback from payers about insufficient alignment with criteria for heart failure cases leads to a shift in how arguments are framed, emphasizing clear documentation of volume status, interventions, and risk factors for decompensation.

Collaborate Across Teams:

- Share insights from overturn rate analysis with case managers, clinical documentation integrity (CDI) teams, and revenue cycle staff to improve collaboration and address gaps holistically.
- **Example Collaboration:**
    - A low success rate for surgical complication denials prompts a joint effort between physician advisors and CDI specialists to ensure that intraoperative and postoperative complications are thoroughly documented.

Actionable Strategies for Improvement

Conduct Regular Overturn Rate Reviews:

- Schedule quarterly reviews to analyze overturn rates at both individual and team levels. Break down data by denial category, payer, and case complexity to gain a comprehensive understanding.
- Use dashboards or analytics tools to visualize trends and identify areas of focus.

Develop Customized Documentation Tools:

- Create condition-specific documentation templates or checklists to ensure that all required clinical data is captured consistently. This reduces variability and strengthens arguments during P2P conferences.
- **Example Tool:**
    - A pneumonia checklist includes fields for oxygen therapy levels, imaging results, CURB-65 scores, and antibiotic response, aligning with common payer criteria.

Integrate Lessons into Prep Sheets:

- Incorporate insights from overturn rate reviews into Prep Sheet templates, ensuring that arguments are tailored to address common payer objections and align with successful strategies.
- **Example Practice:**
    - A Prep Sheet for sepsis denials now includes a dedicated section for documenting serial lactate trends, volume resuscitation, and clinical response.

Encourage Peer-Led Workshops:

- Use overturn rate data to guide peer-led discussions on best practices. Advisors with higher success rates in specific categories can share techniques with colleagues.
- **Example Initiative:**
    - A workshop on overturning heart failure denials focuses on documenting diuretic response, renal function trends, and symptoms of fluid overload.

Practice Tip

Analyze overturn rates by denial category, payer, and case type to uncover trends and refine preparation strategies. Use these insights to create targeted training, improve documentation practices, and enhance collaboration between teams.

Benefits of Using Overturn Rates for Continuous Improvement

Enhanced Advocacy Skills:

- Focusing on data-driven insights helps physician advisors refine their arguments, improving success rates over time.

Improved Documentation Standards:

- Identifying patterns in unsuccessful cases drives improvements in documentation practices, ensuring better alignment with payer criteria.

Streamlined Processes:

- Analyzing overturn rates and implementing changes reduces inefficiencies in denial management workflows.

Stronger Payer Relationships:

- Demonstrating consistent improvement in overturn rates builds credibility with payers, fostering more collaborative relationships over time.

Takeaway

Regularly reviewing overturn rates transforms a simple performance metric into a powerful tool for continuous improvement. By identifying trends, addressing systemic gaps, and tailoring training and documentation practices, physician advisors can strengthen their P2P advocacy, improve success rates, and contribute to the financial and operational health of their organizations.

Key Metrics to Track

Tracking key metrics in peer-to-peer (P2P) advocacy allows physician advisors to assess performance, identify trends, and uncover opportunities for improvement. By monitoring these data points, organizations can optimize denial management processes, enhance the effectiveness of P2P conferences, and improve patient outcomes.

Denial Categories

Understanding success rates across different denial categories provides insight into where physician advisors are excelling and where challenges remain. Common scenarios, such as sepsis, pneumonia, or postoperative care, often require tailored approaches to address specific payer criteria.

- Why It Matters:
    - Certain conditions or denial categories may be more difficult to overturn due to ambiguous payer policies or inconsistent documentation.
    - Tracking success rates by category helps identify areas where additional training, documentation improvement, or argument refinement is needed.
- Examples of Denial Categories:
    - Sepsis Denials:
        - Success in overturning sepsis denials often hinges on clear documentation of hemodynamic instability, lactic acid trends, and fluid resuscitation efforts.
    - Pneumonia Denials:

            - Success rates improve when arguments emphasize oxygen requirements, imaging results, and antibiotic response.
    - Postoperative Complication Denials:
        - Highlighting evidence of instability, such as drops in hemoglobin or signs of infection, strengthens arguments for inpatient care.
- Actionable Steps:
    - Break down denial categories to evaluate success rates and identify trends.
    - Develop category-specific documentation templates to address recurring challenges.
    - Share findings with clinical documentation integrity (CDI) teams to align medical records with payer expectations.

Payer Trends

Monitoring variations in overturn rates among different insurers provides critical insight into payer-specific patterns and challenges. Each insurer may apply criteria differently, creating opportunities to fine-tune advocacy strategies.

- Why It Matters:
    - Understanding payer trends helps physician advisors tailor their arguments to specific insurers' criteria and practices.
    - Identifying insurers with particularly low overturn rates may indicate systemic barriers that require targeted intervention or escalation.
- What to Track:
    - Overturn Rates by Payer:
        - Analyze success rates for each payer to identify patterns, such as frequent denials for specific conditions or consistent challenges with particular insurers.
    - Payer-Specific Criteria:
        - Evaluate differences in how payers interpret standard criteria (e.g., InterQual vs. MCG) or apply proprietary guidelines.
- Examples:
    - High Success Rates: Some insurers may have more lenient criteria for sepsis or heart failure, leading to higher overturn rates.
    - Low Success Rates: Insurers with rigid policies for length-of-stay or observation care may require additional documentation or argumentation strategies.
- Actionable Steps:
    - Maintain a repository of payer-specific insights and guidelines for easy reference during P2P preparation.
    - Develop payer-specific argument frameworks that address common objections unique to each insurer.
    - Share insights with leadership to inform negotiations with insurers or identify opportunities for policy advocacy.

Time to Resolution

The average time from denial to overturn is a crucial metric for identifying inefficiencies in the denial management process. Delays in resolution can impact hospital revenue cycles, administrative workloads, and patient outcomes.

- Why It Matters:
    - Faster resolution times reduce operational costs and improve cash flow for healthcare organizations.
    - Timely overturns ensure that patients receive necessary care without delays caused by protracted appeals.
- What to Assess:
    - Time Per Denial Category:
        - Evaluate whether certain categories, such as length-of-stay denials, take longer to resolve compared to others.
    - Time Per Payer:
        - Identify insurers with prolonged resolution times to prioritize process improvements or escalations.
    - Process Bottlenecks:
        - Pinpoint delays caused by internal workflows, such as late documentation submissions or lack of preparation for P2P conferences.
- Examples:
    - Short Resolution Times: Successful denials resolved within days may indicate strong preparation and alignment with payer expectations.
    - Prolonged Resolution Times: Cases that take weeks to resolve may suggest inefficiencies in documentation, communication, or payer responsiveness.
- Actionable Steps:
    - Implement process reviews to identify and address bottlenecks in denial resolution.
    - Leverage technology, such as AI-driven tools, to automate case tracking and streamline documentation workflows.
    - Share data with administrative teams to ensure alignment across departments, such as revenue cycle management and CDI.

Using Metrics for Strategic Improvement

Tracking these key metrics allows organizations to:

- Identify training needs for physician advisors.
- Standardize best practices for documentation and argumentation.
- Optimize workflows to reduce delays and improve overturn rates.

- Strengthen payer relationships by addressing specific challenges proactively.

By focusing on denial categories, payer trends, and time to resolution, physician advisors can refine their advocacy strategies, improve institutional performance, and ensure patients receive the care they need without unnecessary administrative barriers.

Seeking Feedback and Mentoring Others

The continuous refinement of skills in peer-to-peer (P2P) conferences relies on open communication, shared experiences, and constructive feedback. By actively seeking insights from peers, payers, and colleagues, physician advisors can gain valuable perspectives, enhance their advocacy, and foster a culture of collaboration and growth within their teams.

Solicit Feedback from Payers and Peers

Payer Feedback

Engaging insurance medical directors for feedback after P2P conferences can illuminate areas for improvement and refine future strategies. While some payers may not formally provide feedback, many are open to discussing the strengths and weaknesses of the presented arguments during the conference or in follow-up communications.

- Benefits of Payer Feedback:
    - Identify Gaps: Learn where documentation, clinical reasoning, or alignment with payer criteria could be improved.
    - Understand Payer Perspectives: Gain insights into how payers interpret policies, helping to tailor future arguments more effectively.
    - Strengthen Relationships: Demonstrating openness to feedback fosters a collaborative dynamic with insurance medical directors, increasing the likelihood of successful outcomes over time.
- How to Request Feedback:
    - Politely ask the payer for input at the end of a P2P conference.
        - Example: "Could you share any suggestions for how we might improve documentation or approach similar cases in the future?"
    - Follow up via email when appropriate, particularly for complex or contentious cases.
        - Example: "Thank you for taking the time to review this case with me. If you have any additional feedback on how we can better align with your guidelines, I'd greatly appreciate it."
- Practice Tip:

- Compile payer feedback over time to identify recurring themes, such as documentation weaknesses or common objections. Use these insights to develop targeted training sessions or refine organizational processes.

Peer Reviews

Peer reviews provide a structured opportunity for collaborative learning and skill enhancement. Regularly discussing cases with colleagues fosters a supportive environment where physician advisors can share insights, refine arguments, and collectively address challenges.

- Benefits of Peer Reviews:
  - Diverse Perspectives: Colleagues may offer alternative strategies or identify overlooked aspects of the case.
  - Shared Learning: Reviewing successful and unsuccessful cases collectively enhances the team's overall skillset.
  - Increased Consistency: Collaborative discussions promote alignment in how P2P conferences are approached across the organization.
- How to Conduct Peer Reviews:
  - Schedule regular case review meetings (e.g., weekly or monthly) where team members present recent denials, overturned cases, or particularly challenging scenarios.
    - Example: "In this case, the payer upheld the denial for observation status. How might we reframe the argument to emphasize inpatient criteria?"
  - Use a structured format to guide discussions:
    - Case Summary: Present the clinical scenario, denial rationale, and initial payer objections.
    - Review of Argument: Highlight the points made during the P2P conference.
    - Feedback and Insights: Invite colleagues to share alternative approaches or additional data that could have strengthened the case.
  - Rotate facilitators to ensure diverse leadership and perspectives.
- Practice Tip:
  - Document key takeaways from peer reviews and incorporate them into organizational best practices or training materials. For instance, create templates for common denial scenarios, such as sepsis or heart failure, based on collective input.

Mentoring Junior Physician Advisors

Mentorship plays a pivotal role in fostering the next generation of physician advisors. Experienced advisors can share their expertise, provide practical guidance, and create a supportive environment for learning.

Hands-On Training

Mentors can help junior advisors build foundational skills by sharing real-world insights and practical advice.

- Guidance for Junior Advisors:
  - "Focus on the payer's specific denial rationale and align your argument with their criteria, supported by clear, concise documentation."
  - "Highlight the most critical clinical findings—overloading the discussion with irrelevant details can dilute your message."
- Practice Tip:
  - Pair new physician advisors with seasoned mentors for their first few P2P conferences to provide guidance and real-time feedback.

Shadowing Opportunities

Allow junior physician advisors to observe experienced colleagues during P2P conferences. Shadowing offers invaluable exposure to effective communication techniques, argument framing, and payer dynamics.

- Benefits of Shadowing:
  - Observing seasoned advisors helps junior colleagues learn how to respond to payer objections, use concise language, and maintain a collaborative tone.
  - It builds confidence by providing a real-world demonstration of successful strategies.
- Practice Tip:
  - Follow each shadowing session with a debrief to discuss what the junior advisor observed and answer any questions.

Confidence Building

Encouraging and empowering junior physician advisors is essential for their long-term success. Constructive feedback and positive reinforcement can help them build the confidence needed to navigate challenging P2P conferences.

- Strategies for Confidence Building:
    - Highlight successes, even in cases where the denial was upheld but the argument was well-constructed.
    - Provide actionable feedback, focusing on specific areas for improvement without discouraging overall performance.

Build a Supportive Network

Creating a collaborative environment within the physician advisor team promotes professional growth and enhances team performance.

Establish a Physician Advisor Forum

A structured forum allows physician advisors to share challenges, discuss emerging trends, and celebrate successes.

- Example Practice:
    - Hold monthly meetings where team members rotate as presenters, sharing lessons learned from recent P2P conferences or payer interactions.

Foster Open Communication

Encourage advisors to share both successes and mistakes in a judgment-free setting. Open communication allows the team to collectively learn and improve.

- Practice Tip:
    - Create a shared repository of anonymized case studies and payer feedback for team-wide reference.

Soliciting feedback and mentoring others are vital components of professional development for physician advisors. Engaging with payers and peers to seek constructive input enhances individual skills and builds a culture of continuous learning. By mentoring junior colleagues and fostering collaboration within the team, physician advisors can create an environment where best practices are shared, challenges are addressed collectively, and the entire organization benefits from improved denial management outcomes.

Staying Updated on Payer Policies and Best Practices

In the ever-evolving landscape of healthcare, staying current with payer policies, clinical guidelines, and best practices is crucial for effective denial management and P2P advocacy. Physician advisors who actively engage with up-to-date resources and educational opportunities can craft more compelling arguments, align with payer expectations, and enhance their overall performance.

Review Payer Policies Regularly

Keep Guidelines Updated

Payer policies frequently change to reflect new clinical evidence, regulatory shifts, or operational goals. Physician advisors must regularly review and update their knowledge of these criteria to ensure that their arguments align with current expectations.

- Best Practices:
  - Obtain updated payer-specific criteria annually or as changes are released. Examples include InterQual, MCG, or proprietary payer guidelines.
  - Maintain a centralized, easily accessible digital repository of the latest guidelines for all physician advisors.
  - Assign a team member to monitor and communicate changes in payer policies to the group.
- Example:
  - "A recent update to MCG guidelines expanded the criteria for inpatient admission in patients with moderate COVID-19 who require supplemental oxygen. This change could impact future denial appeals for respiratory cases."

Tailor Arguments to Payer-Specific Requirements

Different payers may apply varying criteria to similar clinical situations. Understanding these nuances allows physician advisors to customize their arguments for maximum impact.

- Strategies:
  - Use a comparative matrix to highlight differences in payer policies, enabling physician advisors to prepare tailored approaches for common denial scenarios.
  - Train the team on identifying and adapting to these variations.
- Example:
  - Payer A may require documentation of failed outpatient treatment for cellulitis, while Payer B emphasizes evidence of systemic symptoms such as fever or tachycardia.

Proactively Engage with Payers

Establishing direct communication channels with payers can help clarify ambiguities in their criteria.

- Tips:
  - During P2P conferences, ask specific questions about unclear or evolving policies.
  - Request educational sessions from major payers to ensure alignment and understanding.

Follow Specialty Organizations

Leverage Authoritative Guidelines

Professional societies offer evidence-based recommendations that can complement or enhance payer-specific criteria. These guidelines are particularly valuable when payer policies lack specificity or do not address complex cases.

- Examples of Key Organizations:
  - American Thoracic Society (ATS): Guidelines on managing respiratory conditions, including pneumonia and chronic obstructive pulmonary disease (COPD).
  - American Heart Association (AHA): Recommendations for cardiac conditions such as heart failure and arrhythmias.
  - American College of Gastroenterology (ACG**):** Guidelines for gastrointestinal bleeding or endoscopic procedures.
- How to Integrate Guidelines:
  - Include references to specialty guidelines in P2P prep sheets to substantiate arguments.
  - Highlight where payer criteria align with or deviate from these guidelines.
- Example:
  - "The American Stroke Association recommends inpatient monitoring for all patients receiving thrombolytic therapy. This aligns with the need for continuous neurological assessment and blood pressure management."

Stay Active in Professional Communities

Engage with professional organizations to access the latest resources, network with peers, and gain insights into emerging trends.

- Opportunities:

- Attend annual conferences or seminars hosted by these organizations.
- Subscribe to newsletters and journals to stay informed about updates and advancements.

Participate in Continuing Education

Engage in Targeted Training

Workshops, webinars, and certification programs tailored to denial management, utilization review, and P2P advocacy provide valuable knowledge and skill-building opportunities.

- Examples of Relevant Training Topics:
  - Navigating updates to the Medicare Two-Midnight Rule.
  - Strategies for addressing payer objections in complex cases like sepsis or heart failure.
  - Advanced documentation techniques to meet payer expectations.
- Suggested Activities:
  - Participate in interactive workshops that simulate P2P scenarios to practice argumentation and objection handling.
  - Attend webinars on emerging trends, such as the role of artificial intelligence in denial prevention.

Pursue Certification Programs

Certifications offer structured training and validation of expertise in denial management and related fields.

- Examples:
  - American College of Physician Advisors (ACPA): Certification in utilization review and denial management.
  - American Board of Quality Assurance and Utilization Review Physicians (ABQAURP): Certification in health care quality and management.
  - Healthcare Financial Management Association (HFMA): Revenue cycle and denial prevention training.
- Benefits of Certification:
  - Enhanced credibility in interactions with payers and institutional stakeholders.
  - Comprehensive understanding of regulatory and payer-specific requirements.

Share Knowledge Across Teams

Encourage team-wide participation in educational initiatives to foster collective growth and maintain consistency.

- Practice Tip:
  - After attending a workshop or conference, present key takeaways during team meetings to disseminate insights and strategies.

Staying updated on payer policies and best practices is an essential component of a physician advisor's role. By regularly reviewing guidelines, leveraging authoritative resources, and engaging in continuous education, physician advisors can remain at the forefront of denial management. These efforts not only enhance individual performance but also strengthen the overall effectiveness of the physician advisor team, contributing to better patient outcomes and organizational success.

Measuring Success Beyond Overturn Rates

While overturn rates are a key metric for evaluating the effectiveness of peer-to-peer (P2P) conferences, they provide only a partial picture of success. A holistic approach considers broader indicators such as team performance, organizational learning, and the promotion of a culture of excellence.[57] By assessing these dimensions, physician advisors and their teams can ensure sustained improvement and align their efforts with institutional goals.

---

1. Evaluate Team Impact

Analyze Collective Performance

The success of a denial management program depends not only on individual physician advisors but also on the collective contributions of the entire team. Regularly analyzing the team's aggregate performance helps identify trends, pinpoint areas for improvement, and recognize shared strengths.

- Strategies:
  - Track overturn rates and trends across the team to evaluate overall effectiveness.
  - Identify patterns in successful appeals to replicate best practices across cases.
  - Assess performance across various denial categories (e.g., sepsis, respiratory failure, or post-surgical complications).

- Example Insight:
    - "Over the past quarter, our team achieved a 70% overturn rate for length-of-stay denials, with notable success in cases involving advanced cardiac monitoring."

Assess Interdisciplinary Collaboration

P2P conference success often hinges on collaboration with other hospital departments, such as case management, clinical documentation integrity (CDI) teams, and revenue cycle staff. Evaluating these relationships can reveal opportunities to improve processes and foster alignment across teams.

- Key Metrics:
    - Documentation Quality: Measure the frequency of denials citing documentation issues to assess the impact of collaboration with CDI teams.
    - Efficiency: Track time spent gathering information for P2P conferences, identifying opportunities to streamline workflows with case managers.
- Example Practice:
    - Establish regular check-ins between physician advisors and CDI specialists to ensure alignment on payer documentation requirements before a denial arises.

---

2. Promote Organizational Learning

Share Insights Across Departments

Insights from P2P conferences can inform broader organizational strategies to prevent denials, reduce inefficiencies, and improve overall performance.

- Practice Tip:
    - Create case studies or brief reports highlighting lessons learned from successful appeals, including examples of documentation best practices or argumentation techniques.
- Example Initiative:
    - A review of overturned pneumonia cases revealed that documenting failed outpatient antibiotic therapy significantly improved outcomes. This insight was shared with case managers to standardize outpatient documentation processes.

Standardize Best Practices

Consistency in denial management improves efficiency and reduces variability in outcomes. Standardizing successful approaches ensures that all physician advisors have the tools and knowledge needed to advocate effectively.

- Implementation:
    - Develop templates for P2P prep sheets tailored to common denial categories, ensuring all key data points are included.
    - Establish workflows that guide physician advisors through critical steps in the denial appeal process, such as preparing concise arguments or addressing common payer objections.
- Example Practice:
    - Introduce a "Denial Response Checklist" that includes items like aligning with payer criteria, citing specialty guidelines, and documenting clinical instability.

---

3. Celebrate Successes

Recognize Individual Contributions

Acknowledging the efforts of individual physician advisors fosters motivation and reinforces the importance of their role in the organization's success.

- Strategies:
    - Highlight top-performing physician advisors during team meetings or in organizational newsletters.
    - Create an "Advocate of the Month" program to recognize significant contributions to overturning challenging denials.
- Example:
    - "Dr. Smith's exceptional advocacy in overturning a complex sepsis denial not only ensured appropriate patient care but also recovered $25,000 in revenue for the hospital."

Celebrate Team Achievements

Showcasing the collective accomplishments of the denial management team reinforces a culture of collaboration and shared success.

- Practice Tip:
    - Share quarterly or annual performance summaries with hospital leadership, emphasizing the team's contributions to financial stability and patient outcomes.
- Example Highlight:
    - "This quarter, the physician advisor team improved the hospital's overall denial overturn rate by 15%, reducing revenue delays and ensuring timely care for patients."

Measuring success in P2P conferences requires a multidimensional approach that extends beyond overturn rates. By evaluating team impact, promoting organizational learning, and celebrating achievements, physician advisors can foster a culture of continuous improvement and excellence. These efforts not only enhance individual and team performance but also contribute to the broader goals of patient care, financial sustainability, and institutional success. Through collaborative strategies and recognition of accomplishments, denial management teams can drive meaningful and lasting progress.

Promoting Professional Growth Through Continuous Improvement

The role of a physician advisor in denial management requires ongoing professional development to navigate the complexities of peer-to-peer (P2P) conferences effectively. By engaging in self-assessment, lifelong learning, and adopting innovative strategies, physician advisors can enhance their skills, adapt to evolving challenges, and contribute more meaningfully to their organizations.

Self-Assessment and Reflection

Regular Performance Reviews

Self-assessment involves critically evaluating individual performance after each P2P conference to identify strengths and areas for improvement. Structured reviews help physician advisors refine their techniques and approach future challenges more effectively.

- Practice Tips:
    - Review the clarity and structure of arguments presented during P2P conferences.
    - Analyze how well payer guidelines were integrated into the discussion.
    - Identify areas where additional documentation or data could have strengthened the case.
- Example Practice:
    - After a P2P conference, assess:
        - Argument Effectiveness: Was the denial rationale addressed directly?
        - Payer Feedback: Were objections resolved satisfactorily?

Analyze Trends in Denied Cases

Tracking patterns in denied cases helps uncover recurring challenges and areas where advocacy techniques may need refinement.

- Actionable Steps:
    - Group denied cases by category (e.g., sepsis, length of stay, or cardiac issues) to identify patterns.

- Note common payer objections and evaluate whether documentation or argumentation gaps contributed to the denial.
- Example Insight:
    - "A review revealed that denials for short inpatient stays often lacked documentation of failed outpatient management, prompting a focus on strengthening preadmission records."

Set Personal Goals

Establishing measurable objectives provides a roadmap for professional growth and ensures continuous improvement.

- Goal Ideas:
    - Improve overturn rates for a specific category of denial.
    - Enhance communication techniques to address payer objections more effectively.
    - Develop expertise in a high-frequency denial area, such as length-of-stay cases.

Engage in Lifelong Learning

Stay Current with Industry Trends

Healthcare is an ever-evolving field, and staying informed about changes in payer policies, clinical guidelines, and best practices is crucial for maintaining relevance and effectiveness.

- Strategies:
    - Subscribe to updates from professional organizations, such as the American College of Physician Advisors or payer-specific newsletters.
    - Attend industry conferences to learn about emerging trends and strategies in denial management.
- Example Practice:
    - Participate in annual workshops focused on updates to payer criteria like InterQual and MCG guidelines.

Participate in Advanced Training Programs

Advanced training programs offer targeted skill development in areas like regulatory compliance, documentation, and payer communication.

- Certification Opportunities:
    - Enroll in utilization review certification programs or workshops offered by professional organizations.
    - Pursue specialized training in negotiation techniques or payer relations.

Seek Mentorship Opportunities

Mentorship from seasoned physician advisors can accelerate skill development and provide valuable insights into effective denial management.

- Action Plan:
    - Shadow experienced colleagues during P2P conferences to observe their argumentation techniques.
    - Schedule regular check-ins with mentors to discuss challenging cases and receive feedback.

Adopt Innovative Strategies

Leverage Technology for Skill Development

Technology, including artificial intelligence (AI) and analytics tools, can streamline processes and identify opportunities for improvement.

- Examples of AI Tools:
    - Use AI-driven software to analyze patterns in denied cases and pinpoint documentation gaps.
    - Automate Prep Sheet creation to ensure consistent inclusion of critical data points.
- Case Insight:
    - "AI analysis revealed that heart failure denials were often upheld due to missing documentation on diuresis outcomes. This insight led to the implementation of a checklist for diuretic therapy notes."

Embrace Simulation Training

Role-playing exercises and simulation training are powerful tools for preparing physician advisors to handle challenging payer objections and refine their communication strategies.

- Practice Ideas:
    - Conduct mock P2P conferences with colleagues acting as payer representatives.
    - Simulate high-pressure scenarios, such as handling unexpected payer objections or limited conference time.
- Example Practice:
    - Role-play a sepsis denial appeal where the "payer" questions the necessity of inpatient admission, preparing advisors to counter such objections effectively.

Participate in Peer-Led Workshops

Workshops organized by peers foster collaboration and shared learning while addressing common challenges in denial management.

- Workshop Topics:
  - Effective strategies for addressing payer-specific objections.
  - Updates on changes in payer policies and clinical guidelines.
  - Techniques for aligning documentation with denial rationale.
- Example Initiative:
  - Quarterly peer workshops focused on refining arguments for high-frequency denial categories like respiratory failure or short-stay admissions.

Foster a Growth Mindset

Embrace Challenges as Learning Opportunities

Difficult cases and upheld denials can provide valuable insights into areas for improvement. Physician advisors should approach these situations as opportunities to refine their skills and expand their knowledge base.

- Practice Reflection:
  - Identify key takeaways from upheld denials and incorporate them into future strategies.
  - Use payer feedback to address documentation gaps or enhance argument clarity.

Cultivate Resilience

Maintaining a positive attitude and commitment to growth, even in the face of setbacks, is essential for long-term success in denial management.

- Motivational Practices:
  - Celebrate small victories, such as mastering a new argument technique.
  - View every denial as a chance to improve advocacy skills.

Celebrate Milestones

Recognizing achievements, whether personal or team-based, reinforces progress and motivates continued professional development.

- Examples:
  - Acknowledge a first-time overturn for a complex denial category.
  - Celebrate contributions to team-wide process improvements or educational initiatives.

Conclusion

Promoting professional growth through continuous improvement is essential for physician advisors to remain effective and adaptable in their roles. By engaging in self-assessment, committing to lifelong learning, and adopting innovative strategies, physician advisors can refine their skills, overcome new challenges, and elevate the standards of denial management. These efforts ensure that physician advisors not only contribute to successful outcomes for patients and organizations but also foster a culture of excellence and continuous learning within their teams.

Measuring success in denial management and peer-to-peer (P2P) conferences is a multifaceted process that goes far beyond tracking overturn rates. While these rates are a critical indicator of performance, true success lies in evaluating the broader impact of P2P advocacy on patient care, financial outcomes, and organizational excellence. By adopting a comprehensive approach, physician advisors can identify strengths, address challenges, and drive continuous improvement in their practices.

This chapter emphasized the importance of contextualizing overturn rates, using them not only as a performance metric but also as a tool for uncovering systemic opportunities for improvement. By analyzing trends in denial categories, payer-specific patterns, and time-to-resolution metrics, physician advisors and their organizations can refine their strategies and create more targeted interventions to reduce denials and enhance outcomes.

Promoting Professional Growth and Collaboration

Key to improving performance is fostering a culture of continuous learning and collaboration. Soliciting feedback from payers and peers, participating in regular case reviews, and mentoring junior physician advisors are vital strategies for professional growth. These practices encourage shared learning and ensure that all team members are equipped with the tools and insights necessary for success. Structured mentorship programs and peer-led workshops help build a supportive environment where advisors can hone their skills, share best practices, and adapt to the dynamic healthcare landscape.

Staying Updated and Innovative

In a rapidly evolving industry, staying current with payer policies, clinical guidelines, and best practices is non-negotiable. Physician advisors must proactively review and integrate updated criteria into their advocacy while leveraging authoritative guidelines from professional societies to strengthen their arguments. Advanced training programs, certifications, and the use of AI-driven tools for data analysis and documentation improvements empower advisors to remain at the forefront of denial management.

Simulation training and technology-driven solutions provide additional opportunities to enhance skills and streamline processes. These innovations not only improve individual performance but also contribute to the overall effectiveness of the physician advisor team.

Holistic Success and Organizational Impact

Success in denial management must also be evaluated through its impact on team performance and organizational goals. Promoting a collaborative approach to denial prevention, improving interdisciplinary workflows, and sharing insights across departments help align the physician advisor team with institutional objectives. Celebrating individual and team achievements reinforces a culture of excellence and motivates advisors to continue refining their skills.

Ultimately, the efforts of physician advisors play a pivotal role in ensuring that patients receive timely, appropriate care while safeguarding the financial health of healthcare organizations. By fostering a mindset of continuous improvement, advisors can address payer concerns with confidence and precision, build stronger relationships with insurers, and optimize outcomes for all stakeholders.

Looking Ahead

As the healthcare landscape continues to evolve, physician advisors must remain adaptable and forward-thinking. Embracing innovation, cultivating resilience, and prioritizing collaboration will enable advisors to navigate the challenges of denial management effectively. Through these practices, physician advisors can ensure their critical role in patient advocacy and institutional success remains impactful and sustainable.

This chapter provided a roadmap for measuring success and improving skills in denial management, offering practical strategies and actionable insights for both individual and organizational growth. By integrating these principles into daily practices, physician advisors can continue to excel in their roles and contribute to a more efficient, equitable, and patient-centered healthcare system.

References

[55] Hammer, David, et al. "UPMC's metric-driven revenue cycle: a commitment to managing by the metrics enabled one healthcare organization to achieve world class revenue cycle operations." Healthcare Financial Management, vol. 61, no. 9, Sept. 2007, pp. 96+. Gale Academic OneFile, link.gale.com/apps/doc/A168587032/AONE?u=anon~2e745e8d&sid=googleScholar&xid=ba5fc986. Accessed 19 Feb. 2025.

[56] Chandawarkar, Rajiv MD, MBA*; Nadkarni, Prakash MD†; Barmash, Elizabeth MD*; Thomas, Stephany*; Capek, Allison PAC*; Casey, Kathleen PAC*; Carradero, Freddy*. Revenue Cycle Management: The Art and the Science. Plastic & Reconstructive Surgery-Global Open 12(7):p e5756, July 2024. | DOI: 10.1097/GOX.0000000000005756

[57] Waymack, Pamela M. "Looking reality straight in the eye: five claim denial facts every CFO should know." Healthcare Financial Management, vol. 60, no. 3, Mar. 2006, pp. 44+. Gale Academic OneFile, link.gale.com/apps/doc/A144152152/AONE?u=anon~3113983e&sid=googleScholar&xid=62d9ae71. Accessed 19 Feb. 2025.

CHAPTER 10

Responding to Upheld Denials – Writing Appeal Letters

Introduction

Even with meticulous preparation, thorough documentation, and compelling arguments during peer-to-peer (P2P) conferences, there will inevitably be instances where insurance denials are upheld. Such outcomes can be frustrating for physician advisors who diligently advocate for appropriate patient care. However, these moments, while disappointing, are also opportunities. They provide a chance to revisit the case, address the payer's objections with greater precision, and build a stronger, more refined argument for appeal.

An upheld denial does not signify the end of the process but rather the beginning of a critical next step: the first-level appeal. Writing a persuasive and comprehensive appeal letter immediately following a denial leverages the momentum of the case. It allows the physician advisor to articulate their clinical reasoning while the details remain fresh, ensuring a well-informed and robust rebuttal. A carefully crafted appeal letter not only strengthens the case but also reinforces the hospital's commitment to patient advocacy and compliance with payer criteria.[58]

This chapter explores practical strategies for crafting effective appeal letters that address payer objections, align with clinical guidelines, and adhere to regulatory standards. Additionally, it emphasizes the importance of timeliness in appeal submission, highlighting how delays can weaken arguments and disrupt revenue cycles. Finally, the chapter examines the transformative role of emerging technologies, particularly artificial intelligence (AI), in enhancing the efficiency, consistency, and impact of appeal letter drafting. By embracing these tools and approaches, physician advisors can navigate the complexities of denial management with greater confidence and effectiveness.

The Physician Advisor's Role in Writing Appeal Letters

The appeal process is a pivotal opportunity to advocate for patients while addressing the payer's rationale for denial. As clinical experts and experienced advocates, physician advisors are uniquely equipped to lead this effort. Their direct involvement ensures that appeal letters are not only thorough and accurate but also strategically aligned with the nuances of the case. Physician advisors bring a depth of understanding and authority that strengthens the appeal's effectiveness and bolsters the healthcare organization's commitment to securing appropriate care.

Key Advantages of Physician Advisor Involvement

1. Clinical Expertise
   Physician advisors possess an in-depth understanding of the patient's clinical presentation, treatment plan, and the rationale for the care provided. This knowledge allows them to construct arguments that are both compelling and evidence-based.
    - Precision in Addressing Denials: Physician advisors can pinpoint and refute the payer's objections by linking clinical data directly to guidelines and medical necessity.
    - Examples of Expertise in Action:
        - For a patient denied inpatient admission for heart failure, the physician advisor can articulate how persistent fluid overload, refractory to outpatient management, necessitated hospitalization.
        - In cases involving complex comorbidities, physician advisors can highlight how overlapping conditions elevated the patient's risk profile, justifying the level of care provided.
    - Advocacy through Evidence: Physician advisors integrate clinical findings, payer criteria, and specialty guidelines to present a well-rounded case that leaves little room for ambiguity.
2. Continuity of Advocacy
   Writing the appeal letter ensures a seamless transition from the peer-to-peer (P2P) conference to the next phase of denial management. This continuity preserves the integrity of the arguments and builds on the foundation established during the initial discussion with the payer.
    - Consistency Across Platforms: By aligning the appeal letter with the P2P conference, physician advisors ensure that key points raised during the conference are reinforced and expanded upon in the written appeal.
    - Enhanced Credibility: When the same physician advisor handles both the P2P and the appeal, it demonstrates a cohesive approach to advocacy, signaling professionalism and commitment to the patient's case.
    - Avoiding Redundancies: Physician advisors can directly address unresolved objections or gaps noted during the P2P, ensuring the appeal provides a comprehensive rebuttal.
3. Timeliness
   Prompt action is essential in the appeal process, as delays can weaken arguments, disrupt the revenue cycle, and prolong patient uncertainty. Physician advisors are well-positioned to act swiftly, leveraging their firsthand knowledge of the case to craft and submit a compelling appeal without unnecessary delays.
    - Maintaining Momentum: Writing the appeal letter immediately after the P2P conference captures the details of the case while they are fresh, preserving the nuances of the discussion and ensuring a focused response.
    - Expedited Resolution: Timely appeals reduce administrative bottlenecks and encourage faster payer reviews, accelerating reimbursement and minimizing potential financial strain on the organization.

- Demonstrating Organizational Efficiency: By submitting appeals promptly, physician advisors convey the hospital's dedication to resolving denials professionally and efficiently, reinforcing trust with payers.

The Impact of Physician Advisor-Led Appeals

Physician advisor involvement in appeal writing elevates the quality and effectiveness of denial management efforts. Their clinical expertise, continuity of advocacy, and timely actions contribute to stronger arguments, improved overturn rates, and enhanced relationships with payers. By leveraging these unique strengths, physician advisors play a critical role in securing appropriate care for patients and protecting the financial health of their organizations.

Crafting the First Appeal Letter

A well-constructed appeal letter is a cornerstone of effective denial management, combining professionalism, clarity, and robust evidence to advocate for the patient. The letter should be concise yet comprehensive, directly addressing the denial rationale and presenting a compelling case for overturning the decision. Below are the essential elements of a successful appeal letter, with practical tips and examples to guide physician advisors in crafting persuasive arguments.[59]

---

1. Start with a Professional and Assertive Tone

The opening of the appeal letter sets the tone for the argument. It should strike a balance between respect for the payer's position and firm advocacy for the medical necessity of the care provided. A confident and professional introduction establishes credibility and emphasizes the urgency of overturning the denial.

- Purpose of the Introduction:
  Clearly state the intent of the letter, referencing the denial decision and highlighting the patient's needs.
- Key Phrases for a Strong Opening:
  - "As a physician, it is my responsibility to advocate for medically necessary care that aligns with clinical and regulatory standards."
  - "This appeal is being submitted to address the incorrect denial of inpatient care, which was necessary to ensure the safety and well-being of the patient."
- Example:
  "I am writing to formally appeal the denial of inpatient care for [Patient Name] (ID: [Patient ID]), following our peer-to-peer conference on [Date]. The care provided was medically necessary and fully consistent with clinical guidelines and regulatory requirements. Denial of this care represents a misapplication of criteria and poses a risk to both patient outcomes and payer compliance with established standards."

---

2. Reiterate the Clinical Case

A concise, well-structured summary of the clinical situation underscores the necessity of the care provided. The summary should include key clinical details, emphasizing the severity of the patient's condition and the appropriateness of the chosen care level.

- Focus Areas:
  - Acute clinical findings (e.g., vital signs, diagnostic results).
  - Specific interventions (e.g., oxygen therapy, IV medications, monitoring).
  - Documented failure of outpatient or lower-acuity treatments, if applicable.
- Key Phrases to Highlight Clinical Necessity:
  - "The patient's condition warranted continuous monitoring and advanced interventions that could only be provided in an inpatient setting."
  - "Outpatient and observation-level care were insufficient to meet the patient's needs due to [specific complications]."
- Example:
  "The patient presented with severe community-acquired pneumonia, characterized by a persistent oxygen requirement of 4L/min to maintain SpO_2 >90%. Despite appropriate outpatient treatment, the patient experienced clinical deterioration, necessitating hospitalization for IV antibiotics, continuous oxygen therapy, and close monitoring of respiratory status. These interventions clearly exceeded the capabilities of observation care and were essential for managing the patient's condition and preventing further complications."

---

3. Address the Payer's Denial Rationale

Directly refuting the payer's denial rationale demonstrates a thorough understanding of the case and payer criteria. This section should focus on the specific reasons cited for the denial and provide evidence-based counterarguments.

- Tips for Addressing Denial Rationale:
  - Restate the payer's reasoning before presenting your rebuttal to demonstrate a clear understanding of their position.
  - Use clinical data and guidelines to dismantle the denial rationale.
- Key Phrases for Refuting Denials:
  - "The payer's assertion that observation care was sufficient overlooks the patient's [specific clinical needs]."
  - "This decision does not account for the risks associated with delaying inpatient interventions, as evidenced by [clinical findings]."

- Example:
"The denial states that observation care would have been sufficient. However, the patient's continuous oxygen dependency and the need for IV antibiotics and close respiratory monitoring clearly meet criteria for inpatient admission. Observation care is designed for patients with less acute needs, and delaying the required inpatient interventions would have placed this patient at significant risk of deterioration."

---

4. Cite Clinical Guidelines

Incorporating references to accepted clinical guidelines strengthens the appeal by grounding the argument in established standards of care. Guidelines from reputable organizations, such as InterQual, MCG, or specialty societies, add authority to the appeal and align it with industry benchmarks.

- Tips for Using Guidelines Effectively:
    - Clearly link clinical data to guideline criteria.
    - Avoid overloading the appeal with unnecessary citations; focus on the most relevant guidelines.
- Key Phrases for Citing Guidelines:
    - "According to widely accepted guidelines for [condition], inpatient care is indicated when [specific criteria]."
    - "This patient's clinical presentation aligns with [guideline] criteria for inpatient admission due to [specific reasons]."
- Example:
"Per MCG guidelines for pneumonia management, inpatient care is recommended for patients requiring oxygen therapy exceeding 2L/min and IV antibiotics. This patient's oxygen dependency of 4L/min, coupled with the need for IV antibiotics, met these criteria unequivocally."

---

5. Close with a Call to Action

The conclusion should reinforce the urgency of overturning the denial and clearly state the desired outcome. A firm but respectful tone helps ensure the payer understands the implications of the denial while encouraging prompt resolution.

- Key Components of a Strong Closing:
    - Request a specific action (e.g., overturning the denial).
    - Provide a timeframe for response to maintain momentum.
    - Emphasize the importance of compliance with clinical and regulatory standards.
- Key Phrases for a Call to Action:

    - "I urge you to promptly overturn this denial to ensure compliance with established care standards and to avoid disruptions in patient care."
    - "Please provide a response to this appeal within [specific timeframe] to expedite resolution."
- Example:
  "I respectfully request that this denial be overturned promptly, given the clear medical necessity of the care provided. This decision is critical to ensuring compliance with clinical guidelines and to prevent further delays in patient care and reimbursement. Please notify me of your decision within five business days."

Putting It All Together: Sample Appeal Letter Outline

1. Introduction
    - Respectful and assertive opening.
    - State the intent of the appeal.
2. Clinical Case Summary
    - Key clinical details supporting the care provided.
    - Highlight the severity of the condition and the interventions required.
3. Rebuttal of Denial Rationale
    - Restate the payer's rationale.
    - Provide a detailed, evidence-based rebuttal.
4. Citations to Clinical Guidelines
    - Reference relevant criteria and align clinical findings with standards.
5. Call to Action
    - Request a specific resolution and provide a response timeframe.

By following this structured approach, physician advisors can craft persuasive appeal letters that effectively advocate for the patient while addressing payer objections with professionalism and precision.

Timely Appeals: Preserving Case Momentum

The importance of timely appeal submissions cannot be overstated in the context of denial management. Acting swiftly after a denial preserves the flow of the case, demonstrates organizational efficiency, and signals a strong commitment to patient advocacy. Promptly addressing denials ensures that critical details remain fresh in the minds of all parties involved, reducing the risk of oversight or delayed resolution.[60]

Benefits of Timely Appeals

1. Preserves Case Momentum
   A prompt appeal ensures that the nuances of the case and the payer's objections are addressed while the details are still top of mind for the physician advisor and other stakeholders. Delays in submission can lead to weakened arguments as important case specifics fade from memory.
   - Key Insight: Timeliness reinforces the continuity of advocacy, ensuring that the energy and effort invested in the peer-to-peer (P2P) conference are not lost.
   - Example Benefit:
     "Submitting an appeal within 48 hours of the denial allows the physician advisor to build on the groundwork laid during the P2P conference, leveraging their familiarity with the payer's objections and maintaining a strong, cohesive argument."
2. Reflects Professionalism
   Timely appeal submissions demonstrate the organization's commitment to resolving denials efficiently and ensuring patients receive the care they need without unnecessary administrative delays. It sends a clear message to payers about the seriousness of the organization's advocacy efforts.
   - Key Indicator of Professionalism:
     - Consistently submitting appeals within defined timeframes (e.g., 48–72 hours) establishes the organization's reputation for efficiency and reliability.
   - Example Perception:
     "A payer may view a hospital with consistently timely appeals as well-organized, increasing the likelihood of productive collaboration in future cases."
3. Reduces Revenue Delays
   Delays in appeal submissions can lead to disruptions in the revenue cycle, creating bottlenecks that strain financial resources and administrative workflows. Submitting appeals promptly minimizes these delays, ensuring a smoother reimbursement process.
   - Example Outcome:
     "Hospitals that prioritize appeal timeliness report faster denial resolutions and improved cash flow, enabling better resource allocation for patient care."

Steps to Ensure Timeliness

1. Establish Clear Protocols
   Defining standardized workflows for denial management is essential to ensure timely appeal submissions. These protocols should outline specific steps and responsibilities for all team members involved.
   - Key Protocol Features:
     - Immediate case review after a denial is received.
     - Scheduled timelines for drafting, reviewing, and submitting appeals.
     - Clear delegation of tasks to ensure accountability.

- Example Workflow:
  "Day 1: Conduct P2P conference and document key insights.
  Day 2: Draft the appeal letter and review with key stakeholders.
  Day 3: Submit the appeal to the payer and log the submission in the denial tracking system."

2. Leverage Prep Sheets
   Prep Sheets prepared for P2P conferences can serve as a valuable resource for crafting appeal letters quickly and effectively. These documents should summarize the clinical case, payer objections, and key arguments, providing a strong foundation for the appeal.
   - Key Elements of Effective Prep Sheets:
     - Patient demographics and clinical details.
     - Summary of the denial rationale.
     - Alignment with payer criteria or clinical guidelines.
   - Example Efficiency Gain:
     "Using Prep Sheets reduced appeal drafting time by 50%, allowing physician advisors to focus on strengthening clinical arguments rather than recreating case summaries from scratch."
3. Delegate Administrative Tasks
   Physician advisors should concentrate on crafting the clinical and strategic components of the appeal letter, while non-clinical tasks such as formatting, submission, and follow-up tracking can be handled by support staff.
   - Examples of Delegated Tasks:
     - Formatting the letter to meet payer submission requirements.
     - Uploading the letter to the appropriate payer portal.
     - Documenting submission dates and tracking responses.
   - Key Benefit:
     "Delegating administrative responsibilities enables physician advisors to focus on high-value activities, such as refining clinical arguments and preparing for future P2P conferences."

Practical Strategies to Enhance Appeal Timeliness

1. Implement Technology Solutions:
   Utilize electronic health record (EHR) systems and denial management software to streamline workflows. Automated reminders and templates can help ensure appeals are submitted on time.
   - Example Application:
     "A denial management system sends automated alerts when a denial is received, prompting immediate action from the physician advisor team."
2. Set Performance Benchmarks:
   Establish and track key performance indicators (KPIs) related to appeal submission times to maintain accountability and identify areas for improvement.

- Example KPI:
  "Maintain an average appeal submission time of less than 72 hours for all denials."

3. Provide Training on Workflow Efficiency:
   Educate physician advisors and support staff on the importance of timeliness and provide practical training on optimizing workflows.
   - Example Training Focus:
     "Streamlining Documentation for Faster Appeals" or "Using Prep Sheets to Expedite Letter Drafting."

Takeaway

Timely appeal submissions are critical for preserving momentum, demonstrating professionalism, and minimizing revenue disruptions. By establishing clear protocols, leveraging Prep Sheets, and delegating administrative tasks, physician advisors and their teams can ensure appeals are submitted efficiently and effectively. This proactive approach not only improves denial resolution times but also enhances the organization's reputation as a committed advocate for patient care.

Using Metrics to Measure Appeal Success

Effectively measuring the outcomes of appeal letters is critical for continuous improvement in denial management. By tracking key performance metrics, organizations can identify strengths, uncover areas for improvement, and refine their strategies to achieve better results. Metrics not only provide a quantitative assessment of the appeal process but also support data-driven decision-making for resource allocation, training, and process enhancements.

---

1. Appeal Success Rate

The appeal success rate measures the percentage of appeal letters that result in overturned denials, providing a direct indicator of the effectiveness of the arguments presented.

- Why It Matters:
  A high success rate reflects the strength of clinical arguments, alignment with payer criteria, and effective advocacy by physician advisors. Conversely, a low success rate may signal gaps in documentation, argument framing, or payer-specific understanding.
- How to Track:
  - Calculate the success rate by dividing the number of successful appeals by the total number of appeals submitted within a given timeframe.

  - Segment data by denial category (e.g., sepsis, length of stay) and payer to uncover patterns.
- Example Insight:
  "The appeal success rate for pneumonia-related denials increased from 65% to 80% after implementing a documentation checklist that ensured oxygen therapy and imaging findings were consistently included in Prep Sheets."
- Actionable Steps:
  - Analyze unsuccessful appeals to identify recurring payer objections or documentation gaps.
  - Provide targeted training to physician advisors on categories with lower success rates.
  - Develop templates for common denial scenarios to standardize the quality of appeal letters.

---

2. Financial Impact

This metric quantifies the revenue recovered through successful appeals, linking denial management efforts to the organization's financial performance.

- Why It Matters:
  Denials represent lost revenue opportunities, and successful appeals have a direct financial impact on the hospital's bottom line. Tracking financial recovery helps justify investments in denial management resources, such as technology and training.
- How to Track:
  - Sum the total dollar amount of overturned denials over a specific period.
  - Break down financial recovery by denial category and payer to identify high-impact areas.
- Example Insight:
  "Overturned denials for short inpatient stays generated an additional $500,000 in revenue in Q3, highlighting the value of targeted advocacy for length-of-stay cases."
- Actionable Steps:
  - Prioritize appeals with the highest potential financial recovery.
  - Use financial impact data to allocate resources, such as assigning experienced physician advisors to high-value cases.
  - Advocate for increased investments in tools, like AI-assisted drafting systems, that enhance success rates and expedite resolution.

---

3. Time to Resolution

The time to resolution measures the duration from appeal submission to final payer decision, offering insights into the efficiency of the appeal process.

- Why It Matters:
  Prolonged resolution times delay reimbursements, disrupt cash flow, and may hinder patient care continuity. Identifying bottlenecks in the appeal process can improve efficiency and reduce delays.
- How to Track:
  - Measure the average time from appeal submission to payer response.
  - Analyze resolution times by payer and denial category to identify trends.
- Example Insight:
  "Appeals submitted within 48 hours of the denial had a 20% faster resolution time compared to those submitted after five days, emphasizing the importance of timeliness."
- Actionable Steps:
  - Implement workflows that prioritize prompt appeal submissions.
  - Address internal bottlenecks, such as delays in gathering documentation or finalizing letters.
  - Use metrics to evaluate payer-specific response times, identifying insurers with consistently prolonged resolution periods.

Expanding Metrics for Comprehensive Analysis

In addition to the core metrics above, consider the following supplementary metrics to gain deeper insights into the appeal process:

1. Appeal Turnaround Time
   - Definition: Measures the time between the initial denial notification and the submission of the appeal.
   - Why It Matters: Short turnaround times reflect operational efficiency and a commitment to resolving denials promptly.
   - Actionable Insight:
     "A decrease in average turnaround time from five days to three days correlated with a 15% improvement in overall appeal success rates."
2. Payer-Specific Success Rates
   - Definition: Tracks success rates by payer to understand variations in criteria interpretation and objection patterns.
   - Why It Matters: Tailoring strategies to payer-specific nuances improves appeal outcomes and fosters stronger payer relationships.
   - Example Insight:
     "Payer A had a 90% success rate for pneumonia-related appeals when arguments included serial oxygen saturation data, compared to a 60% success rate for appeals lacking this information."
3. Training Impact Metrics

- Definition: Evaluates the effectiveness of training programs on denial management by comparing pre- and post-training success rates.
- Why It Matters: Demonstrates the ROI of educational initiatives and identifies areas where further training is needed.
- Example Insight:
  "After training on sepsis documentation, the success rate for sepsis-related appeals increased by 25%."

Using Metrics to Drive Continuous Improvement

1. Identify Trends:
   Regularly review metrics to uncover patterns in denials and appeals. For example, a high success rate for certain payers or denial categories can inform the development of standardized templates or best practices.
2. Set Benchmarks and Goals:
   Use historical data to establish benchmarks for success rates, financial impact, and resolution times. Set incremental goals to drive team performance and track progress.
3. Optimize Workflows:
   Metrics can highlight inefficiencies in the appeal process, prompting targeted interventions. For example, a review of turnaround times might reveal delays in documentation retrieval, suggesting the need for streamlined workflows or automation tools.
4. Focus on High-Impact Areas:
   Allocate resources strategically based on metrics. For instance, prioritize appeals in denial categories with significant financial impact or those with the highest potential for successful overturns.
5. Enhance Collaboration:
   Share metrics with interdisciplinary teams, such as case managers and revenue cycle staff, to foster alignment and collective accountability for improving outcomes.

Takeaway

Tracking and analyzing metrics is essential for evaluating the success of appeal letters and optimizing the denial management process. By focusing on appeal success rates, financial impact, and time to resolution, organizations can refine their strategies, address inefficiencies, and improve patient and financial outcomes. These metrics provide a roadmap for continuous improvement, ensuring that physician advisors remain effective advocates in an increasingly complex healthcare environment.

The Future of Artificial Intelligence in Appeal Letters

Artificial intelligence (AI) is revolutionizing the healthcare industry, and denial management is no exception. AI-powered tools are increasingly being utilized to enhance the efficiency, consistency, and effectiveness of appeal letter drafting. These advancements offer transformative potential for physician advisors and healthcare organizations by streamlining processes, reducing administrative burdens, and improving appeal success rates.[61]

---

1. Automated Letter Drafting

AI's ability to draft appeal letters based on inputted clinical data and denial details is one of its most immediate and impactful applications.

- Efficiency Gains:
  AI tools can extract relevant data from electronic medical records (EMRs) and denial notices to automatically generate a well-structured, evidence-based draft appeal letter. This significantly reduces the time required for manual drafting and allows physician advisors to focus on refining clinical arguments.
- Personalization:
  While automated, these letters are not generic. AI systems can tailor content to specific cases by incorporating clinical details, payer policies, and patient history.
- Example Use Case:
  Input: Clinical data indicating pneumonia requiring oxygen therapy, payer denial citing observation care as sufficient.
  Output: A draft appeal letter that includes a summary of the patient's condition, a rebuttal of the payer's rationale, and references to clinical guidelines.
- Impact:
  Automated drafting has been shown to cut the time required for appeal preparation by up to 50%, enabling quicker submissions and more timely resolutions.

---

2. Consistency

One of AI's greatest strengths lies in ensuring uniformity across appeal letters while preserving the flexibility needed to address individual cases.

- Standardized Formats:
  AI tools can ensure that all appeal letters follow a standardized structure, including critical components such as the clinical summary, rebuttal of the denial rationale, and references to payer criteria or clinical guidelines. This eliminates variability and enhances the professionalism of submissions.

- Minimized Oversights:
  By cross-referencing key data points against established templates and checklists, AI reduces the risk of omitting essential details, such as regulatory citations or payer-specific criteria.
- Reinforcement of Best Practices:
  AI systems can embed institutional best practices into the drafting process, ensuring that every letter reflects the organization's standards for quality and compliance.
- Example Insight:
  "An analysis of past denials revealed that letters missing references to oxygen saturation data were 30% less likely to succeed. AI systems now automatically include these references when relevant."

---

3. Data-Driven Insights

AI's ability to analyze vast amounts of historical data offers unparalleled opportunities to refine denial management strategies and tailor appeal letters for maximum impact.[62]

- Pattern Recognition:
  AI systems can analyze past denials to identify recurring payer objections, such as common reasons for rejecting inpatient admissions for certain conditions. These insights enable physician advisors to anticipate objections and address them preemptively in appeal letters.
- Predictive Modeling:
  Advanced AI tools can predict the likelihood of success for specific denial appeals based on historical trends, guiding advisors to prioritize high-potential cases.
- Tailored Argumentation:
  By identifying payer-specific tendencies, AI can customize appeal arguments to align with the preferences or requirements of individual insurers.
- Example Practice:
  "AI analytics revealed that a particular payer frequently denies inpatient claims for heart failure without explicit documentation of diuretic therapy response. As a result, appeal letters now routinely include this data for relevant cases."
- Continuous Improvement:
  As AI systems learn from successful and unsuccessful appeals, they refine their algorithms, improving the quality and effectiveness of future drafts.

Emerging AI Capabilities in Appeal Letter Drafting

1. Natural Language Processing (NLP)

Advances in NLP allow AI systems to extract and interpret unstructured data from clinical notes, denial letters, and payer policies.[63]

- Practical Applications:
    - Automatically identifying key phrases in a denial rationale and linking them to corresponding clinical data.
    - Generating concise summaries of complex medical histories for inclusion in appeal letters.
- Example Insight:
  "NLP-enabled AI identified a payer's objection to the absence of documented discharge barriers for a length-of-stay denial and included a targeted rebuttal in the appeal letter."

---

2. Real-Time Feedback

AI systems can provide real-time feedback during the drafting process, alerting physician advisors to potential weaknesses in their arguments or missing documentation.

- Practical Applications:
    - Highlighting sections where additional clinical evidence or guideline references are needed.
    - Suggesting alternative phrasing to strengthen the tone or clarity of the argument.
- Example Practice:
  "During drafting, the AI system flagged that the appeal lacked documentation of failed outpatient treatment, prompting the advisor to include this detail before submission."

---

3. Continuous Learning and Adaptation

AI tools evolve through machine learning, becoming increasingly effective over time as they analyze more data and incorporate feedback.

- Future Potential:
    - AI systems could autonomously refine templates and workflows based on the outcomes of past appeals.
    - Continuous learning ensures that AI tools remain aligned with changing payer policies and clinical guidelines.
- Example Evolution:
  "Over time, AI systems adapted to new payer criteria for COVID-19 cases, ensuring appeal letters consistently addressed updated requirements for oxygen therapy documentation."

Challenges and Considerations in AI Integration

1. Data Quality:
   The effectiveness of AI depends on the quality and completeness of the clinical documentation it analyzes. Ensuring accurate and comprehensive records is critical for maximizing AI's potential.
2. Human Oversight:
   While AI can automate and enhance many aspects of appeal letter drafting, physician advisors remain essential for interpreting clinical nuances and ensuring the tone aligns with organizational values.
3. Integration with Existing Systems:
   AI tools must seamlessly integrate with EMRs, payer portals, and other denial management software to avoid workflow disruptions.

The future of AI in appeal letter drafting holds immense promise for transforming denial management. By automating the drafting process, ensuring consistency, and leveraging data-driven insights, AI empowers physician advisors to focus on their clinical expertise while reducing administrative burdens. As technology continues to evolve, AI will play an increasingly central role in enhancing the efficiency, precision, and success of denial appeals, ensuring that healthcare organizations can better advocate for patients and secure appropriate reimbursements. However, the human element—judgment, empathy, and strategic thinking—will remain indispensable, ensuring that AI is used as a tool to amplify, not replace, the expertise of physician advisors.

Conclusion: The Comprehensive Approach to Responding to Upheld Denials

The process of responding to upheld denials, while challenging, represents a critical juncture in the advocacy efforts of physician advisors. This chapter has explored the pivotal role that physician advisors play in crafting effective appeal letters, underscored the importance of timeliness, and highlighted the transformative potential of artificial intelligence (AI) in denial management. By embracing a structured and strategic approach, physician advisors can turn setbacks into opportunities for success, securing appropriate care for patients while reinforcing their organization's financial and operational stability.

Key Takeaways

1. The Physician Advisor's Role: Clinical Expertise and Advocacy

Physician advisors bring unparalleled clinical expertise and continuity to the appeal process, positioning them as uniquely capable advocates. Their direct involvement ensures that appeal letters are:

- Evidence-Based: Drawing directly from patient records and payer criteria.
- Strategically Consistent: Seamlessly transitioning from the peer-to-peer (P2P) conference to the appeal process.
- Timely: Leveraging their firsthand knowledge to act promptly and preserve momentum.

2. Crafting Persuasive Appeal Letters

A successful appeal letter is a meticulous balance of professionalism, clarity, and robust evidence. By following key steps—assertive introductions, concise clinical summaries, targeted rebuttals, and guideline citations—physician advisors can craft arguments that directly address payer objections while aligning with established clinical standards.

- Professional Tone: Demonstrating respect while advocating firmly for patient care.
- Clinical Precision: Summarizing patient conditions in ways that underscore the necessity of care.
- Guideline Alignment: Referencing payer criteria and specialty guidelines to strengthen the argument.

3. Timely Appeals: A Cornerstone of Effective Advocacy

Acting quickly after a denial is essential to maintaining momentum, reducing revenue delays, and demonstrating organizational professionalism. By establishing workflows, leveraging Prep Sheets, and delegating administrative tasks, physician advisors can ensure appeals are submitted promptly without compromising quality.

- Benefits of Timeliness: Faster resolution, preserved case details, and improved payer relationships.
- Strategies for Efficiency: Standardized protocols, technology integration, and team training.

4. Measuring Success Through Metrics

The success of appeal letters is best evaluated through metrics that provide actionable insights:

- Appeal Success Rate: Reflects the effectiveness of arguments and alignment with payer expectations.
- Financial Impact: Demonstrates the tangible value of denial management efforts.
- Time to Resolution: Highlights opportunities to improve efficiency and reduce bottlenecks.

Metrics are not only a measure of current performance but also a guide for future improvements, informing strategies and resource allocation.

5. AI: A Game-Changer in Denial Management

Artificial intelligence is poised to revolutionize the appeal process by automating letter drafting, ensuring consistency, and leveraging data-driven insights. While AI enhances efficiency and precision, the human element—judgment, empathy, and clinical expertise—remains indispensable.

- Capabilities of AI: Automated drafting, real-time feedback, and predictive modeling.
- Future Potential: Continuous learning and adaptation to evolving payer policies.

Broader Implications

The strategies and tools outlined in this chapter extend beyond individual appeal letters, reflecting broader themes of accountability, innovation, and patient-centered care.

1. Advocacy Beyond Appeals

By addressing payer objections with evidence-based arguments, physician advisors not only advocate for patients but also reinforce the importance of compliance with clinical and regulatory standards. Their efforts contribute to a healthcare system that prioritizes medically necessary care and financial fairness.

2. Advancing the Profession

Physician advisors play a pivotal role in evolving the field of denial management. By embracing continuous learning, sharing best practices, and leveraging innovations like AI, they set the standard for excellence and adaptability in an increasingly complex healthcare landscape.

3. Enhancing Organizational Outcomes

Effective denial management has a profound impact on healthcare organizations, improving financial stability, operational efficiency, and patient satisfaction. Physician advisors' contributions are integral to achieving these outcomes, ensuring that denial management efforts align with institutional goals.

Closing Thoughts

Responding to upheld denials is not merely an administrative task; it is a strategic opportunity to advocate for patients, uphold clinical standards, and drive systemic improvements. By combining their clinical expertise with structured workflows and advanced technologies, physician advisors can transform the appeal process into a powerful tool for achieving better outcomes.

Through timely action, data-driven strategies, and a commitment to innovation, physician advisors reinforce their critical role as advocates and leaders in denial management. The future of healthcare will undoubtedly present new challenges, but by staying at the forefront of best practices and emerging technologies, physician advisors will continue to excel in their mission to ensure that every patient receives the care they need and deserve.

References

[58] Akosa AN. Precertification, denials and appeals: reducing the hassles. Fam Pract Manag. 2006 Jun;13(6):45-8. PMID: 16813121.

[59] Baumel, Mark J., and Robert R. Corrato. "Dedicating the right resources to denials management: full-time physician attention helped reduce one health system's denied claims and length of stay." Healthcare Financial Management, vol. 57, no. 12, Dec. 2003, pp. 66+. Gale Academic OneFile, link.gale.com/apps/doc/A111696061/AONE?u=anon~f1a2f011&sid=googleScholar&xid=bbb04065. Accessed 19 Feb. 2025.

[60] Roche, John D. Claim Denied!: How to Appeal a VA Denial of Benefits. Potomac Books, Inc., 2008.

[61] Kendall J. Kiser et al., Large Language Models to Help Appeal Denied Radiotherapy Services. JCO Clin Cancer Inform 8, e2400129(2024).
DOI:10.1200/CCI.24.00129

[62] Cohen, Fred. "AI-Powered Patient Advocacy: Amplifying Health Care Voices." Neurology Live, 5 Nov. 2023. Gale OneFile: Health and Medicine, link.gale.com/apps/doc/A777222379/HRCA?u=anon~3b2556fe&sid=googleScholar&xid=09eaa396. Accessed 19 Feb. 2025.

[63] Michael R. Waters, Sanjay Aneja, Julian C. Hong, Unlocking the Power of ChatGPT, Artificial Intelligence, and Large Language Models: Practical Suggestions for Radiation Oncologists, Practical Radiation Oncology, Volume 13, Issue 6, 2023, Pages e484-e490, ISSN 1879-8500, https://doi.org/10.1016/j.prro.2023.06.011.

CHAPTER 11

The Future of the Physician Advisor Role

Introduction

The physician advisor role is undergoing rapid transformation, mirroring the complexities, challenges, and demands of modern healthcare. Physician advisors serve as pivotal figures in ensuring that hospitals meet regulatory standards, optimize resource utilization, and secure appropriate reimbursements. However, this role has expanded far beyond traditional utilization management, as physician advisors are increasingly called upon to lead initiatives in clinical quality improvement, denial prevention, and strategic planning.

Several forces are driving this transformation. The rise of payer scrutiny, fueled by advanced analytics and artificial intelligence, has heightened the need for robust documentation and precise advocacy. Simultaneously, the shift from fee-for-service to value-based care has redefined how success is measured, emphasizing patient outcomes and cost-efficiency over volume. In this context, physician advisors are tasked not only with managing immediate operational challenges but also with influencing the long-term direction of healthcare delivery.

At the same time, technological advancements, particularly in artificial intelligence (AI) and machine learning, are creating unprecedented opportunities to streamline workflows, enhance decision-making, and improve outcomes. Tools that automate repetitive tasks, predict denial trends, and analyze complex data are enabling physician advisors to focus on strategic and clinical priorities. However, these innovations also bring challenges, such as ensuring data quality, maintaining ethical oversight, and integrating technology seamlessly into established workflows.

As these dynamics evolve, the need for standardized training and professional development for physician advisors has never been more critical. Without a unified framework of core competencies and recognized certifications, the profession risks variability in practice standards and a lack of recognition for its strategic importance within healthcare systems.

This chapter delves into the emerging trends, challenges, and opportunities shaping the future of the physician advisor profession. It highlights the transformative potential of artificial intelligence, examines the shifting landscape of reimbursement models, and underscores the importance of establishing standardized training and mentorship programs. By embracing these developments and positioning themselves as leaders, physician advisors can ensure their role remains indispensable in navigating the complexities of healthcare, advocating for patients, and driving organizational success.

Emerging Trends and Challenges

The physician advisor role is evolving rapidly, shaped by a dynamic healthcare landscape that demands adaptability, innovation, and a proactive approach to problem-solving. Emerging trends and challenges are redefining responsibilities and requiring physician advisors to expand their skill sets while addressing systemic issues with creativity and resilience. Below are the key areas of transformation and their implications for the profession:

---

1. Increased Scrutiny from Payers

Insurance companies are leveraging advanced analytics, machine learning, and automated systems to review claims with unprecedented speed and scale.[64] While these technologies enhance payer efficiency, they often fail to account for clinical subtleties, leading to inappropriate denials that demand robust counterarguments. The increased reliance on algorithm-driven decisions intensifies the challenges faced by physician advisors in advocating for appropriate patient care.

Key Challenges:

- Automated systems prioritize cost containment over individualized clinical needs, increasing the likelihood of denials for complex cases.
- Payers frequently revise criteria and policies, requiring constant vigilance from physician advisors to ensure alignment and compliance.

Strategic Response:

- Proactive Education: Physician advisors must lead efforts to educate clinical teams on payer-specific criteria and denial trends, ensuring documentation anticipates potential objections.
- Enhanced Documentation Practices: Emphasize precise, comprehensive, and timely documentation that aligns with payer expectations while highlighting clinical nuances.
- Collaborative Interventions: Partner with revenue cycle teams and case managers to address systemic gaps in the documentation and denial management processes.

Example Initiative: A physician advisor team implemented monthly training sessions for clinical staff on common payer criteria for inpatient admissions, reducing avoidable denials by 20%.

---

2. Shifting Payment Models

The transition from fee-for-service to value-based care represents a paradigm shift in healthcare reimbursement. New models such as bundled payments and shared savings programs emphasize outcomes, quality, and cost-efficiency, challenging physician advisors to navigate these frameworks while advocating for fair reimbursement.[65]

Key Challenges:

- Balancing cost-containment initiatives with the imperative to deliver high-quality, patient-centered care.
- Integrating quality metrics and financial accountability into everyday workflows.

Strategic Focus:

- Alignment with Quality Benchmarks: Physician advisors must guide care decisions to align with key performance metrics such as readmission rates, length of stay, and patient satisfaction.
- Advocacy for Reimbursement: Advocate for fair reimbursements by ensuring documentation reflects the complexity and necessity of care provided within value-based frameworks.
- Resource Optimization: Collaborate with interdisciplinary teams to streamline care delivery and reduce unnecessary resource utilization.

Example Contribution: In response to a bundled payment initiative for joint replacement surgeries, physician advisors developed standardized care pathways, achieving a 15% reduction in costs while maintaining high patient satisfaction scores.

---

3. Expanding Scope of Responsibilities

The physician advisor role has grown significantly beyond its traditional focus on denial management. Today, physician advisors are integral to a wide array of hospital initiatives, including clinical quality improvement, population health management, and cost containment.[66]

Key Contributions:

- Driving Quality Improvement: Physician advisors identify inefficiencies in care delivery, implement systemic enhancements, and monitor outcomes to ensure continuous improvement.
- Leading Population Health Efforts: Collaborate with care teams to reduce readmissions, manage chronic diseases, and address social determinants of health, improving outcomes across patient populations.

- Cost-Containment Strategies: Evaluate and refine resource utilization to achieve financial sustainability without compromising care quality.

Example in Action: A physician advisor spearheaded a task force to address avoidable readmissions for congestive heart failure, resulting in a 12% reduction within one year by implementing post-discharge follow-up protocols and patient education programs.

---

4. Regulatory Changes

Frequent updates to payer policies, federal regulations, and compliance requirements create a complex and ever-changing environment for physician advisors. These changes demand agility and a thorough understanding of both clinical and administrative impacts.

Key Challenges:

- Staying updated on evolving Medicare, Medicaid, and commercial payer regulations.
- Navigating the implications of policies such as the Medicare Two-Midnight Rule and prior authorization mandates.

Strategic Adaptation:

- Dedicated Regulatory Teams: Establish teams or appoint liaisons to monitor regulatory changes, interpret implications, and disseminate updates to clinical and administrative staff.
- Ongoing Education: Incorporate regulatory training into physician advisor development programs to ensure compliance and mitigate risks.
- Policy Advocacy: Engage with hospital leadership and professional organizations to advocate for transparent and equitable regulatory policies.

Example Practice: A hospital implemented a quarterly compliance review process led by physician advisors, identifying regulatory gaps and reducing noncompliance penalties by 30%.

---

5. Addressing Physician Burnout

The expanding scope of responsibilities, coupled with administrative demands and the high-stakes nature of the role, increases the risk of burnout among physician advisors. Ensuring workforce sustainability requires a multifaceted approach that prioritizes well-being and efficiency.

Key Challenges:

- Balancing clinical, administrative, and strategic responsibilities without overextending resources.
- Navigating the emotional toll of frequent payer denials and complex appeals.

Solutions:

- Streamlined Workflows: Implement advanced technologies, such as AI-powered tools, to automate routine tasks and reduce administrative burdens.
- Resilience and Peer Support: Foster collaborative networks and mentorship programs that promote professional growth and emotional well-being.
- Flexible Scheduling: Advocate for workload management practices that ensure adequate time for recovery and professional development.

Example Initiative: A hospital introduced a wellness program for physician advisors, including peer support groups and access to resilience training, which reduced self-reported burnout levels by 25%.

Opportunities Amid Challenges

While these trends present significant challenges, they also open doors for physician advisors to expand their influence and impact. By embracing innovation, advocating for systemic improvements, and fostering interdisciplinary collaboration, physician advisors can position themselves as indispensable leaders in healthcare transformation.

Key Opportunities:

- Advancing Innovation: Physician advisors can lead efforts to integrate AI and data-driven insights into denial prevention and quality improvement.
- Shaping Policy: Increased involvement in policy advocacy offers opportunities to standardize training and clarify payer criteria.
- Strengthening Collaboration: Building interdisciplinary teams allows for a holistic approach to addressing complex healthcare challenges, improving outcomes for patients and organizations alike.

By navigating these emerging trends and challenges with strategic foresight and adaptability, physician advisors can shape the future of their profession while ensuring sustainable and effective healthcare delivery.

Integrating AI and Technology in Utilization Management

The integration of artificial intelligence (AI) into utilization management represents a paradigm shift in how physician advisors approach denial prevention, clinical documentation, and peer-to-peer (P2P) conferences. By automating repetitive tasks, analyzing complex datasets, and providing actionable insights, AI enhances efficiency and allows physician advisors to focus on higher-level decision-making and advocacy. However, its implementation requires a strategic approach to overcome challenges and maximize its potential.

---

1. AI as a Partner in Utilization Management

AI has become an indispensable tool for physician advisors, revolutionizing their workflows by automating routine administrative tasks and providing insights that support proactive decision-making.[67] This partnership enables physician advisors to prioritize complex cases and dedicate more time to activities that require clinical expertise and strategic judgment.

How AI Supports Utilization Management:

- Automation of Routine Tasks: AI systems can handle time-consuming administrative processes, such as initial documentation review, eligibility verification, and denial categorization. By reducing the manual workload, physician advisors can redirect their efforts toward patient care and advocacy.
- Trend Identification and Analysis: AI tools analyze historical data to uncover patterns in payer behavior, such as common denial reasons or high-risk claims, allowing physician advisors to anticipate and address potential issues before submission.
- Real-Time Alerts: AI-powered platforms send notifications about incomplete or non-compliant documentation, ensuring claims are accurate and meet payer criteria.

Examples of AI Impact:

- Automated Documentation Review: By analyzing patient records for completeness and accuracy, AI reduces the time spent on routine documentation tasks. For instance, it can ensure that diagnostic results, medication administration, and clinical trends are appropriately documented before claims are submitted.
- Predictive Analytics: AI models predict which cases are most likely to face denials based on historical data, enabling physician advisors to take proactive measures, such as refining documentation or obtaining additional pre-authorization.

Result:

Streamlined processes and proactive interventions improve efficiency and reduce the likelihood of denials, allowing physician advisors to focus on complex and high-value cases.

---

2. Enhancing Documentation Accuracy

Accurate and comprehensive clinical documentation is critical for avoiding denials and ensuring timely reimbursements. AI tools assist physician advisors and clinical teams by identifying gaps in documentation and providing actionable recommendations for improvement.[68]

Key Capabilities of AI in Documentation:

- Gap Identification: AI systems analyze clinical records to detect missing or inconsistent information, such as omitted lab trends, incomplete vital signs, or absent treatment details.
- Compliance Checks: AI ensures that documentation aligns with payer-specific criteria, clinical guidelines, and regulatory requirements.
- Real-Time Feedback: AI-powered tools provide instant alerts to clinicians about missing documentation during the charting process, reducing errors and enhancing record accuracy.

Example Use Case:

An AI system reviewing pneumonia cases identifies that 20% of claims lack documented oxygen saturation trends, a key criterion for inpatient admission. By flagging this gap before submission, the system prompts clinicians to include the necessary data, reducing the risk of denial.

Result:

Enhanced documentation accuracy increases the defensibility of claims, reduces the administrative burden of resubmissions, and improves overall efficiency in denial management workflows.

---

3. Streamlining Peer-to-Peer Conferences

P2P conferences require careful preparation, as physician advisors must effectively advocate for appropriate care while addressing payer objections. AI tools simplify this process by organizing and summarizing relevant patient data, enabling physician advisors to present stronger arguments in less time.

AI Contributions to P2P Preparation:

- Structured Summaries: AI consolidates clinical details, denial rationales, and payer-specific criteria into a comprehensive, easy-to-read Prep Sheet for use during P2P conferences.
- Customizable Arguments: AI platforms can suggest tailored rebuttals based on payer policies and the patient's clinical scenario, ensuring physician advisors are well-prepared to address objections.

- Data Visualization: AI systems generate charts and graphs to illustrate clinical trends or demonstrate alignment with guidelines, adding clarity and impact to arguments.

Example Result:

AI-generated Prep Sheets save physician advisors an average of 30 minutes per case, allowing them to allocate more time to crafting persuasive arguments and focusing on patient advocacy.

Outcome:

Streamlined preparation processes enhance the effectiveness of P2P conferences, leading to higher rates of overturned denials and improved payer relationships.

---

4. Challenges of AI Integration

While AI offers transformative potential, its adoption in utilization management comes with challenges that require careful planning, robust oversight, and ongoing support. These challenges include data quality, ethical considerations, and the need for human oversight.[69]

Key Challenges:

1. Dependence on Data Quality: AI systems rely on accurate and comprehensive clinical data to deliver reliable results. Incomplete or inconsistent documentation can undermine AI's effectiveness.
2. Ethical Considerations: AI-generated insights must be used responsibly to ensure equitable decision-making and protect patient privacy. Bias in data or algorithms can lead to unintended disparities.
3. Maintaining Human Oversight: AI tools are not infallible and cannot replace the clinical expertise and judgment of physician advisors. Ensuring that AI recommendations align with clinical nuances and organizational goals is essential.

Mitigation Strategies:

1. Training for Data Quality: Equip clinical teams with the knowledge and tools needed to maintain high-quality documentation, ensuring that AI systems have accurate data to work with.
2. Robust Governance Policies: Implement clear protocols for data security, algorithm transparency, and ethical AI use to address privacy concerns and mitigate bias.
3. Position AI as a Supportive Resource: Use AI as a complementary tool to enhance, rather than replace, the expertise of physician advisors. Encourage collaborative workflows where AI insights inform, but do not dictate, decisions.

The Path Forward: Realizing AI's Full Potential

AI's capabilities in utilization management are continually expanding, promising even greater efficiency and effectiveness in the future. Advancements such as natural language processing (NLP), deep learning algorithms, and real-time decision support systems will further enhance the physician advisor's role.[70]

Future Innovations:

- NLP Integration: Extract actionable insights from unstructured clinical notes, such as discharge summaries or consultation reports.
- Deep Learning Algorithms: Improve predictive accuracy for complex denial patterns, enabling more targeted interventions.
- Real-Time Decision Support: Provide AI-driven recommendations during active P2P conferences, enhancing physician advisors' ability to advocate effectively.

By embracing these advancements and addressing integration challenges proactively, physician advisors can harness AI's full potential to streamline processes, improve outcomes, and enhance their impact on healthcare operations.

Advocacy for Standardizing Physician Advisor Training

As the physician advisor role continues to evolve and expand, standardized training and professional development have become essential for maintaining consistency, credibility, and excellence in this critical field. Ensuring that all physician advisors are equipped with the necessary skills and resources not only enhances their effectiveness but also solidifies their role as indispensable leaders in healthcare operations. This advocacy focuses on defining competencies, establishing certification pathways, fostering mentorship, and elevating the profession through role recognition.

---

1. Establishing Core Competencies

The foundation of standardized training lies in defining core competencies that every physician advisor should master. These competencies ensure uniformity in practice and equip advisors to address the complexities of utilization management, denial prevention, and clinical documentation integrity (CDI). By aligning training programs with these competencies, healthcare organizations can create a cohesive and effective physician advisor workforce.

Key Competencies:

1. Utilization Management:
    - Mastery of payer-specific criteria, denial trends, and appeals processes.
    - Strategies for effective peer-to-peer (P2P) conferences and advocacy.
2. Clinical Documentation Integrity (CDI):
    - Techniques for ensuring accurate and comprehensive documentation that reflects patient acuity and justifies care decisions.
    - Collaboration with clinical teams to address documentation gaps and enhance compliance.
3. Regulatory Compliance:
    - Knowledge of federal and state regulations, such as Medicare's Two-Midnight Rule, Medicaid policies, and commercial payer requirements.
    - Ability to apply evolving regulations to hospital workflows and ensure organizational compliance.

Impact:

Establishing these core competencies ensures that physician advisors are consistently prepared to navigate complex healthcare challenges, advocate effectively for appropriate care, and mitigate compliance risks.

---

2. Certification and Professional Development

Certification programs play a vital role in validating the expertise of physician advisors and enhancing their professional credibility. By obtaining certification, physician advisors demonstrate their commitment to excellence and their proficiency in navigating the multifaceted responsibilities of their role.

Benefits of Certification:

- Enhanced Credibility: Certification reinforces the physician advisor's authority in denial management, utilization review, and clinical documentation.
- Increased Confidence**:** Certification programs provide comprehensive training that empowers advisors to handle complex cases with assurance.
- Professional Growth: Ongoing education as part of the certification process ensures that physician advisors remain updated on industry best practices and regulatory changes.

Notable Certification Programs:

Organizations like the American College of Physician Advisors (ACPA) and the American Board of Quality Assurance and Utilization Review Physicians (ABQAURP) offer certification pathways that encompass core competencies and provide a structured framework for professional development.

Long-Term Benefits:

- Strengthened trust with clinical teams, hospital leadership, and payers.
- Enhanced standing of the physician advisor profession within the broader healthcare system.

---

3. Mentorship and Collaboration

Mentorship is a cornerstone of professional growth and an essential component of standardized training. By pairing new physician advisors with experienced mentors, healthcare organizations can accelerate skill development and foster a collaborative culture that prioritizes continuous improvement.

Benefits of Mentorship:

1. Accelerated Learning:
    - Mentors provide real-time guidance on complex cases, P2P conferences, and appeals, reducing the learning curve for new advisors.
2. Knowledge Sharing:
    - Mentors share institutional knowledge, best practices, and insights from their own experiences, enriching the mentee's understanding of the role.
3. Fostering Collaboration:
    - Mentorship programs encourage teamwork and open communication, creating a supportive environment for addressing challenges and sharing successes.

Examples of Mentorship in Action:

- Monthly Case Reviews:
    - Senior physician advisors lead team discussions on challenging cases, highlighting successful strategies and lessons learned.
- P2P Conference Coaching:
    - Mentors observe and provide feedback during live P2P conferences, offering constructive insights to improve performance.

Outcome:

Mentorship not only accelerates skill acquisition but also strengthens the overall performance of the physician advisor team, contributing to better denial management outcomes and enhanced organizational efficiency.

---

4. Role Recognition and Advocacy

As the physician advisor's responsibilities continue to grow, it is essential to advocate for the recognition of their role as a critical component of healthcare operations. By engaging with professional organizations, contributing to policy discussions, and sharing thought leadership, physician advisors can elevate their profession and influence systemic improvements.

Strategies for Advocacy:

1. Engagement with Professional Organizations:
   - Join groups such as the American College of Physician Advisors (ACPA) to network, share best practices, and promote standardization efforts.
2. Policy Advocacy:
   - Collaborate with hospital leadership and industry stakeholders to influence payer policies, advocate for clearer denial criteria, and support legislative efforts that impact denial management.
3. Thought Leadership:
   - Publish articles, white papers, and case studies that highlight the value of the physician advisor role and share innovative strategies for improving outcomes.

Examples of Advocacy in Action:

- Engaging with Hospital Leadership:
  - Presenting data-driven insights on the financial and operational impact of denial management efforts.
- Contributing to Industry Publications:
  - Sharing successful strategies for overturning denials and improving clinical documentation through widely read journals and platforms.

Result:

By championing their role and advocating for systemic improvements, physician advisors not only enhance their own professional standing but also contribute to the advancement of healthcare operations and patient care.

Standardizing physician advisor training is essential for maintaining consistency, enhancing professional credibility, and preparing for the complexities of modern healthcare. By defining core competencies, supporting certification programs, fostering mentorship, and advocating for role recognition, physician advisors can secure their place as leaders in healthcare innovation and operational excellence. These efforts will not only elevate the profession but also ensure that physician advisors continue to drive positive change in patient care and organizational success.

Conclusion

The physician advisor role stands at a crossroads, where rapid advancements in healthcare delivery intersect with growing challenges in payer relationships, regulatory compliance, and operational efficiency. As this chapter has explored, the profession is undergoing a transformation fueled by the rise of artificial intelligence (AI), shifting reimbursement models, and the need for standardized training. These changes, while daunting, also present extraordinary opportunities for physician advisors to expand their influence, improve healthcare outcomes, and drive systemic improvements.

Embracing Technology and Innovation

Technology, particularly AI, has become a critical ally in managing the increasing complexity of the physician advisor's responsibilities. By automating routine tasks, identifying denial trends, and providing actionable insights, AI enables physician advisors to focus their expertise where it is needed most—on patient advocacy and high-level strategic decision-making.

Key Impacts of AI and Technology:

- Efficiency Gains: Automating documentation review, predictive analytics, and preparation for peer-to-peer (P2P) conferences saves time and streamlines workflows.
- Proactive Denial Prevention: Predictive models enable physician advisors to anticipate and address potential denials before they occur.
- Enhanced Decision-Making: AI-powered dashboards provide real-time insights into denial trends, payer behaviors, and systemic bottlenecks.

Looking Ahead:

The next wave of technological advancements, including natural language processing (NLP) and real-time decision support tools, will further empower physician advisors. To fully realize AI's potential, healthcare organizations must address challenges such as data quality, ethical governance, and the integration of AI into established workflows.

Navigating Value-Based Care Models

The shift from fee-for-service to value-based care has redefined success in healthcare, emphasizing quality and efficiency over volume. This transformation places physician advisors at the forefront of aligning care delivery with evolving payment models while ensuring patients receive appropriate and timely care.

Key Strategies for Physician Advisors:

- Aligning with Quality Benchmarks: Physician advisors must guide care decisions to meet performance metrics such as readmission rates, patient satisfaction, and length of stay.
- Advocating for Fair Reimbursement: Comprehensive and precise documentation ensures that healthcare providers are appropriately compensated within value-based frameworks.
- Optimizing Resource Utilization: Collaboration with interdisciplinary teams can identify inefficiencies and reduce unnecessary costs.

Future Vision:

As value-based care becomes the standard, physician advisors will play a central role in developing care pathways, monitoring outcomes, and advocating for sustainable financial models that prioritize patient well-being.

Expanding Scope of Responsibilities

The evolving healthcare landscape has broadened the physician advisor's role beyond traditional utilization management. Physician advisors are now integral to initiatives in clinical quality improvement, population health management, and cost containment.

Expanded Contributions:

- Clinical Quality Improvement: Physician advisors identify inefficiencies, implement systemic changes, and monitor outcomes to ensure continuous improvement.
- Population Health Management: Reducing readmissions, managing chronic diseases, and addressing social determinants of health are increasingly within the purview of physician advisors.
- Cost Containment: By evaluating resource utilization and optimizing workflows, physician advisors balance financial sustainability with quality care delivery.

Leadership Opportunity:

This expanded scope positions physician advisors as leaders in bridging clinical care and operational strategy, ensuring that healthcare organizations remain adaptive and patient-focused.

Advocating for Professional Standards

Standardized training and certification are critical for ensuring consistency, credibility, and preparedness across the physician advisor profession. Without a unified framework of core competencies, the profession risks variability in practice and diminished recognition for its strategic importance.

Key Areas for Standardization:

- Core Competencies: Establishing mastery in utilization management, clinical documentation integrity (CDI), and regulatory compliance provides a strong foundation for the profession.
- Certification Pathways: Programs like those offered by the American College of Physician Advisors (ACPA) nd the American Board of Quality Assurance and Utilization Review Physicians (ABQAURP) validate expertise and enhance professional standing.
- Mentorship and Collaboration: Experienced physician advisors play a critical role in guiding new professionals, fostering continuous improvement and knowledge-sharing.

Advocacy for Role Recognition:

Physician advisors must engage in policy discussions, contribute to thought leadership, and advocate for the recognition of their role as essential to healthcare operations. By championing these efforts, the profession can secure its place as a vital driver of healthcare innovation and excellence.

Addressing Workforce Challenges

The expanding scope and increasing complexity of the physician advisor role have heightened the risk of burnout. Sustaining a skilled and resilient workforce requires healthcare organizations to prioritize well-being initiatives and streamline administrative demands.

Solutions for Workforce Sustainability:

- Streamlined Workflows: Advanced technologies, such as AI, can reduce administrative burdens and allow physician advisors to focus on high-value activities.
- Collaborative Support Networks: Peer mentorship, case reviews, and wellness programs foster resilience and professional growth.
- Flexible Scheduling: Workload management practices that balance professional demands with personal recovery are essential for long-term retention.

Call to Action:

Healthcare organizations must invest in the tools, resources, and support systems needed to sustain their physician advisor teams, ensuring continued excellence in patient advocacy and operational strategy.

Final Reflections

The physician advisor role is uniquely positioned to shape the future of healthcare, sitting at the nexus of clinical expertise and operational strategy. As this chapter has outlined, the profession is poised to navigate significant challenges while embracing transformative opportunities. By leveraging technology, adapting to value-based care, expanding responsibilities, and advocating for professional standards, physician advisors can elevate their impact and redefine their role in healthcare systems.

Looking Forward:

- Leadership in Innovation: Physician advisors must lead the integration of AI and advanced analytics into denial prevention, quality improvement, and strategic planning.
- Commitment to Excellence: Standardized training, mentorship, and certification are essential for ensuring consistency and preparing physician advisors for future challenges.
- Patient-Centered Advocacy: At its core, the physician advisor role remains rooted in advocating for appropriate care, ensuring that patients receive the services they need while supporting organizational sustainability.

Closing Vision:

The physician advisor of the future will be a catalyst for systemic change, driving improvements in patient outcomes, operational efficiency, and healthcare equity. By championing innovation, collaboration, and continuous improvement, physician advisors will ensure that healthcare systems remain adaptable, effective, and centered on the needs of patients. Their role will not only endure but thrive, defining the future of modern healthcare.

References

[64] Kanagarla, Krishna Prasanth Brahmaji, Improved Claims Processing: Advanced Analytics for Faster and More Accurate Claims Handling In Healthcare or Insurance (January 16, 2025). Available at SSRN: https://ssrn.com/abstract=5099728 or http://dx.doi.org/10.2139/ssrn.5099728

[65] Gupta, Reshma MD, MSHPM; Moriates, Christopher MD. Swimming Upstream: Creating a Culture of High-Value Care. Academic Medicine 92(5):p 598-601, May 2017. | DOI: 10.1097/ACM.0000000000001485

[66] Zander, Karen. "Getting more from physician advisors: building a formal job description to improve effectiveness." Case Management Monthly, vol. 4, no. 12, Dec. 2007, pp. 4+. Gale OneFile: Health and Medicine, link.gale.com/apps/doc/A171623700/HRCA?u=anon~e1bddae7&sid=googleScholar&xid=df107ebd. Accessed 19 Feb. 2025.

[67] Abidi, S.S.R. et al. (2019). AI-Driven Pathology Laboratory Utilization Management via Data- and Knowledge-Based Analytics. In: Riaño, D., Wilk, S., ten Teije, A. (eds) Artificial Intelligence in Medicine. AIME 2019. Lecture Notes in Computer Science(), vol 11526. Springer, Cham. https://doi.org/10.1007/978-3-030-21642-9_30

[68] Lin, Steven Y. et al., Reimagining Clinical Documentation With Artificial Intelligence. Mayo Clinic Proceedings, Volume 93, Issue 5, 563 - 565

[69] Saxena, R., Katage, G., Kumar, C., Pathan, N. M., & Bargir, M. N. (2024). AI redefining healthcare documentation for tomorrow: Exploring the impact of AI on healthcare documentation. In Computational Convergence and Interoperability in Electronic Health Records (EHR) (pp. 51-66). IGI Global.

[70] Le, Q. (2024). AI-driven Clinical Documentation Integrity for Electronic Health Records: Implements AI-driven solutions for clinical documentation integrity in electronic health records, enhancing accuracy and completeness of clinical documentation for billing and q. Journal of Artificial Intelligence Research and Applications, 4(1), 170-181.

APPENDICES

Appendix A: Templates for Preparing for P2P Conferences

1. P2P Conference Preparation Checklist

Use this checklist to ensure all key elements are in place before a P2P conference. The tasks below are designed to provide a structured and organized approach to maximize success during the discussion.

Task	Details	Completed
Review denial letter	Identify specific reasons for denial, including cited payer criteria and potential misinterpretations.	[ ]
Gather clinical documentation	Collect relevant medical records, including diagnostic findings, vital sign trends, treatment history, medication administration records, and any consultations.	[ ]
Analyze clinical relevance	Highlight key portions of the documentation that directly support medical necessity and align with payer criteria.	[ ]
Prepare a Prep Sheet	Create a concise document summarizing the patient's clinical story, focusing on key points aligned with payer guidelines.	[ ]
Research payer guidelines	Review specific InterQual, MCG, or proprietary payer policies. Cross-reference these with documented patient data.	[ ]
Anticipate objections	Based on previous denials or payer patterns, identify potential objections and prepare clear, evidence-based responses.	[ ]
Review clinical outcomes	Summarize projected outcomes or risks if the patient's care level was not escalated (e.g., risk of deterioration, readmission).	[ ]
Confirm P2P logistics	Verify time, contact information, and format (e.g., phone, video call) of the conference. Ensure technology is functioning correctly.	[ ]
Role-play or rehearse	Conduct a brief mock P2P conference with a colleague to refine key arguments and anticipate counterpoints.	[ ]

2. Prep Sheet Template

A Prep Sheet organizes critical information into a clear format to ensure the physician advisor can confidently present the case and address objections during the P2P conference.

Section	Details
Patient Information	Include the patient's name, date of birth (DOB), medical record number (MRN), payer name, and policy ID.
Reason for Admission	Briefly describe the patient's presenting condition and rationale for inpatient admission. Include pertinent background such as comorbidities.
Key Clinical Data	Summarize vital signs, diagnostic results, treatment details (e.g., oxygen support, IV medications), and hospital course. Highlight abnormal findings or trends.
Alignment with Criteria	Cite specific InterQual, MCG, or proprietary payer criteria that support inpatient admission. Provide quotes or references to strengthen the argument.
Risk Analysis	Detail the risks associated with a lower level of care, such as potential deterioration, preventable complications, or delayed recovery.
Anticipated Objections	List likely objections from the payer (e.g., stable vitals, no ICU-level interventions) and prepare counterpoints backed by evidence.
Supporting Studies/References	Include citations of medical literature, clinical guidelines, or specialty society recommendations to bolster your argument.
Conclusion Statement	Craft a concise closing statement emphasizing the necessity of inpatient care based on clinical evidence, payer criteria, and anticipated patient outcomes.

Appendix B: Glossary of Key Terms and Concepts

A

- **Administrative Denials**
  Claim denials resulting from procedural issues, such as incomplete documentation or missed submission deadlines, rather than clinical factors.
- **Advanced Analytics**
  The use of tools such as machine learning and predictive modeling to identify trends and optimize denial management processes.
- **Advocacy Beyond Appeals**
  Efforts by physician advisors to promote systemic improvements in payer policies, ensuring equitable patient access to necessary care.
- **Admissions Criteria**
  Guidelines determining whether a patient qualifies for inpatient admission versus observation or outpatient care.

- **Appeal Success Rate**
  A performance metric that measures the percentage of denied claims successfully overturned.
- **Artificial Intelligence (AI) in Denial Management**
  AI tools that streamline workflows, automate documentation reviews, and identify high-risk claims.
- **Authorization**
  A payer-required process to approve certain healthcare services before they are delivered, ensuring coverage and reimbursement.
- **Automated Prep Sheets**
  AI-generated summaries containing clinical data, used to support effective peer-to-peer conferences.
- **Automated Workflows**
  Technology-driven processes that handle repetitive tasks, such as claim submission and denial tracking, with minimal manual intervention.

B

- **Benchmark Metrics**
  Key performance indicators like denial rates and appeal success rates, used to measure the effectiveness of denial management strategies.
- **Behavioral Health Criteria**
  Guidelines for determining the appropriateness of care for mental health and substance use disorders.
- **Blockchain Technology**
  A decentralized, tamper-proof digital ledger system to enhance transparency and security in data sharing and claims processing.
- **Burnout Mitigation**
  Strategies to address the emotional and physical exhaustion faced by physician advisors, aiming to improve work-life balance and efficiency.

C

- **Care Transitions**
  Processes that ensure continuity of care when patients move between healthcare settings, such as hospital to rehabilitation facilities.
- **Case Mix Index (CMI)**
  A measure of patient complexity and resource intensity, used to adjust hospital reimbursement rates.
- **Clinical Decision Support Systems (CDSS)**
  Technology that provides evidence-based recommendations to improve clinician decision-making.

- **Clinical Documentation Integrity (CDI)**
  Programs aimed at enhancing the quality, accuracy, and specificity of clinical documentation.
- **Coding Compliance**
  Ensuring accurate and appropriate assignment of billing codes to prevent claim denials.
- **Collaborative Workflows**
  Interdisciplinary approaches that integrate clinical and administrative teams to improve denial management.
- **Condition Code 44**
  A billing designation used to change inpatient status to outpatient after a case review, when inpatient criteria are not met.

D

- **Data-Driven Advocacy**
  Using data insights to strengthen arguments during payer negotiations or appeals.
- **Denial Prediction Algorithms**
  Machine learning models that forecast claim denial risks based on historical trends.
- **Denial Rate**
  The percentage of claims initially denied by payers, indicating areas of process inefficiency.
- **Denial Trends Analysis**
  Reviewing denial patterns over time to identify root causes and improve processes.
- **Documentation Gaps**
  Missing or incomplete information in medical records that may lead to claim denials.
- **Dynamic Decision-Making**
  Adapting strategies quickly to changing payer policies or emerging trends in denial management.

E

- **Efficiency Metrics**
  Tools to measure the speed and success of denial resolution efforts, such as average time to appeal closure.
- **Eligibility Verification Processes**
  Confirming a patient's insurance coverage and benefits before providing care.
- **Emerging Innovations in Denial Management**
  Advanced technologies like AI and blockchain that are reshaping denial workflows.
- **Evidence-Based Criteria**
  Standards such as InterQual or MCG used to evaluate medical necessity and justify care decisions.

F

- **Financial Reconciliation**
  Resolving payment discrepancies to recover denied or underpaid claims.
- **Fraud Prevention**
  Ensuring compliance with regulatory and payer requirements to avoid fraudulent claims.
- **Future Innovations in AI for Appeals**
  Tools poised to automate appeal letters, enhancing efficiency in denial overturns.

G

- **Global Payment Models**
  Alternative reimbursement approaches, such as bundled payments, that emphasize value over volume in care delivery.
- **Guideline Alignment**
  Ensuring that clinical care aligns with both payer-specific and evidence-based standards to minimize denials.

H

- **Healthcare Disparities in Denials**
  Differences in denial rates due to inequities in access, documentation quality, or systemic bias.
- **High-Risk Claim Identification**
  Using analytics to flag claims with a high likelihood of being denied based on payer history.

I

- **Integrated Care Pathways**
  Coordinated clinical protocols designed to meet payer expectations while optimizing patient outcomes.
- **Interdisciplinary Collaboration**
  Involving multiple stakeholders, such as physicians, coders, and case managers, to resolve denials effectively.
- **InterQual Guidelines**
  Evidence-based tools used to evaluate the medical necessity of healthcare services.
- **Insurance Medical Director Workflows**
  Processes followed by payers' medical directors during claim reviews and appeals.

L

- **Length of Stay (LOS)**
  A metric that measures the duration of a patient's hospitalization, often scrutinized by payers for efficiency.

- **Longitudinal Patient Data in Denial Appeals**
  Comprehensive health records that support medical necessity arguments in appeals.

M

- **Machine Learning in Analytics**
  AI systems that analyze large datasets to identify trends and optimize denial prevention strategies.
- **Managed Medicaid**
  State Medicaid programs administered by private insurers.
- **Medical Necessity Denials**
  Claims denied due to a determination that the services provided were not essential for the patient's condition.

P

- **Patient Advocacy in Utilization Review**
  Efforts to ensure patients receive appropriate care by challenging denials that may limit necessary treatment.
- **Peer-to-Peer (P2P) Conferences**
  Real-time discussions between healthcare providers and payer medical directors to resolve denials.
- **Predictive Analytics**
  Tools that forecast denial risks and guide proactive interventions.
- **Proprietary Payer Criteria**
  Custom rules developed by payers to evaluate claims, often differing from national standards.

R

- **Revenue Cycle Management (RCM)**
  End-to-end processes involving claim submission, payment collection, and denial management.
- **Root Cause Analysis (RCA)**
  Investigating the fundamental reasons behind claim denials to guide corrective actions.

S

- **Social Determinants of Health (SDOH)**
  Non-clinical factors like housing and income that impact health outcomes and denial rates.
- **Sustainability in Denial Management**
  Practices to maintain high levels of performance and efficiency in denial workflows over time.

T

- **Telehealth-Specific Denials**
  Unique denials associated with virtual care services, often due to coding or documentation issues.
- **Timely Filing Rules**
  Deadlines set by payers for submitting claims, non-compliance with which results in denials.

U

- **Unified Payer Criteria Databases**
  Centralized repositories containing payer-specific rules to streamline the denial prevention process.
- **Utilization Management (UM)**
  Assessment of the necessity and efficiency of healthcare services to ensure alignment with payer standards.

W

- **Workflow Optimization**
  Streamlining processes to improve efficiency and reduce bottlenecks in denial management.

Z

- **Zero-Denial Targeting**
  Strategies aimed at eliminating initial claim denials through improved documentation and adherence to guidelines.

Appendix C: References to Guidelines and Tools

This section provides an overview of key resources and tools physician advisors can access to prepare for P2P conferences and appeals.

- **InterQual**: Evidence-based criteria from Optum for inpatient and outpatient care.
- **MCG Guidelines**: Comprehensive guidelines to evaluate medical necessity for various levels of care.
- **Medicare Resources**: Regulations and policies that guide inpatient admission decisions.
- **Proprietary Payer Criteria**: Accessible through individual payer portals or customer support.
- **Specialty Society Guidelines**:
  - **American Thoracic Society**: Guidelines for managing respiratory conditions.

- **American Heart Association**: Recommendations for cardiac conditions such as heart failure.
- **American College of Surgeons**: Standards for perioperative care and postoperative management.

REFERENCES

[1] Quinn, E (2022). Evolution of the Physician Advisor: A Current Perspective - RACMonitor, https://racmonitor.medlearn.com/evolution-of-the-physician-advisor-a-current-perspective/

[2] Bovbjerg, R. R., Griffin, C. C., & Carroll, C. E. (1993). U.S. Health Care Coverage and Costs: Historical Development and Choices for the 1990s. The Journal of Law, Medicine & Ethics, 21(2), 141–162. doi:10.1111/j.1748-720x.1993.tb01239.x

[3] Kaufman, B. G., Spivack, B. S., Stearns, S. C., Song, P. H., & O'Brien, E. C. (2017). Impact of Accountable Care Organizations on Utilization, Care, and Outcomes: A Systematic Review. Medical Care Research and Review, 107755871774591. doi:10.1177/1077558717745916

[4] Wennberg, J. E. (1984). Dealing With Medical Practice Variations: A Proposal for Action. Health Affairs, 3(2), 6–33. doi:10.1377/hlthaff.3.2.6

[5] Harrington, C., Lynch, M., & Newcomer, R. J. (1993). Medical Services in Social Health Maintenance Organizations. The Gerontologist, 33(6), 790–800. doi:10.1093/geront/33.6.790

[6] Mayes, R. (2006). The Origins, Development, and Passage of Medicare's Revolutionary Prospective Payment System. Journal of the History of Medicine and Allied Sciences, 62(1), 21–55. doi:10.1093/jhmas/jrj038

[7] Terra, S. M. (2009). Regulatory Issues. Professional Case Management, 14(5), 217–223. doi:10.1097/ncm.0b013e3181b094fe

[8] Zander, Karen. "Getting more from physician advisors: building a formal job description to improve effectiveness." Case Management Monthly, vol. 4, no. 12, Dec. 2007, pp. 4+. Gale OneFile: Health and Medicine, link.gale.com/apps/doc/A171623700/HRCA?u=nysl_oweb&sid=googleScholar&xid=a49f3714. Accessed 2 Dec. 2024.

[9] Baumel, Mark J., and Robert R. Corrato. "Dedicating the right resources to denials management: full-time physician attention helped reduce one health system's denied claims and length of stay." Healthcare Financial Management, vol. 57, no. 12, Dec. 2003, pp. 66+. Gale Academic OneFile, link.gale.com/apps/doc/A111696061/AONE?u=nysl_oweb&sid=googleScholar&xid=83446def. Accessed 28 Dec. 2024

[10] Zander, Karen. "Getting more from physician advisors: building a formal job description to improve effectiveness." Case Management Monthly, vol. 4, no. 12, Dec. 2007, pp. 4+. Gale OneFile: Health and Medicine, link.gale.com/apps/doc/A171623700/HRCA?u=nysl_oweb&sid=googleScholar&xid=a49f3714. Accessed 28 Dec. 2024.

[11] Robert J. Becker, Managed care is utilization review, American Journal of Hospital Pharmacy, Volume 47, Issue 10, 1 October 1990, Pages 2274–2276, https://doi.org/10.1093/ajhp/47.10.2274

[12] Mitus, A. Jacqueline. (2008). The Birth of InterQual. Professional Case Management, 13(4), 228–233. doi:10.1097/01.pcama.0000327413.01

[13] Field, M. J., & Gray, B. H. (1989). Should We Regulate "Utilization Management?" Health Affairs, 8(4), 103–112. doi:10.1377/hlthaff.8.4.103

[14] Lawless, Grant D., and Dorothy B. Holt. "Developing an in-house physician advisor program." Physician Executive, vol. 17, no. 3, May-June 1991, pp. 29+. Gale Academic OneFile, link.gale.com/apps/doc/A11008711/AONE?u=nysl_oweb&sid=googleScholar&xid=c2462d1f. Accessed 29 Dec. 2024.

[15] Pahuja D, Chadha P, Sharma U, Bhatia H, Nagpal P, Jordan J, Lannom C, de Los Santos E. PHYSICIAN ADVISERS ADD UP FOR SOME HOSPITALS. Physician Leadersh J. 2017 Mar;4(2):44-48. PMID: 30571884.

[16] Hodges, Jackie. "Effective claims denial management enhances revenue: claims denial management has become a critical component of a hospital's strategic effort to offset the adverse impact of Balanced Budget Act payment reductions." Healthcare Financial Management, vol. 56, no. 8, Aug. 2002, pp. 40+. Gale Academic OneFile, link.gale.com/apps/doc/A90317285/AONE?u=nysl_oweb&sid=googleScholar&xid=579c0167. Accessed 10 Jan. 2025.

[17] Mendez, C. M., Harrington, D. W., Christenson, P., & Spellberg, B. (2014). Impact of Hospital Variables on Case Mix Index as a Marker of Disease Severity. Population Health Management, 17(1), 28–34. doi:10.1089/pop.2013.0002

[18] Oh, J. C., Zheng, Z. E., & Bardhan, I. R. (2017). Sooner or Later? Health Information Technology, Length of Stay, and Readmission Risk. Production and Operations Management. doi:10.1111/poms.12748

[19] Olaniyan, Olakunle, et al. "Managing medical necessity and notification denials: are denials for medical necessity and notification issues eroding your bottom line? Implementing a denials mitigation process can help." Healthcare Financial Management, vol. 63, no. 8, Aug. 2009, pp. 62+. Gale Academic OneFile, link.gale.com/apps/doc/A210170120/AONE?u=nysl_oweb&sid=googleScholar&xid=e3746965. Accessed 10 Jan. 2025.

[20] Rode, Dan. "Medicare audits and the importance of documentation." Healthcare Financial Management, vol. 51, no. 10, Oct. 1997, pp. 82+. Gale Academic OneFile, link.gale.com/apps/doc/A20168613/AONE?u=nysl_oweb&sid=googleScholar&xid=89961abe. Accessed 10 Jan. 2025.

[21] Kilanko, V. (2023). Leveraging Artificial Intelligence for Enhanced Revenue Cycle Management in the United States. International Journal of Scientific Advances, 4(4), 505-14.

[22] Gaines, M. E., Auleta, A. D., & Berwick, D. M. (2020). Changing the Game of Prior Authorization. JAMA. doi:10.1001/jama.2020.0070

[23] Doss, D., Pawar, R., & Maddireddy, R. (2022). Intelligent Automation in Health Insurance. Bimaquest, 22(1). Retrieved from http://www.bimaquest.niapune.org.in/index.php/bimaquest/article/view/120

[24] PMLA , Volume 127 , Issue 4: Special Topic Work , October 2012 , pp. 983 - 986
DOI: https://doi.org/10.1632/pmla.2012.127.4.983

[25] Pal, S. et al. (2022). Driving Impact in Claims Denial Management Using Artificial Intelligence. In: Singh, M., Tyagi, V., Gupta, P.K., Flusser, J., Ören, T. (eds) Advances in Computing and Data Sciences. ICACDS 2022. Communications in Computer and Information Science, vol 1613. Springer, Cham. https://doi.org/10.1007/978-3-031-12638-3_10

[26] Lee C, Britto S, Diwan K. Evaluating the Impact of Artificial Intelligence (AI) on Clinical Documentation Efficiency and Accuracy Across Clinical Settings: A Scoping Review. Cureus. 2024 Nov 19;16(11):e73994. doi: 10.7759/cureus.73994. PMID: 39703286; PMCID: PMC11658896.

[27] Nimkar P, Kanyal D, Sabale SR. Increasing Trends of Artificial Intelligence With Robotic Process Automation in Health Care: A Narrative Review. Cureus. 2024 Sep 18;16(9):e69680. doi: 10.7759/cureus.69680. PMID: 39429258; PMCID: PMC11489308.

[28] Quan, Zhiyu; Valdez, Emiliano A. . (2018). Predictive analytics of insurance claims using multivariate decision trees. Dependence Modeling, 6(1), 377–407. doi:10.1515/demo-2018-0022

[29] Robinson, E. . (2024). Predictive Modeling of Health Insurance Claims in the United States. American Journal of Statistics and Actuarial Sciences, 5(1), 35–46. https://doi.org/10.47672/ajsas.1994

[30] Arun Chandramouli, "Leveraging Predictive Analytics to Minimize Claim Denials in Healthcare Revenue CycleManagement", J. Tech. Innovations, vol. 2, no. 4, Jan. 2025, doi: 10.93153/3aexs190.

[31] Hölbl, Marko; Kompara, Marko; Kamišalić, Aida; Nemec Zlatolas, Lili . (2018). A Systematic Review of the Use of Blockchain in Healthcare. Symmetry, 10(10), 470–. doi:10.3390/sym10100470

[32] Sanderson, Brian, et al. "3 pillars of a successful denials management program: with a strong denials management process in place, hospitals can reduce their denial rate and increase cash recoveries." Healthcare Financial Management, vol. 62, no. 9, Sept. 2008, pp. 74+. Gale Academic OneFile, link.gale.com/apps/doc/A188898722/AONE?u=nysl_oweb&sid=googleScholar&xid=bcef7653. Accessed 12 Jan. 2025.

[33] Russell, S., Montes Suros, F., & Kumar, A. (2024). Exploiting Machine Learning Bias: Predicting Medical Denials. Proceedings of the AAAI Symposium Series, 3(1), 58-63. https://doi.org/10.1609/aaaiss.v3i1.31181

[34] Bajowala, S.S., Milosch, J. & Bansal, C. Telemedicine Pays: Billing and Coding Update. Curr Allergy Asthma Rep 20, 60 (2020). https://doi.org/10.1007/s11882-020-00956-y

[35] Zouo, S. J. C., & Olamijuwon, J. (2024). Financial data analytics in healthcare: A review of approaches to improve efficiency and reduce costs.

36 Paz, Anthony J., "Online collaboration platforms: Communication implications for workplace virtual teams" (2020). Senior Honors Theses and Projects. 678. https://commons.emich.edu/honors/67

[37] Medicare Benefit Policy Manual, Chapter 1: Inpatient Hospital Services Covered Under Part A. Section 10: "Covered Inpatient Hospital Services Covered Under Part A."

[38] CMS Final Rule for Inpatient Prospective Payment System (IPPS). 78 Fed. Reg. 50495 (August 19, 2013), pages 50905–50909

[39] Centers for Medicare & Medicaid Services. Medicare Program; Hospital Inpatient Prospective Payment Systems for Acute Care Hospitals and the Long Term Care; Hospital Prospective

Payment System and Fiscal Year 2014 Rates; Quality Reporting Requirements for Specific Providers; Hospital Conditions of Participation; Payment Policies Related to Patient Status; Final Rule. August 19, 2013. http://www.gpo.gov/fdsys/pkg/FR-2013-08-19/pdf/2013-18956.pdf.

[40] Dadana S, Bornovali SM. HCQM Two-Midnight Rule. [Updated 2023 Jul 30]. In: StatPearls [Internet]. Treasure Island (FL): StatPearls Publishing; 2025 Jan-. Available from: https://www.ncbi.nlm.nih.gov/books/NBK594265/

[41] Sommers, B. D., Tomasi, M. R., Swartz, K., & Epstein, A. M. (2012). Reasons For The Wide Variation In Medicaid Participation Rates Among States Hold Lessons For Coverage Expansion In 2014. Health Affairs, 31(5), 909–919. doi:10.1377/hlthaff.2011.0977

[42] Chisolm, D.J., Brook, D.L., Applegate, M.S. et al. Social determinants of health priorities of state Medicaid programs. BMC Health Serv Res 19, 167 (2019). https://doi.org/10.1186/s12913-019-3977-5

[43] Daxa P. Clarke, MD, FAAP, Katherine Johnson, MD, FAAP, "Utilization Management", Caring for the Hospitalized Child: A Handbook of Inpatient Pediatrics, American Academy of Pediatrics Section on Hospital Medicine, Jeffrey C. Gershel, MD, FAAP, Daniel A. Rauch, MD, FAAP, SFHM

[44] Pasetsky J, Garcia-Young JA, Beatley E, Yu JB. Peer-to-Peer Phone Calls and Letters Appealing Insurance Denials of Service: Practical Tips and Resources. Pract Radiat Oncol. 2024 Nov-Dec;14(6):e434-e437. doi: 10.1016/j.prro.2024.06.015. Epub 2024 Jul 30. PMID: 39089624.

[45] Fairchild KD, Duncan SD, Stroustrup A, McNamara PJ. Insurance denials: a peer-to-peer problem in neonatology. J Perinatol. 2024 Aug;44(8):1210-1211. doi: 10.1038/s41372-024-01991-7. Epub 2024 May 11. Erratum in: J Perinatol. 2024 Aug;44(8):1233. doi: 10.1038/s41372-024-02028-9. PMID: 38734803; PMCID: PMC11300295.

[46] Overton AR, Lowry AC. Conflict management: difficult conversations with difficult people. Clin Colon Rectal Surg. 2013 Dec;26(4):259-64. doi: 10.1055/s-0033-1356728. PMID: 24436688; PMCID: PMC3835442.

[47] Stephens, E., William, L., Lim, LL. et al. Complex conversations in a healthcare setting: experiences from an interprofessional workshop on clinician-patient communication skills. BMC Med Educ 21, 343 (2021). https://doi.org/10.1186/s12909-021-02785-7

[48] Enabling difficult conversations in the Australian health sector . Aust J Adv Nurs [Internet]. 2021 Aug. 26 [cited 2025 Feb. 19];38(3). Available from: https://www.ajan.com.au/index.php/AJAN/article/view/310

[49] Dorela Priftanji, John D Hill, Daniel M Ashby, Managing difficult conversations, American Journal of Health-System Pharmacy, Volume 77, Issue 21, 1 November 2020, Pages 1723–1726, https://doi.org/10.1093/ajhp/zxaa149

[50] Prober CG, Grousbeck HI, Meehan WF 3rd. Managing Difficult Conversations: An Essential Communication Skill for All Professionals and Leaders. Acad Med. 2022 Jul 1;97(7):973-976. doi: 10.1097/ACM.0000000000004692. Epub 2022 Jun 23. PMID: 35767404.

[51] Stephens, E., William, L., Lim, LL. et al. Complex conversations in a healthcare setting: experiences from an interprofessional workshop on clinician-patient communication skills. BMC Med Educ 21, 343 (2021). https://doi.org/10.1186/s12909-021-02785-7

[52] West, M.A. (2020). Compassionate and Collective Leadership for Cultures of High-Quality Care. In: Montgomery, A., van der Doef, M., Panagopoulou, E., Leiter, M.P. (eds) Connecting Healthcare Worker Well-Being, Patient Safety and Organisational Change. Aligning Perspectives on Health, Safety and Well-Being. Springer, Cham. https://doi.org/10.1007/978-3-030-60998-6_13

[53] O'Daniel M, Rosenstein AH. Professional Communication and Team Collaboration. In: Hughes RG, editor. Patient Safety and Quality: An Evidence-Based Handbook for Nurses. Rockville (MD): Agency for Healthcare Research and Quality (US); 2008 Apr. Chapter 33. Available from: https://www.ncbi.nlm.nih.gov/books/NBK2637/?utm_source=chatgpt.com

[54] Mata, Á.N.S., de Azevedo, K.P.M., Braga, L.P. et al. Training in communication skills for self-efficacy of health professionals: a systematic review. Hum Resour Health 19, 30 (2021). https://doi.org/10.1186/s12960-021-00574-3

[55] Hammer, David, et al. "UPMC's metric-driven revenue cycle: a commitment to managing by the metrics enabled one healthcare organization to achieve world class revenue cycle operations." Healthcare Financial Management, vol. 61, no. 9, Sept. 2007, pp. 96+. Gale Academic OneFile, link.gale.com/apps/doc/A168587032/AONE?u=anon~2e745e8d&sid=googleScholar&xid=ba5fc986. Accessed 19 Feb. 2025.

[56] Chandawarkar, Rajiv MD, MBA*; Nadkarni, Prakash MD†; Barmash, Elizabeth MD*; Thomas, Stephany*; Capek, Allison PAC*; Casey, Kathleen PAC*; Carradero, Freddy*. Revenue Cycle

Management: The Art and the Science. Plastic & Reconstructive Surgery-Global Open 12(7):p e5756, July 2024. | DOI: 10.1097/GOX.0000000000005756

[57] Waymack, Pamela M. "Looking reality straight in the eye: five claim denial facts every CFO should know." Healthcare Financial Management, vol. 60, no. 3, Mar. 2006, pp. 44+. Gale Academic OneFile, link.gale.com/apps/doc/A144152152/AONE?u=anon~3113983e&sid=googleScholar&xid=62d9ae71. Accessed 19 Feb. 2025.

[58] Akosa AN. Precertification, denials and appeals: reducing the hassles. Fam Pract Manag. 2006 Jun;13(6):45-8. PMID: 16813121.

[59] Baumel, Mark J., and Robert R. Corrato. "Dedicating the right resources to denials management: full-time physician attention helped reduce one health system's denied claims and length of stay." Healthcare Financial Management, vol. 57, no. 12, Dec. 2003, pp. 66+. Gale Academic OneFile, link.gale.com/apps/doc/A111696061/AONE?u=anon~f1a2f011&sid=googleScholar&xid=bbb04065. Accessed 19 Feb. 2025.

[60] Roche, John D. Claim Denied!: How to Appeal a VA Denial of Benefits. Potomac Books, Inc., 2008.

[61] Kendall J. Kiser et al., Large Language Models to Help Appeal Denied Radiotherapy Services. JCO Clin Cancer Inform 8, e2400129(2024). DOI:10.1200/CCI.24.00129

[62] Cohen, Fred. "AI-Powered Patient Advocacy: Amplifying Health Care Voices." Neurology Live, 5 Nov. 2023. Gale OneFile: Health and Medicine, link.gale.com/apps/doc/A777222379/HRCA?u=anon~3b2556fe&sid=googleScholar&xid=09eaa396. Accessed 19 Feb. 2025.

[63] Michael R. Waters, Sanjay Aneja, Julian C. Hong, Unlocking the Power of ChatGPT, Artificial Intelligence, and Large Language Models: Practical Suggestions for Radiation Oncologists, Practical Radiation Oncology, Volume 13, Issue 6, 2023, Pages e484-e490, ISSN 1879-8500, https://doi.org/10.1016/j.prro.2023.06.011.

[64] Kanagarla, Krishna Prasanth Brahmaji, Improved Claims Processing: Advanced Analytics for Faster and More Accurate Claims Handling In Healthcare or Insurance (January 16, 2025). Available at SSRN: https://ssrn.com/abstract=5099728 or http://dx.doi.org/10.2139/ssrn.5099728

[65] Gupta, Reshma MD, MSHPM; Moriates, Christopher MD. Swimming Upstream: Creating a Culture of High-Value Care. Academic Medicine 92(5):p 598-601, May 2017. | DOI: 10.1097/ACM.0000000000001485

[66] Zander, Karen. "Getting more from physician advisors: building a formal job description to improve effectiveness." Case Management Monthly, vol. 4, no. 12, Dec. 2007, pp. 4+. Gale OneFile: Health and Medicine, link.gale.com/apps/doc/A171623700/HRCA?u=anon~e1bddae7&sid=googleScholar&xid=df107ebd. Accessed 19 Feb. 2025.

[67] Abidi, S.S.R. et al. (2019). AI-Driven Pathology Laboratory Utilization Management via Data- and Knowledge-Based Analytics. In: Riaño, D., Wilk, S., ten Teije, A. (eds) Artificial Intelligence in Medicine. AIME 2019. Lecture Notes in Computer Science(), vol 11526. Springer, Cham. https://doi.org/10.1007/978-3-030-21642-9_30

[68] Lin, Steven Y. et al., Reimagining Clinical Documentation With Artificial Intelligence. Mayo Clinic Proceedings, Volume 93, Issue 5, 563 - 565

[69] Saxena, R., Katage, G., Kumar, C., Pathan, N. M., & Bargir, M. N. (2024). AI redefining healthcare documentation for tomorrow: Exploring the impact of AI on healthcare documentation. In Computational Convergence and Interoperability in Electronic Health Records (EHR) (pp. 51-66). IGI Global.

[70] Le, Q. (2024). AI-driven Clinical Documentation Integrity for Electronic Health Records: Implements AI-driven solutions for clinical documentation integrity in electronic health records, enhancing accuracy and completeness of clinical documentation for billing and q. Journal of Artificial Intelligence Research and Applications, 4(1), 170-181.

www.ingramcontent.com/pod-product-compliance
Lightning Source LLC
LaVergne TN
LVHW080043210926
848481LV00011B/3272